An Introduction to Data-Driven Control Systems

An Introduction to Data-Driven Control Systems

Ali Khaki-Sedigh
K.N. Toosi University of Technology
Iran

IEEE PRESS
WILEY

Published by John Wiley & Sons, Inc., Hoboken, New Jersey.
Published simultaneously in Canada.

Library of Congress Cataloging-in-Publication Data Applied for:

Hardback ISBN: 9781394196401

Cover Design: Wiley
Cover Image: © zf L/Getty Images

Set in 9.5/12.5pt STIXTwoText by Straive, Chennai, India

Contents

Preface

Model-based control systems. Model-based control system analysis and design approaches have been the dominant paradigm in control system education and the cornerstone of control system design for decades. These methodologies rely on accurate mathematical models and assumptions to achieve the desired system behaviour. In the early decades of the last century, despite the tremendous interest in model-based control approaches, many PID controllers in the industry were designed based on the data-driven technique of Ziegler–Nichols PID parameter tuning, which is considered the first data-driven control approach. Later, the advanced adaptive and robust model-based control techniques evolved to combat the uncertainty challenge in the established model-based techniques. These advanced control techniques successfully controlled many real-world and industrial plants. Yet, both strategies require mathematical models and prior plant assumptions mandated by the theory.

Data-driven control methodologies. The limitations and uncertainties associated with models and assumptions, on the one hand, and the emergence of progressively complex systems, on the other hand, have sparked a paradigm shift towards data-driven control methodologies. The exponentially increasing number of research papers in this field and the growing number of courses offered in universities worldwide on the subject clearly show this trend. The new data-driven control system design paradigm has re-emerged to circumvent the necessity of deriving offline or online plant models. Many plants regularly generate and store huge amounts of operating data at specific instants of time. Such data encompasses all the relevant plant information required for control, estimation, performance assessment, decision-making and fault diagnosis. This data availability has facilitated the design of data-driven control systems.

Intended audience. This book is an introduction to data-driven control systems and attempts to provide an overview of the mainstream design approaches in the field. The selected approaches may be called with caution the conventional approaches, not including the approaches based on soft computing techniques.

A unique chapter is devoted to philosophical–historical issues regarding the emergence of data-driven control systems as the future dominant control design paradigm. This chapter will be particularly appealing to readers interested in gaining insights into the philosophical and historical aspects of control system design methodologies. Concepts from the philosophy of science and historical discussions are presented to show the inevitable prevalence of data-driven techniques in the face of emerging complex adaptive systems. This book can cover a graduate course on data-driven control and can also be used by any student or researcher who wishes to start working in the field of data-driven control systems. This book will present the primary material, and the reader can perceive a general overview of the developing data-driven control theory. The book presentation avoids detailed mathematical relations and derivations that are available in the cited technical papers on each subject. However, algorithms for easy implementation of the methods with numerical and simulation examples are provided. The software codes are available upon request from the author. Data-driven control is also a hot research topic; many final-year undergraduate and postgraduate students are interested in starting a research project in its different areas. The available reading sources are the technical papers and the limited number of research monographs and books on the subject. However, the technical papers are very specialised and involve deep mathematical derivations. The limited number of published monographs and books also specialise in specific subject areas and do not provide a general introduction and overview of different methodologies for a first-time reader in data-driven control. The selected topics in this book can be individually taught in many different courses on advanced control theory. Also, for an interested researcher in any of the covered fields, it would be beneficial to learn about the basics of other alternative methodologies to plan a research programme.

Prerequisites. The book is designed for graduate-level courses and researchers specialising in control systems across various engineering disciplines. The book assumes that the reader possesses a solid understanding of feedback control systems as well as familiarity with the principles of discrete-time control systems and optimisation problems. Moreover, a basic understanding of system identification, adaptive control and robust control can enhance the reader's comprehension and appreciation of data-driven control methodologies.

Overview of the book. The book is organised as follows. Chapter 1 introduces both the model-based and data-driven control system design approaches. It discusses the early developments and the current status quo of model-based control systems, as well as the challenges they face. The chapter also explores adaptive and robust control methodologies as a means to overcome some of these challenges. Subsequently, the data-driven control system design approach is presented, and the technical aspects of different data-driven control schemes are discussed.

Chapter 2 takes a philosophical perspective to analyse the paradigm shifts in control system design. It presents scientific theory, revolutions and paradigm shifts, drawing parallels to the evolution of control system design methodologies. The historical development of control systems design paradigms and their philosophical foundations is introduced, and a general classification of control systems is given. The chapter concludes with an exploration of the paradigm shifts towards data-driven control methodologies, with a focus on the influence of the unfalsified philosophy.

Chapters 3 and 4 present data-driven adaptive switching supervisory control and multi-model adaptive switching supervisory control, respectively. The philosophical backbone of the presented methodologies is Popper's falsification theory, which is introduced in a data-driven control context by Safonov. It is shown in Chapter 3 that the unfalsified adaptive switching supervisory control can effectively control unknown plants with guaranteed closed-loop stability under the minimum assumption of the existence of a stabilising controller. Although several closed-loop transient improvement techniques are presented in Chapter 3, the multi-model unfalsified adaptive switching control is introduced in Chapter 4 to ensure a superior closed-loop transient performance. It is shown that performance improvement is achieved by utilising a model set to select the appropriate controller based on the falsifying theory. The adaptive memory concept, input-constrained design problems and quadratic inverse optimal control notion are also discussed.

Chapter 5 presents the virtual reference feedback tuning approach. It is shown that by formulating the controller tuning problem as a controller parameter identification problem, a data-based controller design methodology is derived. In this approach, a virtual reference signal is introduced, and it is assumed that the controller structure is known a priori. After introducing the basic concepts and methodology, the problems of appropriate filter design, measurement noise, non-minimum phase zero challenges, closed-loop stability and extensions to multivariable plants are addressed in this chapter.

The simultaneous perturbation stochastic approximation optimisation technique is introduced and utilised in Chapter 6 for the design of data-driven control systems. It is shown that this circumvents the necessity of an analytical closed-form solution to the control optimisation problems that require almost exact mathematical models of the true plant. The essentials of the technique are presented for plants with unknown exact mathematical models. Then, after selecting a controller with a fixed known structure but unknown parameters, by minimising a cost function, the controller parameters are derived. The presented data-driven control methodology is then applied to unknown, under-actuated systems as a practical case study.

Chapter 7 presents a class of data-driven controllers based on Willem's Fundamental Lemma. It is initially shown that persistently exciting data can be used to represent the input–output behaviour of a linear system without the need to identify the linear system's matrices. The derived so-called equivalent data-based representations of a linear time invariant (LTI) system are subsequently utilised to design data-driven state-feedback stabilisers and predictive controllers called Data-enabled Predictive Control, or DeePC for short. Results with measurement noise and nonlinear systems are also given in this chapter.

Chapter 8 presents data-driven controllers based on Koopman's theory and the Fundamental Lemma presented in Chapter 7. The fundamentals of Koopman's theory are briefly reviewed for data-driven control. It is shown that nonlinear dynamical systems are presented by higher dimensional linear approximations. The main notions of *lifting* or *embedding* and the effective tools of (extended) dynamic mode decompositions are introduced and a data-driven Koopman-based predictive control scheme is presented by incorporating Willem's Fundamental Lemma of Chapter 7. A robust stability analysis is provided, and the results are finally applied to the ACUREX solar collector field.

The model-free adaptive control design is a data-driven control design approach based on dynamic linearisation methodologies and is presented in Chapter 9. The three main dynamic linearisations discussed in this chapter are shown to capture the system's behaviour by investigating the output variations resulting from input signals. These data models are utilised for controller design and their virtual nature makes them inappropriate for other system analysis purposes. Also, in Chapter 9, the virtual data model results and their corresponding model-free adaptive controllers are extended to multivariable plants.

Some preliminary concepts that are useful for the chapters are presented in the Appendix. The chapters are accompanied by problem sets that provide readers with the opportunity to reinforce their understanding and apply the concepts discussed. A solution manual is also provided for instructors teaching a class on data-driven control using this book by contacting the author.

October 2023

Ali Khaki-Sedigh
Department of Electrical Engineering
K.N. Toosi University of Technology

Acknowledgements

The preparation of this book has greatly benefited from the invaluable contributions and support of numerous postgraduate students and colleagues who generously dedicated their time to share expertise, valuable suggestions and corrections throughout the process. I extend my sincere gratitude to the following individuals, whose significant efforts have greatly enriched the content of this book:

- Amirehsan Karbasizadeh from the Department of Philosophy at the University of Isfahan, for his insightful discussions on the philosophy of science and his invaluable comments that notably enhanced Chapter 2.
- Mojtaba Nooi Manzar from Faculty of Electrical and Computer Engineering, Shahid Beheshti University, for his contributions to the initial draft of Chapters 3 and 4, as well as corrections and the simulation results for these chapters.
- Mohammad Moghadasi, Mehran Soleymani and Maedeh Alimohamadi, my master's students in the Advanced Control Laboratory, for their diligent proofreading of Chapter 3.
- Bahman Sadeghi and Maedeh Alimohamadi, my master's students in the Advanced Control Laboratory, for their valuable contributions to Chapter 4.
- Mohammad Jeddi and Fatemeh Hemati Kheirabadi, my master's students in the Advanced Control Laboratory, for their insightful contributions to Chapter 5.
- Sepideh Nasrollahi, my PhD student in the Advanced Control Laboratory, for her contributions to Chapter 6, as well as her valuable contributions to other chapters and the creation of numerous figures throughout the book.
- Tahereh Gholaminejad, my PhD student in the Advanced Control Laboratory, for her significant contributions to Chapters 7 and 8.
- Sara Iman, PhD student from Iran University of Science and Technology, for her meticulous proofreading and useful comments on Chapters 7 and 8.

- Ali Rezaei, my master's student in the Advanced Control Laboratory, for his valuable contributions to Chapter 9 and the effort he dedicated to simulations throughout the book.

Finally, I would like to express my sincere appreciation to the anonymous reviewers who provided invaluable feedback during the review process of this book, and special thanks to Wiley-IEEE Press for their exceptional professionalism, dedication, industry knowledge and seamless coordination that exceeded my expectations. Last but not least, I am also grateful to my family for their collaboration and support, allowing me to dedicate most of my holidays, weekends and evenings to completing this book.

List of Acronyms

ASSC	Adaptive switching supervisory control
BIBO	Bounded-input bounded-output
CFDL	Compact-form dynamic linearisation
CSP	Concentrated solar power
DAL	Dehghani–Anderson–Lanzon
DDKPC	Data-driven Koopman predictive control
DeePC	**D**ata-**e**nable**d** **P**redictive **C**ontrol
DFT	Discrete Fourier transform
DMD	Dynamic mode decomposition
EDMD	Extended dynamic mode decomposition
ETFE	Empirical transfer function estimate
FFDL	Full-form dynamic linearisation
GLA	Generalised Laplace analysis
GPC	Generalised predictive control
ICLA	Increasing cost level algorithm
IFAC	International Federation of Automatic Control
LFT	Linear fractional transformation
LICLA	Linearly increasing cost level algorithm
LLC	Linearisation length constant
LMI	Linear matrix inequality
LQG	Linear quadratic Gaussian
LQR	Linear quadratic regulator
LSTM	Long short-term memory
LTI	Linear time-invariant
MFAC	Model-free adaptive control
MMUASC	Multi-model unfalsified adaptive switching control
MMUASC-R	MMUASC with reset time
MMUCGPC	Multi-model unfalsified constrained GPC
MPC	Model predictive control

MPUM	Most powerful unfalsified model
PE	Persistence of excitation, persistently exciting
PFDL	Partial-form dynamic linearisation
PG	Pseudo-gradient
PID	Proportional integral derivative
PJM	Pseudo-Jacobian matrix
PPD	Pseudo-partial derivative
PPJM	Pseudo-partitioned-Jacobian matrix
PSO	Particle swarm optimisation
QFT	Quantitative feedback theory
SA	Stochastic approximation
SCLI	Stably causally left invertible
SICE	Society of Instrument and Control Engineering
SIHSA	Scale-independent hysteresis algorithm
SISO	Single-input-single-output
SNR	Signal-to-noise ratio
SPSA	Simultaneous perturbation stochastic approximation
SVD	Singular value decomposition
THSA	Threshold hysteresis algorithm
UASC	Unfalsified adaptive switching control
UASC-R	UASC with reset-time
UASSC	Unfalsified adaptive switching supervisory control
VRFT	Virtual reference feedback tuning

1

Introduction

1.1 Model-Based Control System Design Approach

1.1.1 The Early Developments

The advent of models in control systems theory and design is rooted in the seminal paper of Maxwell *On Governers* [1]. Norbert Wiener, in introducing the word cybernetics in Ref. [2] describes the Maxwell paper as '*… the first significant paper on feedback mechanisms is an article on governors, which was published by Clerk Maxwell in 1868*' and in Ref. [3], Maxwell is recognised as the 'father of control theory'. The Maxwell magic was to introduce differential equations in modelling the behaviour of the flyball governor feedback control system invented by James Watt in 1788. This ground-breaking contribution by Maxwell introduced the concept of mathematical modelling in the stability analysis of a closed-loop control system, an idea that soon found many applications and advocates and solved many until then unsolved stability analysis problems. The differential equations encountered in the flyball governor model were nonlinear. By linearising these nonlinear equations, Maxwell managed to introduce the notions of what is today called real poles, imaginary poles and the significance of pole position in the right half plane. This model-based approach to the analysis of a control system through the differential equations of motion was performed for the first time in the history of control theory. Hence, it is plausible to introduce Maxwell as the pioneer of the model-based control theory.

In the early twentieth century, control system design methodologies such as the *classical control* techniques initiated by Bode, Nyquist, Evans and Nichols were all model-based approaches to control design since the transfer function knowledge of the controlled system is required. The transfer function can be derived from a set of algebraic and differential equations that analytically relate inputs and outputs, or it could be obtained from simple tests performed on the plant with the assumptions of linearity and time invariance. Later in the 1960s,

An Introduction to Data-Driven Control Systems, First Edition. Ali Khaki-Sedigh.

Kalman introduced the model-based state-space approach that was more detailed and mathematical.

The only notable data-driven technique of the first half of the last century is the Ziegler–Nichols proportional-integral-derivative (PID) parameter tuning proposed in Ref. [4], which became a widely used control technique in the industry [5]. It is stated in Ref. [4] that "*A purely mathematical approach to the study of automatic control is certainly the most desirable course from a standpoint of accuracy and brevity. Unfortunately, however, the mathematics of control involves such a bewildering assortment of exponential and trigonometric functions that the average engineer cannot afford the time necessary to plow through them to a solution of his current problem.*" This statement from the eminent control engineers of that time shows the long-lasting influence of mathematical model-based design techniques on the control systems design community. In describing their work, they immediately state that '*the purpose of this paper is to examine the action of the three principal control effects found in present-day instruments, assign practical values to each effect, see what adjustment of each does to the final control, and give a method for arriving quickly at the optimum settings of each control effect. The paper will thus first endeavor to answer the question: "How can the proper controller adjustments be quickly determined on any control application?"*' This statement can enlighten aspects of the philosophy of the data-driven control systems that evolved in the late twentieth century onwards.

1.1.2 Model-based Control System Status Quo

Model-based control system design is the dominant paradigm in control system education and design. This approach is based on derived analytical models from basic physical laws and equations or models from an identification process. Models are only approximations of reality and cannot capture all the features and characteristics of a plant under control. High-frequency un-modelled dynamics are an example, as in robotic and spacecraft applications where the residual vibration modes are not included in the model [6]. The structure of a model-based control system is shown in Figure 1.1. In the case of adaptive control strategies, the approximate plant model is updated using the input–output data.

As is shown in Figure 1.1, the plant model, derived from first principles or identified from plant-measured data, is used to design a fixed-order controller satisfying the specified closed-loop requirements. However, the designed controller does not necessarily satisfy the pre-defined requirements when connected to the real plant, and the closed-loop performance is limited by the *modelling errors*. Modelling errors can have many root causes, such as un-modelled dynamics, unknown or varying plant parameters resulting from changing operating points, equipment ageing or faults and inappropriate model structures.

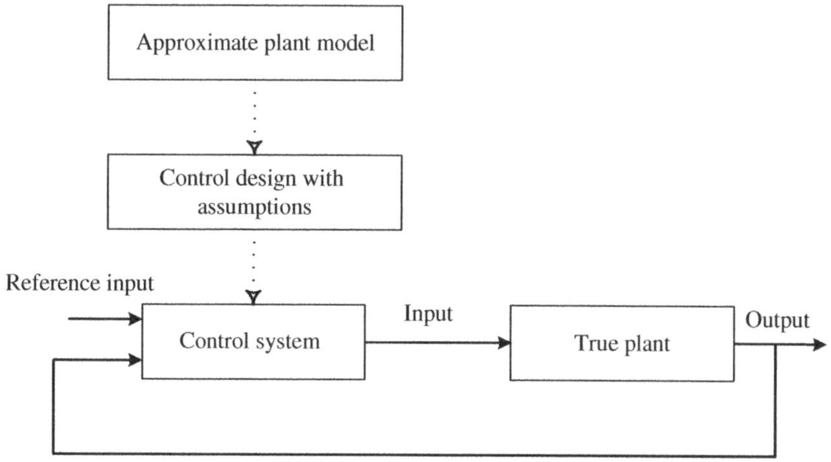

Figure 1.1 The structure of a model-based control system design.

Modelling errors due to un-modelled dynamics are justified in the standard practice of model-based control design when the system is complex and is of a *high* order, and a low-order model is employed to facilitate the control design. On the other hand, there can be a tendency to increase the model order to find a suitable model. It is shown in Ref. [7] that this is not generally true if the model has to be used for control design. In fact, the *order* of a real system is a badly defined concept, and even the most accurate models are only an approximation of the real plant. In the real world, a *full-order model* does not exist, and any description is, by definition, an approximation [7]. Model-based control design can only be employed with confidence in real-world applications if the model structure is perfectly known.

The issue of model-based control system design and the paradigm shifts to and from model-based approaches is further elaborated in Chapter 2.

1.1.3 Challenges of Models in Control Systems Design

The introduction of the state-space concept by Kalman in 1960, together with the newly established notion of optimality, resulted in a remarkable development of model-based control design methods. Before Kalman's state-space theory, most of the control design was based on transfer function models, as is in the Bode and Nyquist plots or the root-locus method and the Nichols charts for lead–lag compensator design.

In the cases where reliable models were unavailable, or in the case of varying parameters and changing operating conditions, the application of the model-based control was severely limited. In the mid-1960s, the system identification strategy evolved. The proposed Maximum Likelihood framework for the identification of input–output models resulted in the prediction error-type identifiers. The advent of identification theory solved the problem of controlling complex time-varying plants using model-based control design methodologies.

Initially, control scientists working on the identification methods aimed at developing sophisticated models and methodologies with the elusive goal of converging to the *true system*, under the assumption that the true system was in the defined model set. Later, they realised that the theory could best achieve an approximation of the true system and characterise this approximation in terms of bias and variance error on the identified models. Finally, system identification was guided towards a control-oriented identification. In all the modelling strategies, modelling by first principles or by identification from data, *modelling errors* are unescapable, and *explicit quantification* of modelling errors is practically impossible. Hence, the modelling strategies cannot provide adequate practical uncertainty descriptions for control design purposes. Therefore, the first modelling principle given in Ref. [8], that arbitrarily small modelling errors can lead to arbitrarily bad closed-loop performance, is seriously alarming for control systems designers.

Application of the *certainty equivalence principle* (see Chapter 2) was based on the early optimistic assumption that it is possible to almost perfectly model the actual plant and the mathematical model obtained from the first principles or by identification from input–output data is valid enough to represent the true system. However, applications in real-world problems did not meet the expectations of the control scientists and designers. Therefore, an obvious need prevailed to guarantee closed-loop stability and performance in the model-based control design approaches. This led to the development of the model-based approaches of fixed-parameter robust control and adaptive control system design [9].

The mathematical models derived from the physical laws have been effectively used in practical applications, provided that the following assumptions hold:

- Accurately model the actual plant.
- Priori bounds on the noise and modelling errors are available.

Also, identification models have been employed in many practical applications. The identified model can capture the main features of the plant, provided that

- Compatibility of the selected model structure and parameterisation with the actual plant's characteristics is assumed.
- The experiment design is appropriate; that is, for control problems, the selection of the input signal is in accordance with the actual plant's characteristics or the persistence of excitation (PE) condition.

It is important to note that even in the case of an accurately modelled plant, if the assumptions about the plant characteristics are not met, the mathematical theorems rigorously proving closed-loop robust stability and performance and parameter convergence are not of practical value.

Hence, in summary, if

- An accurate model is unavailable, or
- The assumptions regarding the plant do not hold,

the designed model-based controller, validated by simulations, can lead to an unstable closed-loop plant or poor closed-loop performance.

1.1.4 Adaptive and Robust Control Methodologies

Adaptive and robust control systems have successfully controlled many real-world and industrial plants. However, both strategies require many prior plant assumptions to be mandated by the theory. The key questions are the closed-loop robust stability and robust performance issues in practical implementations. The assessment of these specifications is not possible a priori, as unforeseen events may occur in practice. Hence, the control engineer must resort to *ad hoc* methods for a safe and reliable closed-loop operation. This is often done by performing many tests for many different variations of uncertainties and operating scenarios in the Monte Carlo simulations. However, with the growing plant complexity and the possible test situations, the cost of these heuristic tests increases. Hence, the limitations inherent in the adaptive and robust controllers are clearly observed. Parameter adjustments and robust control and their synergistic design packages are the ultimate solutions of the model-based control scientists for the utmost guarantee of safe and reliable closed-loop control.

A closed-loop system's performance degradation and even instability are almost inevitable when the plant uncertainty is too large or when the parameter changes or structural variations are too large or occur abruptly. The adaptive switching control was introduced as one of the robust adaptive control techniques to handle such situations and lessen the required prior assumptions. This led to the switching supervisory control methods, where a supervisor controller selects the appropriate controller from a controller bank, similar to their ancestor, the gain-scheduling methodology, which has been and still is widely used in many applications. This mindset and the recently developed selection process based on the falsification theory are regarded as the first attempt towards truly data-driven, almost plant-independent adaptive control algorithms [10].

1.2 Data-driven Control System Design Approach

To circumvent the necessity of deriving offline or online plant models, an alternative approach to control system design is to use the plant data to directly design the controller. This is the *data-driven* approach, which appeared at the end of the 1990s. Many plants regularly generate and store huge amounts of operating data at specific instants of time. Such data encompass all the relevant plant information required for control, estimation, performance assessment, decision makings and fault diagnosis. This facilitates the design of data-driven control systems. The term data-driven was initially used in computer science and has entered the control system science literature in the past two decades. Although data-driven control was actually introduced in the first decades of the twentieth century (see Chapter 2), the approach was not called data-driven at that time. The data-driven control and data-based control concepts are differentiated in Ref. [11]. Also, Ref. [12] has elaborated on the difference between data-based and data-driven control. It is stated in Ref. [12] that '*data-driven control only refers to a closed loop control that starting point and destination are both data. Data-based control is then a more general term that controllers are designed without directly making use of parametric models, but based on knowledge of the plant input-output data. Sorted according to the relationship between the control strategy and the measurements, data based control can be summarized as four types: post-identification control, direct data-driven control, learning control, and observer integrated control.*'

The main features of the data-driven control approaches can be categorised as follows:

- Control system design and analysis employ only the measured plant input–output data. Such data are the controller design's starting point and end criteria for control system performance.
- No priori information and assumptions on the plant's dynamics or structure are required.
- The controller structure can be predetermined.
- The closed-loop stability, convergence and safe operation issues should be addressed in a data-driven context.
- A designer-specified cost function is minimised using the measured data to derive the controller parameters.

The structure of a data-driven control system design is shown in Figure 1.2.

Several definitions for data-driven control are proposed in the literature. The following definition from Ref. [11] is presented.

Definition 1.1 Data-driven control includes all control theories and methods in which the controller is designed by directly using online or offline input–output data of the controlled system or knowledge from the data processing but not any explicit information from a mathematical model of the controlled process and whose stability, convergence and robustness can be guaranteed by rigorous mathematical analysis under certain reasonable assumptions.

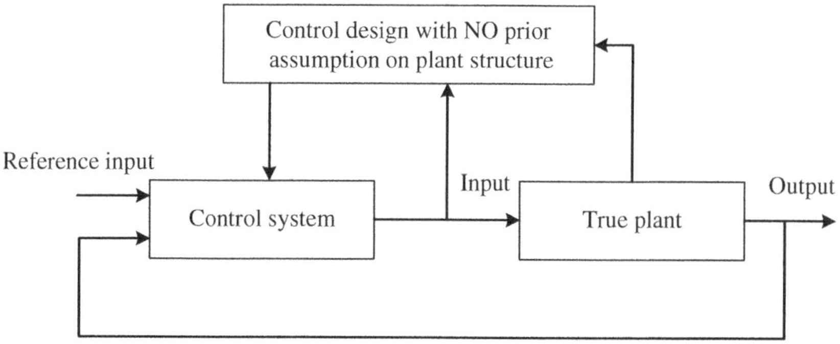

Figure 1.2 The structure of a data-driven control system design.

The three key points of this definition are the direct use of the measured input–output data, data modelling rather than first principles modelling or identified modelling, and the guarantee of the results by theoretical analysis.

1.2.1 The Designer Choice: Model-based or Data-driven Control?

In general, systems encountered in control systems design can be categorised as simple, complicated, complex and complex adaptive (see Chapter 2 for definitions and more details). In real-world applications, the controlled plants and all the conditions that they may confront in terms of models and assumptions can be categorised into the following classes:

Class 1: In this class, it is possible to derive accurate mathematical models from the first principles or the identification-based schemes, and it can be anticipated that the theoretically indispensable plant assumptions hold. This class includes simple plants and certain well-modelled complicated systems.

Class 2: In this class, for some even simple plants, many complicated systems, and a few complex systems, models derived from the first principles or the identification-based schemes are crudely accurate, but uncertainties can be used to compensate for the modelling error with known bounds, and it can be anticipated that the theoretically indispensable plant assumptions hold.

Class 3: In this class, conditions are similar to those of class 2, but with the difference that the theoretically indispensable plant assumptions may not be guaranteed to hold.

Class 4: In this class, for some complicated systems and most complex systems, models derived from the first principles or the identification-based schemes models are crudely accurate, and the uncertainties used to describe the modelling errors are difficult to obtain accurately, and it can be anticipated that the theoretically indispensable plant assumptions may not hold.

Class 5: In this class, for a few complicated, some complex, and complex adaptive systems, derivation of models from the first principles or the identification-based schemes, and reliable uncertainty descriptions are difficult or practically unavailable, and it can be anticipated that the theoretically indispensable plant assumptions do not hold.

The plants falling into the class 1 category have been successfully controlled by the well-established and well-documented model-based control strategies from the classical and state-space schools of thought. For the plants falling into the class 2 category, both adaptive and robust control schools are well developed and have been successfully implemented in practice. Although there are still many open problems in the adaptive and robust control approaches to reliably control all such plants, solutions are conceivable in the future with the present theoretical tools or some extensions and modifications. Adaptive and robust control methodologies must be selected with much hesitation for the control of the real-world plants falling into the class 3 category. In such cases, a data-driven approach would be the recommended choice. For the plants falling into the class 4 category, the data-driven approach is the strongly recommended choice. Although some of the present adaptive and robust control techniques may be employed in a few class 4 category plants, their application is difficult, time-consuming and with no guaranteed safety and reliability. In the case of plants falling into the class 5 category, data-driven control is the sole choice. Many of the future real-world plants are going in this direction [13], and the control scientist's community must be equipped with a well-established and strong sufficient theoretical background of data-driven control theory to handle these control problems. The final point to note is that practical controllers should not be too complex, difficult or non-economical to use.

To summarise, the main characteristics of data-driven control systems that make them appealing to selection by a designer are given as follows:

- In the data-driven control approaches, the design methodologies do not explicitly include any parts or the whole of the plant model or are not restricted by the assumptions following the traditional modelling processes. Hence, they are basically *model-free* designs.
- The stability and convergence derivations of the data-driven approach do not depend on the model and uncertainty modelling accuracy.
- In the data-driven control framework, the inherently born concepts of un-modelled dynamics and robustness in the model-based control methodologies are non-applicable.

1.2.2 Technical Remarks on the Data-Driven Control Methodologies

The following remarks are important to clarify certain ambiguities and concepts in the current data-driven control literature:

Remark 1: In the control literature, the control design techniques that *implicitly* utilise the plant model, such as the direct adaptive control and the subspace-identification-based predictive control methods, are sometimes categorised as data-driven control. However, their controller design, stability and convergence analysis are fundamentally model-based and also require fulfilling strong assumptions on different model characteristics such as the model order, relative degree, time delay, noise, uncertainty characteristics and bounds. Hence, this book categorises such techniques as model-based rather than data-driven.

Remark 2: In dealing with mathematical models, issues such as nonlinearity, time-varying parameters and time-varying model structures cause serious limitations and require complex theoretical handling. However, such issues at the input–output data level are non-existent. In fact, a truly data-driven control approach should be able to deal with the above control problems.

Remark 3: The concepts of robustness and persistency of excitation that appear in adaptive and robust control methodologies are general notions that must also be dealt with in the data-driven control approach. However, new definitions and frameworks are necessary to pursue these concepts in the data-driven control context.

Remark 4: The theory–practice gap in the model-based approaches is greatly alleviated in the data-driven approach as the implementations are directly field-based.

Remark 5: A very rich literature on the mathematical system theory and immensely valuable experiences in the implementation of model-based control techniques is available. It would not be desirable or wise to ignore such valuable information. The fact is that plant models can play a vital role in the design of control systems. One aspect is the application of model-based controller design techniques, if possible. The other aspect would be the cooperation of data-driven control with other control theories and methods. The relationship between data-driven and model-based control should be complementary, and data-driven approaches can learn and benefit from the established model-based concepts. The effective employment of existing accurate information about the plant by the data-driven approach is an open problem for further research.

Remark 6: Data-driven control is predicted to be the dominant paradigm of control design science, complementing and substituting the present model-based paradigm.

1.3 Data-Driven Control Schemes

In this section, six different data-driven control schemes are briefly introduced. A classification and a brief survey on the available data-driven approaches are given in Ref. [11].

1.3.1 Unfalsified Adaptive Control

Unfalsified control was proposed by Safonov in 1995 [14]. The underlying philosophy of unfalsified control is Popper's falsification theory proposed for the demarcation problem in the philosophy of science (see Chapter 2).

Unfalsified control is a data-driven control theory that utilises physical data to learn or select the appropriate controller via a falsification or elimination process. In the unfalsified feedback control configuration, the goal is to determine a control law C for plant P such that the closed-loop system T satisfies the desired specifications, where the plant is either unknown or only partially known, and the input–output data are utilised in selecting the control law C. In the unfalsified control, the control system *learns* when new input–output information enables it to eliminate the candidate controllers from the control bank. The three elements that form the unfalsified control problem are as follows:

- Plant input–output data.
- The bank of candidate controllers.
- Desired closed-loop performance specification denoted by T_{spec} consisting of the 3-tuples of the reference input, output and input signals (r, y, u).

Definition 1.2 [15] A controller C is said to be *falsified* by measurement information if this information is sufficient to deduce that the performance specification $(r, y, u) \in T_{\text{spec}}\ \forall r \in \mathbb{R}$ would be violated if that controller were in the feedback loop. Otherwise, the control law C is said to be *unfalsified*.

Figure 1.3 shows the general structure of the closed-loop unfalsified control system. The inputs to the falsification logic and algorithm are the plant input–output data, the set of candidate controllers in a control bank or set, and the desired closed-loop performance. The controllers are verified using a falsification logic and algorithm with performance goals and physical data as its inputs. Note that no

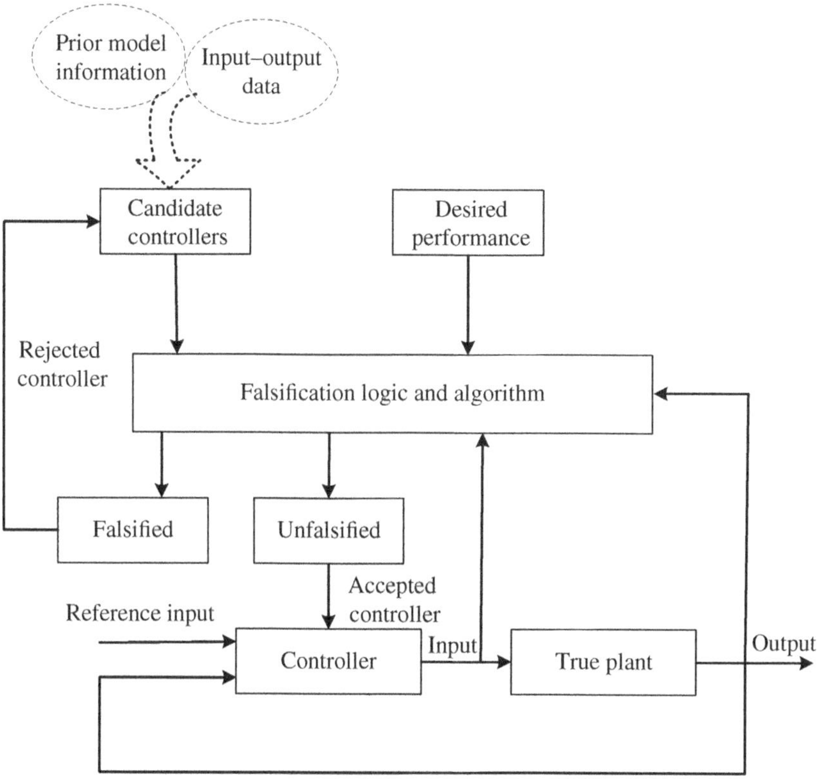

Figure 1.3 The unfalsified control structure.

plant models are required in the verification process. However, in the design of the controller members of the set of candidate controllers, a plant model can be beneficial. This is, of course, before the selection of the candidate controllers.

The selection process recursively falsifies candidate controllers that fail to satisfy the desired performance specification. The whole process is designed by the input–output data rather than the mathematical model of controlled plants. The candidate controllers set can be generated utilising priori model information or can be generated or updated based on the plant input–output data.

1.3.1.1 Unfalsified Control: Selected Applications

Model-based control techniques are the dominant control design methodologies utilised in marine systems. However, as stated in Ref. [16], the emerging field of smart sensors has provided a unique opportunity to have access to online plant measurements at a low price. Data-driven control algorithms are applied to marine systems by presenting the results of the application of unfalsified control to the problem of dynamic positioning of marine vessels subjected to environmental forces in Ref. [16]. Also, in Ref. [17], robust switching controllers are developed by applying the unfalsified control concept to the autopilot design of extremely manoeuvrable missiles. It is claimed that this can overcome challenging problems of highly nonlinear dynamics, wide variations in plant parameters and strong performance requirements in missiles.

Controlling the flight of micro-aerial vehicles is a highly challenging task due to the inherent nonlinearities and highly varying longitudinal and lateral derivatives. In Ref. [18], the application of unfalsified control theory is presented for such systems. Also, the unfalsified control concept is employed for the automatic adaptation of linear single-loop controllers and is successfully implemented in a continuous non-minimum phase stirred tank reactor model with the Van de Vusse reaction scheme in Ref. [19]. Finally, in Ref. [20], an application of the unfalsified control theory to the design of an adaptive switching controller for a nonlinear robot manipulator is presented, where the manipulator has many factors that are not accurately characterised in a model, such as link flexibility and the effects of actuator dynamics, saturation, friction and mechanical backlash.

1.3.2 Virtual Reference Feedback Tuning

Virtual Reference Feedback Tuning (VRFT) is a general methodology for the controller design of a plant with no available mathematical models. It is a non-iterative or one-shot direct method that minimises the control cost in the 2-norm type by utilising a batch of input–output data collected from the plant, where the optimisation variables are the controller parameters.

VRFT formulates a controller tuning problem as a controller parameter identification problem by introducing a virtual reference signal. It is therefore called a *data-based* controller design method in Ref. [21], indicating it is an adaptive method where the adaptation is performed offline. In VRFT, a controller class is selected, and a specific controller is chosen based on the collected data. The selection is performed offline. Initially, a batch of input–output data is collected from the plant, and utilising this data and the virtual reference concept, a controller is selected. The selected controller is then placed in the loop without any further adaptation. Note that the designed controller is verified for stability and performance requirements before its placement in the loop.

To further elaborate on the one-shot direct data-based nature of VRFT, it returns a controller without requiring iterations and/or further access to the plant for experiments after the batch input–output plant data collection. This is possible because the *design engine* inside VRFT is intrinsically global, and no gradient-descent techniques are involved. The input signals must be exciting enough, and in the case of poor excitation, the selected controller can be inappropriate with non-satisfactory closed-loop responses. The results obtained by VRFT are related to the information content present in the given batch of input–output data. The one-shot feature of VRFT is practically attractive because [21]:

- It is low-demanding, i.e. access to the plant for multiple experiments is not necessary, and therefore the normal operation of the plant is not halted.
- It does not suffer from local minima and initialisation problems.

The virtual reference concept is fundamental to VRFT; its basic idea is presented in Ref. [22]. Let the controller $C(z;\theta)$ be such that the closed-loop system transfer function is as a desired reference model $M(z)$. Consider any reference signal $r(t)$, the model reference output will be $M(z)r(t)$. Hence, a necessary condition for the closed-loop system to have the same transfer function as the reference model is that the output of the two systems is the same for a given $\bar{r}(t)$. In typical model reference controllers, such as the model reference adaptive control, for a given reference input and reference model, the controller $C(z;\theta)$ is designed to satisfy the above condition. However, in the model reference designs, it is assumed that the plant model is available or its structure is known with certain given assumptions.

The basic idea of the virtual reference approach is an intelligent selection of $\bar{r}(t)$ so that the determination of the controller becomes plausible. Suppose only the input–output data of the plant are given, the plant transfer function $P(z)$ is not available. Given the measured output $y(t)$, consider a reference input $\bar{r}(t)$ such that $M(z)\bar{r}(t) = y(t)$. $\bar{r}(t)$ is called a *virtual reference input* because it is not, in reality, implemented to generate $y(t)$. Notice that $y(t)$ is the desired output of the closed-loop system when the reference signal is $\bar{r}(t)$. Then, the corresponding tracking error is $e(t) = \bar{r}(t) - y(t)$, and it is known that when $P(z)$, that is truly unknown, has $u(t)$ as the input signal, its output is $y(t)$. Therefore, the desired controller to be designed must generate $u(t)$ when fed by $e(t)$. Since both signals $u(t)$ and $e(t)$ are known, and they are the output and input of the controller, respectively, the design task reduces to the identification problem of a dynamical system describing the $e(t)$ and $u(t)$ relationship. VRFT is extended to nonlinear plants in Ref. [21], and it is shown that the one-shot approach must be replaced with a multi-pass, iterative procedure. Also, other extensions, such as multivariable plants, are cited in Ref. [11]. Finally, the general structure of a VRFT is shown in Figure 1.4. A general model reference scheme is depicted in Figure 1.4a, where it is desired for the designed controller $C(z;\theta)$ to achieve a closed-loop performance close to the reference model $M(z)$. That is, the responses of the closed-loop system and the reference model to the reference input $r(t)$ be similar. Figure 1.4b depicts the VRFT design scheme. Note that by defining the virtual reference input, the responses of the reference model and plant for this

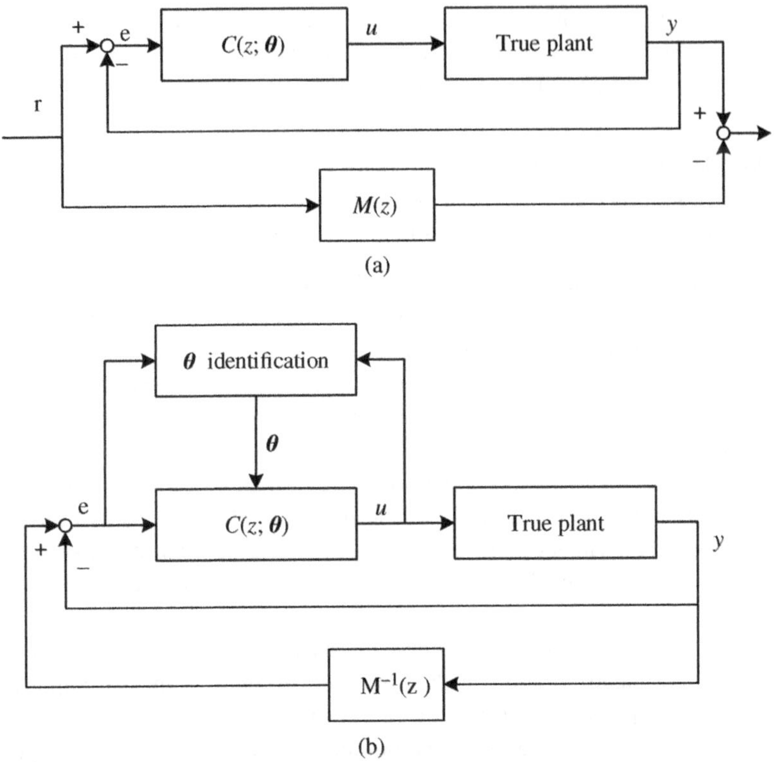

Figure 1.4 The VRFT general structure. (a) Model reference scheme of VRFT, (b) schematic diagram of VRFT.

input would be similar. A controller that gives the plant input for the given error is identified as shown in Figure 1.4b.

1.3.2.1 VRFT: Selected Applications

VRFT was utilised for the designing of closed-loop controllers for functional electrical stimulation (FES)-supported systems in Ref. [23], where feedback controllers for knee joint movement of people with paraplegia using FES of the paralyzed quadriceps muscle group are designed using VRFT. Open-loop measured data are collected, and the controller is identified in such a way that the closed loop meets the desired model reference. This is extended for FES-supported standing up and sitting down in Ref. [24], which is a difficult task due to the lack of stability and its nonlinear dynamic behaviour. The control strategy is based on two control loops, where the inner loop is a linear stabilising PID controller and the outer loop is designed using the VRFT strategy for controller tuning.

A switching control system for a vertical-type one-link arm using VRFT is reported in Ref. [24], and VRFT is applied to self-balancing manual manipulators that countervail the weight of a load that must be manually handled and moved by a human operator in Ref. [25].

1.3.3 Simultaneous Perturbation Stochastic Approximation

Simultaneous perturbation stochastic approximation (SPSA) was introduced in Ref. [26] as a stochastic perturbation (SA) algorithm that is based on a simultaneous perturbation gradient approximation. SPSA has recently attracted considerable attention in data-driven control designs. The essential feature of SPSA that makes it attractive for data-driven control is the underlying gradient approximation that requires only input–output measurements of the plant. This is an important feature as, in many practical situations, it may not be possible to obtain reliable knowledge of the plant input–output relationships. The gradient-based optimisation algorithms that are widely employed in the model-based control theory may be infeasible if no plant model is available, or its control results may be unreliable if the plant model is not accurate enough [27].

Consider a general nonlinear control problem where the plant model is not available, and the applied controller has a fixed structure, such as a PID or a neural network controller. A cost function $J(\boldsymbol{\theta})$ is defined that entails weighted square error of the outputs and the plant inputs, where the vector $\boldsymbol{\theta}$ contains the controller parameters. The design problem is, therefore, the minimisation of $J(\boldsymbol{\theta})$ with respect to $\boldsymbol{\theta}$. SPSA is utilised for this optimisation problem since the classical techniques would require a plant model for the mathematical calculations. The general structure of an SPSA-based data-driven control is shown in Figure 1.5.

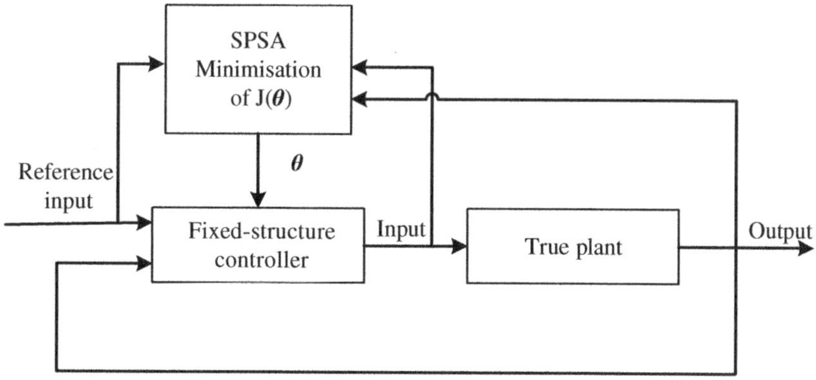

Figure 1.5 A SPSA-based data-driven control system.

1.3.3.1 SPSA: Selected Applications

A list of applications of SPSA can be found in Ref. [28]. The general areas for the application of SPSA, as stated in Ref. [28], include statistical parameter estimation, simulation-based optimisation, pattern recognition, nonlinear regression, signal processing, neural network training, adaptive feedback control and experimental design. The specific control system application includes aircraft modelling and control, process control and robot control [28]. Also, in Ref. [29], a data-driven control algorithm for a variable-speed wind turbine in the partial load region is designed and optimised online to capture the maximum energy from the wind. The online optimisation is based on the SPSA methodology.

1.3.4 The Willems' Fundamental Lemma

In the model-based control systems analysis and design techniques, a dynamical system is associated with a mathematical input–output or state-space model. However, the *behavioural approach* to system theory proposed by Willems in the early 1980s provided a system theoretic platform that separated the plant from its many representations by defining it as a set of trajectories. The behavioural approach has been recently employed for data-driven analysis, signal processing and control. A review of these ideas and techniques and some of the methods that originated from them are presented in Ref. [30]. These techniques are unsupervised; that is, they directly utilise measured plant data without any further processing. They also do not involve a parametric transfer function or state-space-type representation of the plant and are therefore classified as non-parametric methods.

System identification is a practical solution for obtaining plant models from a given set of measured input–output data. For a given trajectory of a linear time invariant (LTI) plant, it is possible to derive a representation of that plant. In theory, as the data are an exact trajectory of the plant, it can be recovered exactly. The key objective would therefore be to derive a desirable representation of the plant by employing a sufficiently rich trajectory. This objective is achieved through *Willems Fundamental Lemma* or *Fundamental Lemma* in short. The Fundamental Lemma provides a framework to identify the data-generating plant from the measured data and was initially presented in Ref. [31].

The Fundamental Lemma was overlooked by the control systems community for about 13 years since Willems first introduced it. The approach was considered a purely theoretical system identification result that provided identifiability conditions for deterministic linear time-invariant systems. However, with the increasing interest in data-driven control methodologies and the importance of non-parametric representations in data-driven control design problems, the Fundamental Lemma has become an effective tool in many data-driven control methodologies [30].

Consider an unknown deterministic linear time-invariant system with a controlled input vector $\mathbf{u}_k \in \mathbb{R}^m$ and a measured output vector is given by $\mathbf{y}_k \in \mathbb{R}^p$. Let the sequence of inputs and the corresponding system outputs for $k = 0, 1, \ldots, T-1$ be denoted by $\mathbf{u}_{[0,\,T-1]} := [\mathbf{u}_0{}^T\ \mathbf{u}_1{}^T \ldots \mathbf{u}_{T-1}{}^T]^T$ and $\mathbf{y}_{[0,\,T-1]} := [\mathbf{y}_0{}^T\ \mathbf{y}_1{}^T \cdots \mathbf{y}_{T-1}{}^T]^T$, respectively. The input–output is restricted to the interval $[0, T-1]$, and this provides an input–output *trajectory* of the unknown system for this interval. The system input sequences $\mathbf{u}_{[0,\,T-1]}$ is *persistently exciting of order L*, if the following Hankel matrix has full row rank:

$$\mathcal{H}_L(\mathbf{u}_{[0,T-1]}) = \begin{bmatrix} \mathbf{u}_0 & \mathbf{u}_1 & \cdots & \mathbf{u}_{T-L} \\ \mathbf{u}_1 & \mathbf{u}_2 & \cdots & \mathbf{u}_{T-L+1} \\ \vdots & \vdots & & \vdots \\ \mathbf{u}_{L-1} & \mathbf{u}_L & \cdots & \mathbf{u}_{T-1} \end{bmatrix}$$

Also, the Hankel matrix of input and output sequences is defined as follows:

$$\begin{bmatrix} \mathcal{H}_L(\mathbf{u}_{[0,T-1]}) \\ \mathcal{H}_L(\mathbf{y}_{[0,T-1]}) \end{bmatrix} = \begin{bmatrix} \mathbf{u}_0 & \mathbf{u}_1 & \cdots & \mathbf{u}_{T-L} \\ \vdots & \vdots & & \vdots \\ \mathbf{u}_{L-1} & \mathbf{u}_L & \ldots & \mathbf{u}_{T-1} \\ \mathbf{y}_0 & \mathbf{y}_1 & \cdots & \mathbf{y}_{T-L} \\ \vdots & \vdots & & \vdots \\ \mathbf{y}_{L-1} & \mathbf{y}_L & \cdots & \mathbf{y}_{T-1} \end{bmatrix}$$

where $L \geq 1$ and the columns of the above matrix are the measured inputs–outputs of the system. The system's linearity implies that every linear combination of the columns of this matrix, that is, $\begin{bmatrix} \mathcal{H}_L(\mathbf{u}_{[0,T-1]}) \\ \mathcal{H}_L(\mathbf{y}_{[0,T-1]}) \end{bmatrix} \mathbf{g}$ for any real vector $\mathbf{g} \in \mathbb{R}^{T-L+1}$, is also an input–output trajectory of the system under control. The Fundamental Lemma states that for a persistently exciting input $\mathbf{u}_{[0,\,T-1]}$, any other input–output plant trajectory of length L such as $\left[\bar{\mathbf{u}}_{[0,L-1]}^T\ \bar{\mathbf{y}}_{[0,L-1]}^T\right]^T$ can be expressed in terms of $\left[\mathbf{u}_{[0,L-1]}^T\ \mathbf{y}_{[0,L-1]}^T\right]^T$ for an appropriate vector $\mathbf{g}$, provided that the system is controllable. And, any linear combination $\begin{bmatrix} \mathcal{H}_L(\mathbf{u}_{[0,T-1]}) \\ \mathcal{H}_L(\mathbf{y}_{[0,T-1]}) \end{bmatrix} \mathbf{g}$ for any real vector $\mathbf{g} \in \mathbb{R}^{T-L+1}$ is also a trajectory of the unknown system.

The Fundamental Lemma provides a platform for an open-loop and closed-loop data-based representation of systems that can be used in data-driven control strategies. In the closed-loop identification results, controller parameter vectors are parameterised using input–output data.

The design of data-driven control methodologies based on the Fundamental Lemma can be summarised in the following general steps:

- Sufficiently rich input–output data are collected from the unknown plant.

- State estimation, input–output predictions, or necessary algebraic equations (whichever is applicable) are derived or solved by utilising non-parametric representations based on the Fundamental Lemma.
- Control signals are produced by solving the required equations or an optimisation problem (whichever is applicable).

1.3.4.1 Fundamental Lemma: Selected Applications

The Fundamental Lemma has been used in developing algorithms for system identification, data-driven simulation and data-driven control, which are well-reviewed in Ref. [30]. The design of data-driven state feedback, robust control, and predictive control based on the Fundamental Lemma is presented in more detail in Chapter 7. The methods were initially formulated for deterministic, linear and time-invariant systems. They were further extended for application in stochastic, nonlinear and time-varying systems through robustification and adaptation of the optimisation methods. The application of the fundamental lemma for the control of general unknown nonlinear systems in the context of the Koopman operator is discussed in Chapter 8. The most notable data-driven control design based on the Fundamental Lemma that has also been practically applied to real systems is the **D**ata-**e**nabl**e**d **P**redictive **C**ontrol or DeePC (see Chapter 7). There are reports on the practical and experimental application of DeePC in different engineering fields. These applications are summarised as follows [30]:

- In the process control field, DeePC is applied to a nonlinear laboratory four-tank process [32].
- In power systems, DeePC is applied for power system oscillation damping in a large-scale numerical case study, which R&D groups have successfully replicated on industrial simulators [33]. Also, DeePC has been successfully experimentally implemented on grid-connected power converters [34] and synchronous motor drives [35].
- In the robotics and mechatronics field, [36] used an aerial robotics simulation case study via the DeePC method. Reference [37] experimentally implemented this case study and, for the first time, demonstrated the performance, robustness and real-time implementability of DeePC. DeePC has also been experimentally implemented to control a 12-ton autonomous walking excavator [38].

1.3.5 Data-Driven Control System Design Based on Koopman Theory

Linear control system analysis and design methodologies are widely studied and employed in many practical control problems. They are applied to nonlinear systems mainly by linearising the nonlinear system at its operating points. However, the Koopman and Perron Frobenius operators have provided a fundamentally different approach to nonlinear dynamical systems by providing linear representations for even strongly nonlinear dynamics. Although the outlook of these approaches was promising, their application in control design was not taken seriously due to their infinite-dimensional nature, making them difficult to work with in practice. This has changed for Koopman operators by obtaining low-dimensional matrix approximations of the Koopman operators through dynamic mode decomposition techniques. Koopman operators are now considered for the practical design of data-driven control schemes of unknown nonlinear plants due to the increasing availability of data and the effectiveness of numerical methods for handling Koopman operators. In principle, the Koopman operator provides a linear embedding of nonlinear dynamics and enables the application of linear control methods to the nonlinear system [39]. Koopman has recently gained attention with its application in data-driven control [40]. The data-driven application is facilitated by the application of Willems' Fundamental Lemma, and a class of data-driven control methods are developed for unknown systems.

Let a general discrete-time nonlinear system be described as $\mathbf{x}_{k+1} = \mathbf{F}(\mathbf{x}_k)$, where $\mathbf{x}_k$ is the system state and $\mathbf{F}$ describes the nonlinear dynamics. A new vector of coordinates $\mathbf{z} = \boldsymbol{\varphi}(\mathbf{x})$ can be defined such that the nonlinear system is described in the new coordinates as the linear representation $\mathbf{z}_{k+1} = \boldsymbol{K}\mathbf{z}_k$, and the linear dynamics is entirely determined by the matrix $\boldsymbol{K}$. Also, Let $\psi : \boldsymbol{X} \rightarrow \mathbb{R}$ belong to the set of *measurement* or *output functions,* also called an *observable function* in Koopman theory. The discrete-time infinite-dimensional Koopman operator $\mathcal{K}_F$ associated with the nonlinear system for an *observable* function $\psi(\mathbf{x}_k)$ is defined through the following composition that advances the measurement functions $\mathcal{K}_F\psi(\mathbf{x}_k) := \psi(\mathbf{F}(\mathbf{x}_k)) = \psi(\mathbf{x}_{k+1})$; it is shown that the Koopman operator is a linear operator. The linearity property makes the Koopman operator attractive for control systems design of nonlinear systems.

The Koopman operator results in a linear dynamical system described by $\psi_{k+1} = \mathcal{K}_F\psi_k$ which is analogous to the original nonlinear system, except that the resulting Koopman-based dynamical representation is infinite-dimensional. However, the infinite dimensionality makes the Koopman linear model impractical for engineering and, in particular, control systems design purposes. A practical strategy to overcome this problem is to identify key measurement functions that linearly evolve with the system's dynamics. The so-called Koopmans's eigenfunctions φ of $\mathcal{K}_F$ and the corresponding eigenvalue λ are practical measurements that behave linearly in time and are suited for further control design purposes.

The Koopman operator is infinite-dimensional, so there will be an infinite number of Koopman eigenfunctions and eigenvalues for a system. However, to construct the Koopman finite-dimensional linear state-space model, only a subset of the eigenfunctions φ_1 to φ_p with a finite p is sufficient. Characterising the Koopman operator and approximating its eigenvalues and eigenfunctions, or spectral decomposition from measurement data, is a significant research interest in the Koopman theory for control systems design.

A p-dimensional linear subspace of the space of real-valued observables spanned by the Koopman eigenfunctions is $\{\varphi_1(\mathbf{x})\ \varphi_2(\mathbf{x})\ \dots\ \varphi_p(\mathbf{x})\}$, where $p \gg n$ are employed to construct a finite-dimensional approximation of the Koopman operator $\mathcal{K}_F$ be denoted by $\mathcal{K}_{F_p}$. Then, it can be shown that there exists a matrix representation of $\mathcal{K}_{F_p}$, called the *Koopman Matrix* of the system, $\boldsymbol{K}$, which corresponds to the action of the Koopman operator on observables in the considered finite-dimensional subspace. The Koopman *lifted state* is defined as follows:

$$\mathbf{z} = \Phi(\mathbf{x}) = \begin{bmatrix} \varphi_1^T(\mathbf{x}) \\ \vdots \\ \varphi_{n_z}^T(\mathbf{x}) \end{bmatrix}$$

and the approximated Koopman *lifted dynamics* is $\mathbf{z}_{k+1} = \mathbf{A}\mathbf{z}_k$, where it can be shown that $\mathbf{A} = \boldsymbol{K}^T$. The l system outputs $\mathbf{y}_k$ in the control design applications are equivalent to the observables for some matrix $\mathbf{C} \in \mathbb{R}^{l \times p}$, as $\mathbf{y}_k = \mathbf{C}\mathbf{z}_k$. The general model-based cycle of the Koopman-based linear controller design is shown in Figure 1.6.

The Koopman finite-dimensional linear model is an approximated model. The computation of finite-dimensional Koopman linear state-space model matrices

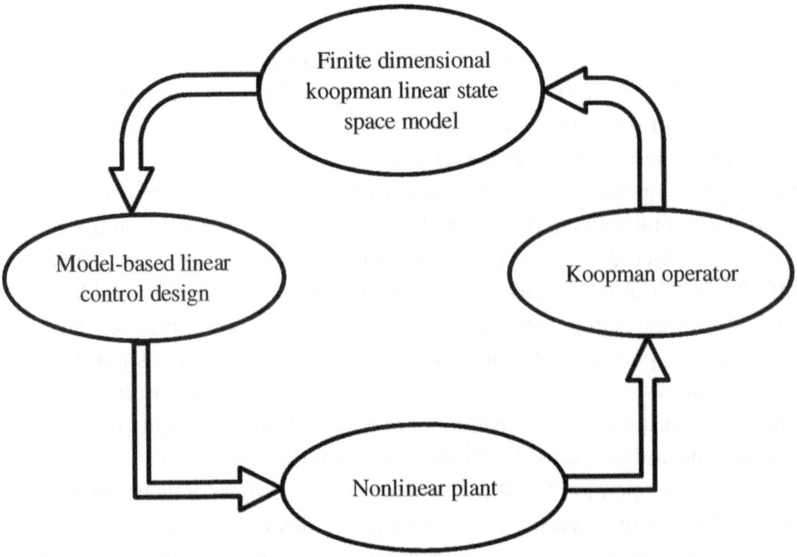

Figure 1.6 The model-based cycle of the Koopman-based linear controller design for general nonlinear plants.

A, **B** and **D** for a nonlinear system is not unique. In the data-driven approach, without identifying a Koopman linear model, it is possible to directly design a data-driven Koopman-based predictive controller from a finite-length data set of a nonlinear system. Willems lemma provides a practical solution to this problem. Indeed, all other sequences of the Koopman state variable vectors of the nonlinear system can be obtained using an available sequence of the Koopman state variable vector with a pre-selected structure of the Koopman eigenfunctions.

In the Koopman model-based control design, the plant's linear state space is directly derived from the Koopman operator representation of the original nonlinear system, and in Koopman data-driven control, the underlying model is derived from the Fundamental Lemma that is directly written from the measured plant data set. Figure 1.7 shows the general data-driven cycle of the Koopman-based linear controller design.

1.3.5.1 Koopman-based Design: Selected Applications

Koopman-based controllers have been applied in a variety of applications [39]. The applications in Ref. [39] include active learning and experimental applications where the environment is unknown or changing; Koopman-based model predictive control (Koopman-MPC) is applied to a nanometric positioning control problem; a methodology for heat transfer modelling based on Koopman mode decomposition is presented that estimates the parameters of the advection and diffusion equations, such as the effective velocity and diffusivity, and is used with real data in the case of a building with an atrium; the data-driven analysis of voltage dynamics in power grids using extended dynamic mode decomposition with delay embedding, the full-state dynamics of a power grid using voltage–amplitude measurements of a single bus is reconstructed; performance of nonlinear consensus

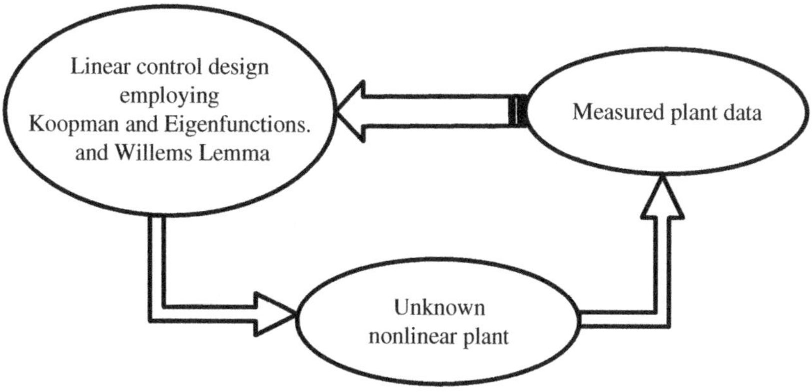

Figure 1.7 The data-driven design cycle of the Koopman-based linear controller design for general nonlinear plants.

networks is estimated based on Koopman mode decomposition and interpolation methods, and the performance estimate of nonlinear consensus networks is expressed in terms of Koopman modes and eigenvalues.

In Ref. [41], a deep MPC based on the Koopman operator is applied to control the heat transfer fluid temperature of a distributed-parameter model of the ACUREX solar collector field. It is shown that a linear equivalent of the highly nonlinear model of ACUREX is obtained and successfully used for the MPC design of the solar collector field. A data-driven Koopman predictive control with robust stability proof is applied to the ACUREX solar collector field comprising 480 single-axis parabolic trough collectors arranged in 10 parallel loops, with 48 collectors in each loop [42].

Koopman's theory is utilised and experimentally validated in the modelling and control of a tail-actuated robotic fish [43]. A data-driven approach for constructing a linear representation in terms of higher-order derivatives of the underlying nonlinear dynamics is presented, and the nonlinear system is then controlled with a linear quadratic regulator (LQR) feedback scheme.

Fluidised bed spray granulation (FBSG) processes allow for the generation of dust-less and free-flowing particles of desired dimensions and are, therefore, widely used in chemical, pharmaceutical and food industries [44]. However, the process dynamics are often unknown, and in Ref. [44], a data-driven Koopman-based control system is proposed that relies only on input–output measurements. The linearisation achieved by the Koopman embedding allows for linear-quadratic control design, which does not depend on the operating point.

In Ref. [45], a data-driven Koopman model-based predictive control for automatic train operation systems is proposed. Automatic train operation systems of high-speed trains are critical to guarantee operational safety, comfort and parking accuracy. Due to the train's uncertain dynamics and actuator saturation, model-based design techniques confront difficulties. In Ref. [45], an explicit linear Koopman model is employed to reflect the complex dynamics of a train, and a linear MPC is designed to satisfy the comfort and actuator constraints. Also, an online update strategy for the Koopman model of the train and closed-loop stability proofs is presented.

As a final example of the Koopman theory in control systems design applications, the decentralised control of interconnected systems based only on local and interconnection time series in a Koopman operator framework is considered [46]. Interconnected systems with an increasing presence in many areas of science and technology are complex adaptive systems that are intrinsically difficult to control. Interconnected systems are present, from robotics to power grids and social systems. There are often several subsystems that can be highly interconnected, and handling the interconnections in the modelling and control of such systems is a difficult or even impossible task in complex situations. Koopman-based lifting techniques are developed to control coupled nonlinear systems in Ref. [46], employing a decentralised control design strategy for interconnected systems based on local and interconnection data time series. Initially, a local control-oriented model is developed based only on the measured data, then an optimisation problem associated with the predictive control design is formulated, and finally, an identification procedure for systems, which are coupled through their inputs, is proposed

1.3.6 Model-free Adaptive Control

Model-free adaptive control (MFAC) is a data-driven control method initially proposed in 1994 for a class of general discrete-time nonlinear systems [11]. Contrary to the classical adaptive control techniques that identify a linear or nonlinear model of the plant, a *virtual equivalent dynamical linearisation data model* is derived at each closed-loop operating point.

There are several classical linearisation techniques, all of which require plant models; some are complex with a very high computational burden. For example, the feedback linearisation technique requires an accurate mathematical model of the controlled plant, and Taylor's linearisation approach uses Taylor expansion around the operating point and neglects higher-order terms to obtain a linearised model. Its accuracy can be (practically slightly) improved in the piecewise linearisation extension. The orthogonal function-approximation-based linearisation is based on a carefully selected set of orthogonal basis functions to approximate the nonlinear model of a controlled plant and has a heavy computational burden.

To overcome the problems of model assumption complexity, the dynamic linearisation technique is based on the concepts of pseudo-partial derivative (PPD), the pseudo-gradient (PG), or the pseudo-Jacobian matrix (PJM), using the plant input–output data. The various dynamic linearisation data models are the compact-form dynamic linearisation (CFDL) data model, the partial-form dynamic linearisation (PFDL) data model and the full-form dynamic linearisation (FFDL) data model.

The MFAC is formed using the equivalent virtual data model and a proposed controller, as shown in Figure 1.8. The resulting MFAC approach is a data-driven control method, and no a priori model information is required. Also, the MFAC does not require external testing signals or training processes, as in the case of neural adaptive controllers. Finally, under certain practical assumptions, the monotonic convergence and bounded-input bounded-output (BIBO) stability of the CFDL data-driven MFAC and PFDL data-driven MFAC can be guaranteed [47], and the BIBO stability of the FFDL data-driven MFAC is also proved in Ref. [48].

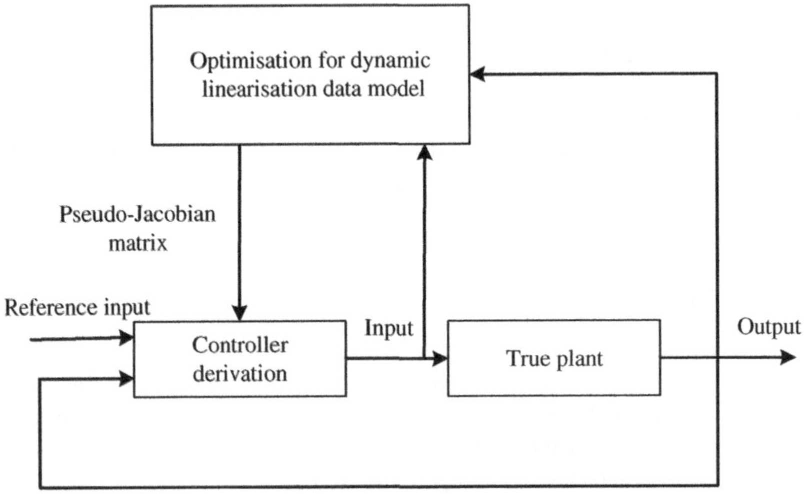

Figure 1.8 The model-free adaptive control.

To summarise the basic principles of the basic MFAC design, as shown in Figure 1.8, consider the general nonlinear multivariable true plant, under certain not restricted and plant-model free dependent assumptions, where it is shown that there exists a time-varying matrix $\mathbf{\Phi}(k)$, called the Pseudo-Jacobian Matrix (PJM), such that the unknown general system can be described by an equivalent CFDL data model as $\Delta\mathbf{y}(k+1) = \mathbf{\Phi}(k)\Delta\mathbf{u}(k)$, where $\Delta\mathbf{y}(k+1) = \mathbf{y}(k+1)-\mathbf{y}(k)$ and $\Delta\mathbf{u}(k) = \mathbf{u}(k)-\mathbf{u}(k-1)$. The PJM is a time-varying matrix dependent on the input–output data up to the current time, and all the system's complex characteristics are compacted in this matrix. The mathematical description of $\mathbf{\Phi}(k)$ is not analytically formulated in general, but its numerical behaviour is simple and can be approximated. Note that the existence of the PJM is proved theoretically.

The system is assumed output-controllable from a data viewpoint, and a general cost function $J(\mathbf{u}(k))$ in terms of the reference input and the plant input–output data is defined. Minimisation of $J(\mathbf{u}(k))$ utilising $\mathbf{\Phi}(k)$ gives the control signal.

1.3.6.1 MFAC: Selected Applications

In Ref. [49], it is asserted that, at the time of publication, more than 500 researchers worldwide have actively engaged in the field, directly employing the controller design and analysis methods of the MFAC theory. These researchers have successfully applied MFAC approaches to address a diverse range of control problems encountered in practical applications. It is stated that MFAC and its improved forms have been applied in over 150 different fields. In Ref. [49], the applications of prototype MFAC are reviewed and the typical applications considered are motion control systems, industrial control systems, power control systems and many other more specific applications. The interested reader is referred to Ref. [49] for further details.

As an example, a MFAC approach for density control of freeway traffic flow via ramp metering is presented in Refs. [50, 51]. Reference [51] presents a model-free periodic adaptive control approach by taking advantage of the periodicity feature of the freeway traffic flow. Ramp metering is an effective tool for freeway congestion control, where various model-based control methods have been studied to regulate the amount of traffic entering a freeway from entry ramps during certain time periods so that the flow on the freeway does not exceed its capacity. The freeway traffic flow system is nonlinear, with couplings and uncertainties, and an accurate model is hardly available in practice. It is shown in Refs. [50, 51] that the proposed MFAC approaches can provide acceptable results with the given input–output data of the freeway traffic system without requiring any other priori of the control system.

1.4 Outline of the Book

This book comprises nine chapters. We attempted to present some mainstream techniques in the data-driven control design school of thought. This is done by studying and presenting the fundamentals of the respective algorithms, simulation results and examples to clarify the main issues. Most chapters can be studied independently, and necessary prerequisites are specified in the following chapter descriptions. The detailed mathematical derivations are avoided, and the reader is referred to the available cited papers and related references. Problems are provided for Chapters 3–9 to further extend the knowledge of the particular methods. Chapter 1 provides an introduction to data-driven control methodologies. It briefly reviews the inherent characteristics of all the model-based control systems designs. The drawback and main flaw of the model-based controller designs are summarised as requiring an accurate plant model and the associated assumptions to be satisfied for its mathematical robustness, stability, convergence and performance proofs. It is concluded that an inevitable alternative and complement to the model-based approach is the data-driven control systems design methodology. The main features and the general structure of the data-driven control approaches are presented, and a definition of data-driven control is given. Based on a broad classification of systems as simple, complicated, complex and complex adaptive, guidelines for selecting between the model-based or the data-driven control from a control system designer's viewpoint are proposed. Several remarks are given to clarify certain ambiguities and concepts in the current data-driven control literature. The data-driven control techniques that are reviewed are the techniques discussed in the subsequent chapters. Also, several applications of the above techniques are presented.

Chapter 2 presents the philosophical perspectives of the paradigm shift in control systems design and attempts to show from a philosophical viewpoint that data-driven control is inevitable for the control systems design solution of the present and future complex systems. The historical–philosophical discussions can further enlighten the future direction of control system design methodologies and deepen our look into the different control design paradigms. Also, in this chapter, the systems encountered in engineering design are classified and defined into four simple, complicated, complex and complex adaptive classes. Although the author believes that a historical–philosophical perspective can tremendously help for a deeper understanding of the control design paradigm shifts, this chapter can be skipped without loss of continuity and affecting the remaining chapters.

Chapter 3 deals with the unfalsified adaptive switching supervisory control (UASC) approach. This approach has a strong philosophical background in the problem of demarcation in the philosophical views of Popper. Adaptive switching supervisory control is briefly reviewed, and its ability to cope with the incompetencies of classical adaptive control is highlighted. After introducing the basic definitions and concepts, it is shown that UASC as a performance-based supervisory control approach circumvents the need for model estimations and does not require the general model-based restrictive assumptions. Also, closed-loop stability theorems are given, and discussions on closed-loop performance are provided. The Dehghani-Anderson-Lanzon (DAL) phenomenon is introduced, and modifications to handle closed-loop performance degradations are presented. This chapter is a prerequisite for Chapter 4.

Chapter 4 presents the multi-model unfalsified adaptive switching control (MMUASC) methodology. Initially, the multi-model ASSC approach is presented that divides large plant uncertainty sets into smaller subsets and results in a finite number of identified models $M_1, M_2, \cdots, M_N$. Each model M_i has a substantially reduced uncertainty, and correspondingly, a controller C_i is designed to ensure closed-loop stability and desired performance. This greatly improves the closed-loop performance of highly uncertain plants. However, an appropriate switching criterion must be defined at each instant of time, and it is assumed that the exact match condition holds for the plant, which is quite challenging when the system has large uncertainties. Based on the results developed in Chapter 3, it is shown that UASC is employed to circumvent the need for the exact match condition. The combination of the two approaches gives the MMUASC, and it is shown that the benefits of both the multi-model adaptive control approach and the unfalsified control strategy are achievable. Various features of MMUASC are studied in this chapter, and in particular, closed-loop stability and the important concept of input constraints are handled. Finally, the popular predictive controllers are considered the underlying designed controllers, and the inverse optimal control approach is utilised for the virtual signal calculations.

Chapter 5 presents the virtual reference feedback tuning (VRFT) approach for the design of data-driven control systems. This chapter can be studied independently. It is shown that VRFT can be considered a one-shot direct adaptive control design methodology where adaptation is performed offline. In VRFT, a controller class is selected, and the specific controller is determined for closed-loop control by minimising a controlled cost of the 2-norm type by using a batch of input–output data collected from the plant. It is indicated that in VRFT, the optimisation variables are the controller parameters. The basic VRFT concept is initially presented. Then, its extensions are presented to deal with the model reference discrepancy issue in terms of a specific filter design, measurement noise with the introduction of instrument variables, and open-loop non-minimum phase zeros challenges. Multivariable plant extensions and optimal model reference selection are also treated. Finally, closed-loop stability with VRFT design is handled.

Chapter 6 presents the simultaneous perturbation stochastic approximation (SPSA)-based data-driven control design approach. SPSA is an optimisation technique that does not require direct gradient information from the plant, which is derived from the plant governing mathematical relations. It is discussed that SPSA iteratively searches for the optimal solution relying only on measurements of the objective function and not on measurements of the gradient of the objective function. The essential features of SPSA are presented, and it is shown that this technique can be readily applied to the tuning parameters of a given controller with the available measured input–output data from a general nonlinear plant. The main theoretical result of the SPSA algorithm, that is, the algorithm convergence is provided. The SPSA data-driven control scheme is given, and its application for the parameter tuning of PID controllers and MPC is discussed. Finally, data-driven control of under-actuated systems is presented as a difficult control design case study to show the effectiveness of the SPSA-based data-driven control of such systems. This chapter can be studied independently.

Chapter 7 presents the data-driven control system design schemes based on Willem's Fundamental Lemma. The Fundamental Lemma formulated in a behavioural system theory framework is discussed and certain conditions are derived under which the state trajectory of a state-space representation spans the whole state-space. In this chapter, a state-space systems approach is adopted for describing the principal issues and design algorithms. It is essentially shown that if the length T input trajectory of a controllable (in a behavioural setting) LTI system is persistently exciting of a sufficiently high order, then the length T measured input–output data spans all possible input–output solutions of a smaller length. Then, an alternative data-based open-loop and closed-loop representation of LTI systems that can replace the state-space and transfer function models are provided. This equivalent data-based representation is further employed for the design of data-driven control systems. The specific designs considered are the data-driven state-feedback stabilisation schemes both in deterministic and noisy environments, and the data-driven predictive control represented as a data-enabled predictive control or DeePC algorithm. Finally, a data-driven predictive control for nonlinear systems is presented. This chapter can be studied independently, and it is a prerequisite for the data-driven part of the next chapter.

Chapter 8 presents a class of data-driven controllers based on Koopman theory and the Fundamental Lemma. Initially, it is discussed that Koopman's theory has provided an effective tool for handling nonlinear dynamical systems by what is called lifting or embedding the nonlinear dynamics into a higher-dimensional space where its evolution is approximately linear. A Koopman-based platform is presented for linear analysis and design of strongly nonlinear dynamical systems via the infinite-dimensional linear Koopman operator. The basic concepts and definitions of Koopman's theory are presented without indulging in detailed mathematical concepts and derivations. To implement the Koopman operator for practical control systems design, it is shown that certain key measurement functions are derived that linearly evolve with the system's dynamics and capture its dominant dynamical behaviour. It is further shown that data-driven methods can be used to construct the Koopman required functions. Dynamic mode decomposition (DMD) and extended dynamic mode decomposition (EDMD) are the two primarily presented methods that use data-driven strategies to approximate the Koopman linear model for a nonlinear system. To apply the Koopman theory for data-driven control purposes, it is shown that Willems' Fundamental Lemma can be an effective tool. In this chapter, the application of the Koopman operator in Willems' Fundamental Lemma context is presented. Given that Willems' Fundamental Lemma applies to LTI systems, and the Koopman operator attempts to find a universal LTI model for a nonlinear system, the Koopman operator is therefore used to apply Willems' Fundamental Lemma to unknown nonlinear systems. In this respect, two lemmas are presented to provide a data-driven representation of nonlinear systems utilising Koopman and Willem's theories. Also, a data-driven Koopman predictive control strategy is presented, and its robust stability analysis is given. Finally, a parabolic solar collector field practical plant is considered for evaluating the presented theories. The Koopman theory part of this chapter can be studied independently, but for its application in data-driven control strategies, Chapter 7 is a prerequisite.

Chapter 9 presents the MFAC strategy which is an effective data-driven control systems approach. MFAC is theoretically founded on the so-called pseudo-partial-derivative (PPD) concept which is discussed in this chapter.

Common linearisation techniques such as Taylor's linearisation, piecewise linearisation and polynomial function approximation require a mathematical model of the plant. On the other hand, it is argued that a linearised model to be employed in an adaptive data-driven control application should have a simple structure, few adjustable parameters and be readily applicable to measured input–output data. It is shown that this objective is achieved by MFAC through different forms of dynamic linearisation models. These models capture the dynamical system's behaviour by the variations of the output signal resulting from the changes in the current input signal, towards a control-design-oriented model that cannot be used for other system analysis purposes. It is shown in this chapter that a virtual description of the closed-loop input–output data is provided that enables the derivation of completely data-driven controller structures in MFAC. The three dynamic linearisation techniques and their corresponding MFAC designs discussed are compact-form dynamic linearisation (CFDL), partial-form dynamic linearisation (PFDL) and full-form dynamic linearisation (FFDL). Extensions to multivariable control systems are also given. The presented results in this chapter are in the form of several theorems and corresponding algorithms for practical implementations. This chapter can be studied independently. However, a combined SPSA-MFAC strategy is also presented to widen the applicability of MFAC techniques. Chapter 6 would be a prerequisite for this section.

References

1 J. C. Maxwell, "I. On governors," *Proceedings of the Royal Society of London,* vol. 16, pp. 270–283, 1868.

2 N. Wiener, *Cybernetics or Control and Communication in the Animal and the Machine*, second edition. MIT Press, 1961.

3 C.-G. Kang, "Origin of stability analysis:\" On Governors\" by JC Maxwell [Historical Perspectives]," *IEEE Control Systems Magazine,* vol. 36, no. 5, pp. 77–88, 2016.

4 J. G. Ziegler and N. B. Nichols, "Optimum settings for automatic controllers," *Trans. ASME,* vol. 64, no. 11, pp. 759–765, 1942.

5 K. Åström and T. Hägglund, *PID Controllers: Theory, Design, and Tuning*, second edition. Research Triangle Park, NC, USA: Instrument Society of America 1995.

6 H. Elci, R. W. Longman, M. Phan, J.-N. Juang, and R. Ugoletti, "Discrete frequency based learning control for precision motion control," in *Proceedings of IEEE International Conference on Systems, Man and Cybernetics*, 1994, vol. 3: IEEE, pp. 2767–2773.

7 S. Formentin, K. Van Heusden, and A. Karimi, "Model-based and data-driven model-reference control: a comparative analysis," in *2013 European Control Conference (ECC)*, 2013: IEEE, pp. 1410–1415.

8 R. Skelton, "Model error concepts in control design," *International Journal of Control*, vol. 49, no. 5, pp. 1725–1753, 1989.

9 M. Gevers, "Modelling, identification and control," in *Iterative Identification and Control*: Springer, 2002, pp. 3–16.

10 M. Stefanovic and M. G. Safonov, *Safe Adaptive Control: Data-driven Stability Analysis and Robust Synthesis*. Springer, 2011.

11 Z.-S. Hou and Z. Wang, "From model-based control to data-driven control: survey, classification and perspective," *Information Sciences*, vol. 235, pp. 3–35, 2013.

12 J.-W. Huang and J.-W. Gao, "How could data integrate with control? A review on data-based control strategy," *International Journal of Dynamics and Control*, vol. 8, no. 4, pp. 1189–1199, 2020.

13 R. M. Murray, K. J. Astrom, S. P. Boyd, R. W. Brockett, and G. Stein, "Future directions in control in an information-rich world," *IEEE Control Systems Magazine*, vol. 23, no. 2, pp. 20–33, 2003.

14 M. G. Safonov and T.-C. Tsao, "The unfalsified control concept: a direct path from experiment to controller," in *Feedback Control, Nonlinear Systems, and Complexity*: Springer, 1995, pp. 196–214.

15 M. G. Safonov and T.-C. Tsao, "The unfalsified control concept and learning," *IEEE Transactions on Automatic Control*, vol. 42, no. 6, pp. 843–847, 1997.

16 V. Hassani, A. M. Pascoal, and T. F. Onstein, "Data-driven control in marine systems," *Annual Reviews in Control*, vol. 46, pp. 343–349, 2018.

17 P. B. Brugarolas, V. Fromion, and M. G. Safonov, "Robust switching missile autopilot," in *Proceedings of the 1998 American Control Conference. ACC (IEEE Cat. No. 98CH36207)*, 1998, vol. 6: IEEE, pp. 3665–3669.

18 A. Wolniakowski and A. Mystkowski, "Application of unfalsified control theory in controlling MAV," in *Solid State Phenomena*, 2013, vol. 198: Trans Tech Publications Ltd, pp. 171–175.

19 T. Wonghong and S. Engell, "Automatic controller tuning via unfalsified control," *Journal of Process Control*, vol. 22, no. 10, pp. 2008–2025, 2012.

20 T. C. Tsao and M. G. Safonov, "Unfalsified direct adaptive control of a two-link robot arm," *International Journal of Adaptive Control and Signal Processing*, vol. 15, no. 3, pp. 319–334, 2001.

21 M. C. Campi and S. M. Savaresi, "Direct nonlinear control design: the virtual reference feedback tuning (VRFT) approach," *IEEE Transactions on Automatic Control*, vol. 51, no. 1, pp. 14–27, 2006.

22 M. C. Campi, A. Lecchini, and S. M. Savaresi, "Virtual reference feedback tuning: a direct method for the design of feedback controllers," *Automatica,* vol. 38, no. 8, pp. 1337–1346, 2002.

23 F. Previdi, T. Schauer, S. M. Savaresi, and K. J. Hunt, "Data-driven control design for neuroprotheses: a virtual reference feedback tuning (VRFT) approach," *IEEE Transactions on Control Systems Technology,* vol. 12, no. 1, pp. 176–182, 2004.

24 F. Previdi, M. Ferrarin, S. M. Savaresi, and S. Bittanti, "Closed-loop control of FES supported standing up and sitting down using Virtual Reference Feedback Tuning," *Control Engineering Practice,* vol. 13, no. 9, pp. 1173–1182, 2005.

25 F. Previdi, F. Fico, D. Belloli, S. M. Savaresi, I. Pesenti, and C. Spelta, "Virtual Reference Feedback Tuning (VRFT) of velocity controller in self-balancing industrial manual manipulators," in *Proceedings of the 2010 American Control Conference,* 2010: IEEE, pp. 1956–1961.

26 J. C. Spall, "Multivariate stochastic approximation using a simultaneous perturbation gradient approximation," *IEEE Transactions on Automatic Control,* vol. 37, no. 3, pp. 332–341, 1992.

27 J. C. Spall, "An overview of the simultaneous perturbation method for efficient optimization," *Johns Hopkins APL Technical Digest,* vol. 19, no. 4, pp. 482–492, 1998.

28 James C. Spall, Johns Hopkins University, Applied Physics Laboratory. https://www.jhuapl.edu/spsa/ (accessed 03/12/2021).

29 M. Nouri Manzar and A. Khaki-Sedigh, "Online data-driven control of variable speed wind turbines using the simultaneous perturbation stochastic approximation approach," *Optimal Control Applications and Methods,* vol. 44, no. 4, pp. 2082–2092, 2022.

30 I. Markovsky and F. Dörfler, "Behavioral systems theory in data-driven analysis, signal processing, and control," *Annual Reviews in Control,* vol. 52, pp. 42–64, 2021.

31 J. C. Willems, P. Rapisarda, I. Markovsky, and B. L. De Moor, "A note on persistency of excitation," *Systems & Control Letters,* vol. 54, no. 4, pp. 325–329, 2005.

32 J. Berberich, J. Köhler, M. A. Müller, and F. Allgöwer, "Data-driven model predictive control: closed-loop guarantees and experimental results," *at-Automatisierungstechnik,* vol. 69, no. 7, pp. 608–618, 2021.

33 L. Huang, J. Coulson, J. Lygeros, and F. Dörfler, "Decentralized data-enabled predictive control for power system oscillation damping," *IEEE Transactions on Control Systems Technology,* vol. 30, no. 3, pp. 1065–1077, 2021.

34 L. Huang, J. Zhen, J. Lygeros, and F. Dörfler, "Quadratic regularization of data-enabled predictive control: theory and application to power converter experiments," *IFAC-PapersOnLine,* vol. 54, no. 7, pp. 192–197, 2021.

35 P. G. Carlet, A. Favato, S. Bolognani, and F. Dörfler, "Data-driven continuous-set predictive current control for synchronous motor drives," *IEEE Transactions on Power Electronics*, vol. 37, no. 6, pp. 6637–6646, 2022.

36 J. Coulson, J. Lygeros, and F. Dörfler, "Distributionally robust chance constrained data-enabled predictive control," *IEEE Transactions on Automatic Control*, vol. 67, no. 7, pp. 3289–3304, 2021.

37 E. Elokda, J. Coulson, P. N. Beuchat, J. Lygeros, and F. Dörfler, "Data-enabled predictive control for quadcopters," *International Journal of Robust and Nonlinear Control*, vol. 31, no. 18, pp. 8916–8936, 2021.

38 F. Wegner, *Data-enabled Predictive Control of Robotic Systems*, ETH Zurich, 2021.

39 A. Mauroy, Y. Susuki, and I. Mezić, *Koopman Operator in Systems and Control*. Springer, 2020.

40 S. L. Brunton, M. Budišić, E. Kaiser, and J. N. Kutz, "Modern Koopman theory for dynamical systems," *arXiv preprint arXiv:2102.12086*, 2021.

41 T. Gholaminejad and A. Khaki-Sedigh, "Stable deep Koopman model predictive control for solar parabolic-trough collector field," *Renewable Energy*, vol. 198, pp. 492–504, 2022.

42 T. Gholaminejad and A. Khaki-Sedigh, "Stable data-driven Koopman predictive control: concentrated solar collector field case study," *IET Control Theory & Applications*, vol. 17, no. 9, pp. 1116–1131, 2023.

43 G. Mamakoukas, M. Castano, X. Tan, and T. Murphey, "Local Koopman operators for data-driven control of robotic systems," in *Robotics: Science and Systems*, 2019.

44 A. Maksakov and S. Palis, "Koopman-based data-driven control for continuous fluidized bed spray granulation," *IFAC-PapersOnLine*, vol. 54, no. 3, pp. 372–377, 2021.

45 B. Chen, Z. Huang, R. Zhang, W. Liu, H. Li, J. Wang, Y. Fan, and J. Peng, "Data-driven Koopman model predictive control for optimal operation of high-speed trains," *IEEE Access*, vol. 9, pp. 82233–82248, 2021.

46 D. Tellez-Castro, C. Garcia-Tenorio, E. Mojica-Nava, J. Sofrony, and A. Vande Wouwer, "Data-driven predictive control of interconnected systems using the Koopman operator," *Actuators*, 2022, vol. 11, no. 6: MDPI, p. 151.

47 Z. Hou and S. Jin, *Model Free Adaptive Control: Theory and Applications*. CRC Press, 2019.

48 Z. Hou and S. Xiong, "On model-free adaptive control and its stability analysis," *IEEE Transactions on Automatic Control*, vol. 64, no. 11, pp. 4555–4569, 2019.

49 Z. Hou, R. Chi, and H. Gao, "An overview of dynamic-linearization-based data-driven control and applications," *IEEE Transactions on Industrial Electronics*, vol. 64, no. 5, pp. 4076–4090, 2016.

50 R. Chi and Z. Hou, "A model-free adaptive control approach for freeway traffic density via ramp metering," *International Journal of Innovative Computing, Information and Control,* vol. 4, no. 11, pp. 2823–2832, 2008.

51 C. Rong-Hu and H. Zhong-Sheng, "A model-free periodic adaptive control for freeway traffic density via ramp metering," *Acta Automatica Sinica,* vol. 36, no. 7, pp. 1029–1033, 2010.

2

Philosophical Perspectives of the Paradigm Shift in Control Systems Design and the Re-Emergence of Data-Driven Control

2.1 Introduction

In this chapter, the paradigm shifts in control systems design from the philosophy of science frameworks of Popper and Kuhn are studied. The concepts of Popper's falsification theory and definition of scientific work, and also Kuhn's scientific revolutions and paradigm shifts, are employed in the control system design context. A historical perspective is given to acquire a deeper understanding of these paradigm shifts. Based on the historical periods, philosophical arguments about the paradigm shifts in control system design are highlighted. Kuhn's scientific revolutions and paradigm shifts theory are adopted to model the paradigm shifts in control systems design.

The historical timeline of paradigm shifts in automatic control is divided into the following six main periods:

- The pre-history and primitive control era, where automatic control can be considered an art.
- The pre-classical period that marks the introduction of mathematics in systems analysis and, in particular, stability analysis.
- The classical period and the first data-driven control technique that is a revolution and paradigm shift in control system sciences.
- The modern control systems analysis and design that is a paradigm shift to complete model-based control design.
- The uncertainty combat that is the other major paradigm shift to resolve the anomalous results build-up from the modern model-based methodologies.
- The re-emergence of data-driven techniques.

Analysing the paradigm shifts in control system design from a philosophical–historical perspective and based upon the fact that for the control of the present and future complex and complex adaptive systems, the present model-based

An Introduction to Data-Driven Control Systems, First Edition. Ali Khaki-Sedigh.

control system design techniques are not competent, it is argued that a fourth paradigm shift is necessary and is on its formation edge. This paradigm shift will be the re-emergence of data-driven control methodologies with new tools and accessories.

2.2 Background Materials

2.2.1 Scientific Theory

There is a distinction between the terms observational and theoretical in the philosophy of science. Observations in science and engineering are mainly undertaken by tests and experiments. Terms and predicates such as faster, stable, overshoot and oscillation in control system design are claimed to be observational because they get their meaning directly from experience, as James Watt observed these phenomena from observing the performance of the fly-ball governor prior to any mathematical formulations [1]. Indeed, any observer could have had a sense of these terms or their equivalents without any technical-mathematical knowledge. The more obvious observational examples are chair, loud, is green, is straight, is lighter than. The conditions under which assertions involving observational terms are verified in experience coincide with the conditions that are true. On the other hand, the theoretical terms such as state-controllability, internal stability and state-space realisation in control systems design get their meaning via a specific mathematical theory of systems. Scientists, including control systems scientists, have tried to account for the meaning of theoretical terms based on the meaning of observational counterparts whenever applicable [2].

Karl Popper is one of the most eminent philosophers of science of the twentieth century. The central problem in the Popperian philosophy of science is differentiating between science and non-science, that is of *demarcation* [3]. Popper believes that observation is *theory-laden*. As stated in [4], 'Roughly speaking, this means that apparently neutral statements about observational data are invariably shot through with assumptions about scientific theory'. Popper argues that limited numbers of observations cannot prove general theoretical claims and recognises that statements about data rely on general theoretical claims. Hence, all observation statements are laden with theory [4]. Popper rejects the view that science and non-science are distinguishable on an induction basis and further argues that induction is not a method of scientific analysis and inference. Instead, he proposes the notion of *falsifiability* for scientific judgment. Popper's criterion of demarcation is falsifiability and claims that a theory is scientific if and only if it is *falsifiable*. In the falsifiability logic, if a test result is positive, the theory is *unfalsified* or *corroborated*, as described by Popper [5]. Corroboration is a technical term introduced by Popper to indicate that test results cannot confirm a hypothesis. It is important to note that corroboration does not mean proven true, and even a regularly corroborated theory cannot be viewed as highly probable or even as more probable.

Moreover, even perfect corroboration carries no evidential weight, as stated in [5]. 'Corroborated is a hypothesis that (1) has not yet been refuted and (2) has stood up to severe tests (i.e. attempts at refutation)' [2]. If a theory is unfalsified by observations, it does not become probable but becomes corroborated. Finally, if a test result is negative, the theory is *refuted* or *falsified*.

In summary, contrary to the common belief of many scientists that a theory is provisionally true if it is repeatedly corroborated, corroboration does not lead to proof of truth. This shows Popper's perception of the *asymmetry*, which holds between verification and falsification. That is, a single actual counterexample falsifies the theory, but it is impossible to verify or prove a theory by reference to experience. However, Caldwell [5] argues that just as corroboration does not prove a theory true, falsification does not prove it false. It is unnecessary to reject a falsified theory, and it may be decided not to reject a falsified theory if there is no alternative theory to replace it.

Also, Popper reasons that science values theories with high informative content, and informative content is inversely proportional to probability and is directly proportional to testability. Hence, to falsify or corroborate the theory, the degree of aggressiveness of the experimental and test efforts by which a theory is put through is of fundamental importance. Popper states that 'There can be no ultimate statements in science: there can be no statements in science which cannot be tested, and therefore none which cannot in principle be refuted, by falsifying some of the conclusions which can be deduced from them'.

2.2.2 Scientific Revolutions and Paradigm Shifts

Thomas Samuel Kuhn is an influential philosopher of science of the last century, his book *The Structure of Scientific Revolutions* is claimed to be the most cited academic book in the social sciences, where he introduced the term *paradigm shift* to the scientific community [6, 7]. Kuhn proposed the notion of paradigm shifts in the progress of scientific knowledge rather than merely progressing linearly and continuously. A paradigm can be defined as the set of concepts and practices that define a scientific discipline at any particular period [6], and it is characterised by established objectives and principles, theories and methodologies, values and indices and action guidelines and rules.

The concept of a paradigm can also be substituted by the notions of *disciplinary matrix* and *exemplars*. The disciplinary matrix includes the generally accepted set of laws or basic equations governing the dynamics of the underlying systems studied by a scientific community and the set of models derived from the underlying theories that can describe the phenomena. Finally, the values or indices are used to evaluate scientific theories. Exemplars are model solutions to problems and are expressive of the fundamental concepts of the paradigm [2].

Kuhn states that 'scientific advancement is not evolutionary, but rather a series of peaceful interludes punctuated by intellectually violent revolutions and in such revolutions one conceptual world view is replaced by another'. He further claims that 'Well-established theories collapse under the weight of new facts and observations which cannot be explained, and then accumulate to the point where the once useful theory is obsolete' [6]. This was, of course, in contrast to the previously accepted regime of scientific developments. It was believed that science develops by the accumulation of facts and truths and, in some cases, by correcting and modifying previous theories.

In Kuhn's philosophical view, science development has *normal* and *revolutionary* phases. Where the revolutionary phases are qualitatively different from normal science. In normal science, scientists do their routine job within a given discipline. They agree on principles and theories within the discipline and have common past achievements with exemplars. It is stated in [4] that 'normal science is meant to bring out the idea that this type of science is business as usual'.

The concepts of puzzles and problems are crucial to distinguishing between normal science and revolutionary phases. Puzzles are problems that are solvable by the settled theories within the paradigm and are solved by scientists. In contrast, problems are unsolvable by the settled theories regardless of the amount of creative work and effort put into solving them.

Normal science is *puzzle-solving* and not *problem-solving*, as Kuhn defines it. Also, Kuhn argues that in periods of normal science where the paradigm is stable and assumed valid, the scientific theories are involved in a routine puzzle-solving phase. However, once new problems are launched, and the scientific community cannot solve them by the settled paradigm doctrine, anomalous results build up. Then, the frustrated solution-seeking force starts a revolution and initiates a paradigm change. Kuhn states: 'when the paradigm is successful, the profession will have solved problems that its members could scarcely have imagined and would never have undertaken without commitment to the paradigm'. The five phases of the process of scientific change include both normal science and revolution. In the first phase or the pre-paradigm phase, there is no consensus on any specific theory. Theories are inharmonic and incomplete. Towards the end of this phase, the scientific community converges into a unified conceptual framework. The second phase is the period of normal science, in which puzzles are solved within the context of the dominant paradigm. However, anomalies and many unsolved puzzles by the theories emerge. The third phase is the phase of the scientific crisis. Normal science within the settled paradigm fails, and science inevitably enters the next phase to overcome the crises. The fourth phase is the period of paradigm shift or scientific revolution in which the underlying assumptions are critically questioned and radically revised. The scientific revolution gives birth to a new paradigm. The fifth phase is the return of the scientific community to normal science within the new paradigm.

In normal science, paradigms and the theories within them are settled and taken for granted and not questioned to the limit of change; they can successfully solve scientific puzzles. In revolutions, they are seriously questioned and changed in revolutionary science. Problems unsolvable within the framework of normal science are tackled and solved by the revolutionists. Scientific movement in normal science is continuous, whereas conceptual discontinuities are the prominent feature of the revolutions and leading incommensurability in them. Incommensurability was introduced by Kuhn to apprehend the mapping from one paradigm to another paradigm in a scientific revolution. The pre-revolutionary and the post-revolutionary paradigms are said to be *incommensurable* in that there are no strict translations of the terms and predicates of the old paradigm into those of the new [2].

Kuhn states, 'The normal-scientific tradition that emerges from a scientific revolution is not only incompatible but often actually incommensurable with that which has gone before' [7].

2.2.3 Revolutions in Control Systems Design from Kuhn's Perspective

In Kuhn's perspective, during the period of normal science, the efforts of the scientific community, in this case, the control systems designers and scientists, are exerted to solve puzzles defined by the paradigm, in this case, to design control systems for closed-loop stability and performance. Significant achievements are achieved in solving the practical control problems in the settled control paradigm.

The automatic control approaches are divided into six main periods. These are, as will be further elaborated in the following sections, the pre-history and the primitive control, the pre-classical period, the classical period, the modern control period, the uncertainty combat period and the re-emergence of data-driven techniques.

The vagueness of the notion of a paradigm expounded in [8] leads to the conclusion that clearly defined periods of normal science are not feasible. In control systems design, the paradigms are observed from a control design *practicability* viewpoint. Although overlaps can be drawn between the paradigms, the periods of normal control science can be well defined. In these paradigms, there is a high degree of agreement, both on theoretical assumptions and on the problems to be solved within the framework provided by those assumptions [8]. During the six mentioned periods, the practical or engineering competency and the underlying theory are such that the control design puzzles are solved, and the emerging anomalies are not initially serious enough to refute the techniques and theories. However, with the growing number of unsolved puzzles and anomalies arising from technological advancement, the control system designers' confidence in the settled paradigm is eroded. The crisis of confidence and the technological requirements have resulted in control system design paradigm shifts.

The notion of Kuhn's scientific revolutions is seen in these paradigm shifts. The terms *paradigm* and *paradigm shifts* are loosely used in control theory literature. Often, the claimed paradigm shifts do not comply with Kuhn's notion. For example, in [9], a paradigm for fuzzy logic control, the term paradigm is considered a general class of fuzzy logic controllers, or in [10], the term paradigm is employed to distinguish different computational intelligence approaches. Similarly, the term paradigm is employed in the context of development [11], where the modelling and control design phases have objective functions related to the ultimate achieved performance.

In [12, 13], the term paradigm is used closest to Kuhn's notion. In [12], it is claimed that an alternative framework for control design is presented that compliments the existing theory in actively and systematically exploring the use of nonlinear control mechanisms for better performance and representing a control strategy that is *rather* independent of mathematical models of the plants, thus achieving inherent robustness and reducing design complexity. However, it can be argued that the presented methodology is a novel evolution of the previously existing nonlinear control methodologies mentioned in the paper. Also in [13], it is claimed that the proposed methodology is in the *spirit of Thomas Kuhn* to reduce the *inherent tension* between engineering practice and modern control theory, where mathematical rigour and precision are prized over utility. It is also claimed that a paradigm shift is proposed through reflection on the current paradigms in both theory and practice, and the necessity of the paradigm shift is demonstrated. However, even the referred papers and similar research papers can be considered in strict terms innovative methods within a settled normal science to overcome certain practicability problems. It is argued in this chapter that paradigm shifts and scientific revolutions in Kuhn's notion of control theory have occurred in very general frameworks. This will be elaborated on in Section 2.3. The keywords in Kuhn's notion are revolutionary science, exemplary instances, puzzles, normal science, anomalies and incommensurability. Often, the claimed paradigm shifts in control theory literature are *evolutionary* steps in design thinking or methodology, as in the two previously cited papers. In paradigm shifts, the control scientists and engineers no longer agree on control system design principles. Note that from an individual-artistic approach to control design to a mathematical approach, mathematics was the main analysis and design tool, and the individual-artistic approaches became obsolete. Model-based control design paradigm made the analysis and design techniques utterly dependent on specific external or internal mathematical models and assumptions. The paradigm shifts to tackle uncertainties in models and assumptions formed a new class of control design techniques. However, in this case, not all of the model-based analysis and design accomplishments were obsolete. Still, attempts were made to restructure the mindsets to create a new framework for designs and develop new mathematical tools. The status of a control system designer during a period of normality, and the employer's view, are well described in Kuhn's terms [7], 'If it fails the test, only his own ability, not the corpus of current science, is impugned. In short, though tests frequently occur in normal science, these tests are of a peculiar sort, for, in the final analysis, it is the individual scientist rather than the current theory which is tested'.

Finally, the good-making qualities of a control theory, following the general characteristics given by Kuhn [7], which are called the five ways [8], are as follows:

1. *Accuracy*: A theory should be accurate within its domain. In the case of control systems design, the resulting analysis and designed controller from a control theory should be observed in the experimental and practical implementations.
2. *Consistency*: A theory should be consistent. In the case of control systems analysis and design, the application results from a specific plant should be close to those achieved by other control methodologies in the paradigm.
3. *Applicability*: A theory should have a broad scope. In the case of control systems analysis and design methodology, it should be able to handle a wide range of plants and be robust in the face of unforeseen conditions.
4. *Simplicity*: A theory should be simple. In the case of control systems analysis and design, in applying the methodology from a simple system to a complex adaptive system, the cost, structure and assumptions should not be fundamentally different. No conditions or prior assumptions apart from those associated with the observable data are necessary.
5. *Fruitfulness*: A theory should be fruitful for new research findings. In the case of control systems analysis and design, it should have the capability of integrating the results and technical achievements of other methodologies to enhance its performance or to do so by theoretical and technical self-modifications.

Kuhn mainly holds that these factors provide *the shared basis for theory choice* [8]. These factors are paradigm neutral and can be employed to select rival available control methodologies.

2.2.4 Philosophical Issues in Control Engineering and Control Systems Design

Control systems are employed to regulate the output or, in general, the behaviour of systems in the desired way. The basic principle in a control system involves measuring an error signal and adjusting the system input to derive the system behaviour towards the desired performance. Control systems are implemented using automation technology in a wide range of plants. Analysing the system's operation characteristics and making the control systems operational through such technologies are the tasks of control systems engineering. The control engineer links the numerous elements of a complex system, as in system engineering.

At the heart of a control engineering solution package, including hardware and software components, to monitor and improve the performance of a system is the control system design process. The control system design process consists of three significant steps as follows. Initially, the objectives and controlled-measured variables must be established. Then, the system must be defined and modelled, as in the model-based control design techniques. Finally, the control system must be designed using the recognised design techniques. Hence, control engineering uses the knowledge provided by control systems science (design), but control engineering is not simply the control systems science. At the design stage, the control engineer or control scientist is merely interested in correcting the system behaviour, and the surrounding environment and issues are not considered in the design study. Except if it produces effects on the system that are needed to be considered in the design, which is mainly modelled as external disturbances, noises or interactions. Thus, the control system designer only views systems from the input and output points. However, in control engineering practice and the implementation process of control system designs, environmental, economical, human interface and many other outside factors can be extremely important.

From a philosophical perspective, the issues concerning control systems engineering practice and control systems design are different. Some of the philosophical arguments regarding the control engineering design and implementations are the overall design optimality, the notion of uncertainty, control engineering philosophy, emergence in control engineering, innovation, values, sustainable control design, maintenance, modelling and simulation-related issues, verification and validation of a control system design, risk, safety, resilience in control systems, social responsibility and ethical issues in control engineering, professional code of ethics for control engineers and the concept of autonomy in control engineering [14]. The convergence of sciences brings new philosophical agenda, and new models for knowledge creation and integration become necessary. Control systems engineering is deeply involved in the convergence of many sciences from different disciplines and the evolution of cyber-physical and human systems. This has philosophical, social, behavioural and economic implications.

On the other hand, philosophy further develops and revolutionises the control system design thinking and can shape the background thoughts and beliefs in a new control design methodology. The evolution of a general system theory, the emergence of artificial intelligence and soft computing concepts, and their role in control systems design, data-driven concepts to handle uncertainties, and falsification theory in the philosophy of science are examples of the philosophical concepts that entered the control systems design realm and resulted in design paradigm shifts. The final point to note is that in the current control theory and practice literature, the term *philosophy* is often used to show the underlying design concept and implementation, for example, refer to Bickley and White [15]. That is, the philosophy of control system design is used to provide the motives and philosophy behind the design and implementation of different control strategies such as feedback control, feedforward control, cascade control, ratio control and centralised or decentralised control. Control design philosophy in this context would be an outlook or approach used by control system engineers in everyday work, which does not involve some formal philosophy. The term philosophy in this chapter has a dedication to utilising philosophical perspectives to explore the practice of control systems design.

2.2.5 A General System Classification

In general, systems encountered in engineering design can be defined into four classes, *simple*, *complicated*, *complex* and *complex adaptive* [14]. Note that there is often a case-based overlap class type in systems. Simple systems have one input and one output with internal linear dynamics that govern system behaviour. Systems with multi inputs and multi outputs that consist of several simple subsystems can be a simple system if it exhibits weak or no interconnectedness or interactions between the subsystems. Complicated systems are systems with multi inputs and multi outputs and strong interactions between subsystems. However, these interactions should not lead to emergence. Emergence occurs when a system property is manifested, and its subsystems do not have that property on their own. However, some complicated systems can exhibit *domino behaviour*, where a failure in any of the subsystems leads to total or partial system failure. The failure degree would depend on the interaction measure. Complex systems are systems with strong interaction among their subsystems, where the interactions or interconnectedness can lead to emergence. Note that single input and single output complex systems exist with emerging system behaviours. Finally, complex adaptive systems are systems with strong interactions among their subsystems, and their subsystems (agents) can have time-varying characteristics, or they can adapt their behaviour to changes in the system, be it changes in the environment or the behaviour of other agents, as stated in [14].

2.3 Paradigm Shifts in Control Systems Design

The history of automatic control can be divided into six main periods. These periods are the pre-history and primitive control up to 1900, the pre-classical period from 1900 to 1935, the classical period and the first data-driven control methodologies from 1935s to 1950s, the modern control period from 1950s to 1960s, the uncertainty combat period from 1960s onward and the re-emergence of data-driven techniques during the past two decades and onward. In what follows, these paradigm shifts are briefly reviewed from a historical–philosophical viewpoint.

2.3.1 Pre-history and Primitive Control

Feedback control systems have been known and used for more than 2000 years [16]. The earliest examples are water clocks and the feedback mechanisms described by Heron of Alexandria in the first century AD. Later, from 800 to 1200, Muslim engineers designed feedback mechanisms for their mechanical tools and toys. They managed to design and implement the first on-off controllers [17]. In *The Book of Ingenious Devices* and *The Book of Tricks*, written in Arabic around 850, many mechanical devices with innovative feedback mechanisms were introduced. Up to the time of the Industrial Revolution, the designs were a combination of trial and error, artistic nature, creativity and innovation of the design engineers. Gradually and after the seminal work of Maxwell in introducing the mathematical relations governing the James Watt governor and analysing the system behaviour, a new era in control theory was commenced. However, it was not until the end of the nineteenth century and early twentieth century that one could claim the application of fundamental mathematical theories in control systems design [18, 19].

2.3.2 Pre-classical Control Paradigm

The *first paradigm shift* in control theory occurred with the introduction of mathematics to replace the personalised-based designs of the previous era. This can be called a *scientific revolution* in control system theory and design by all standards. All the previously applied techniques were obsolete, and the new attitude and mathematically oriented mindsets replaced the old traditions. We call these periods the pre-history to the primitive period of control system design [20]. Although mathematics was introduced in the mid-1800s, it was only used for system stability analysis, and it was not used for control design techniques. Along with the mathematical developments, the automatic control theory was to flourish after a perspective of *system* entered the scientific literature that included 'a dynamical

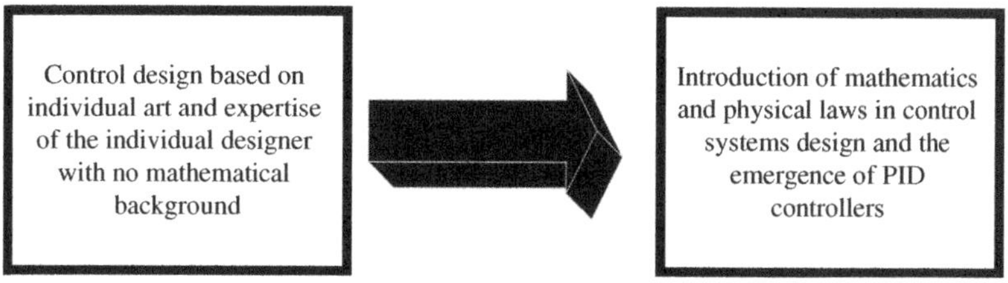

Figure 2.1 The first paradigm shift.

entity with definite "inputs" and "outputs" joining it to other systems and to the environment' [17]. The control scientific community appreciated the importance of the general system theory in the transition from the control of simple systems of the nineteenth century to the evolution of the complicated systems of the first decades of the last century, namely the Bell Telephone System and mass communication over long distances, and the guided and controlled systems deployed in the world wars. Figure 2.1 depicts the first paradigm shift.

2.3.3 General System Theory and the Philosophical Foundations of Model-Based Control

Before the general system concept and employing transfer functions as black boxes or state variables and state-space models, the dominant philosophical concept in control systems design was a *mechanicism* approach. The main features of this approach are as follows. *Determinism*, where the behaviour of a plant is univocally describable by its past outputs. *Reductionism*, where the behaviour of the plant under control is derivable from the laws and principles governing the behaviour of its parts. *Primary reactivity*, where the changes of the plant behaviour can be described by the sole action of its environment [21]. This foundation led to tailor-made control designs for plants, and a general design platform did not exist. In philosophical circles, anti-mechanicism grew at the end of the nineteenth century to advocate the concepts of *wholeness* and its consequence, *system*. This has been interpreted in the history of modern science as the history of the triumph of the categories of 'function' and 'relation' over the ones of 'substance' and 'thing' [21], which had significant implications in control theory. With the function or black-box approach, control theory could study a system by avoiding a nomothetic knowledge of its inner interactions and solely working with the relationships between inputs and outputs. Later, inner system relationships were abstractly modelled in terms of state variables from a general system perspective.

In the first two decades of the twentieth century, mathematician and science philosopher A.N. Whitehead who was a mathematics graduate from Trinity

College, Cambridge, rejected the idea that reality is essentially constructed of totally independent matters with no interconnections. Instead, he proposed an event-based or process ontology, where events are fully interconnected, and strong interactions exist. This led to Whitehead's philosophy of organisms, and it is now widely recognised as the *process philosophy* [22]. Another development was due to Karl Ludwig von Bertalanffy, a biologist known as one of the key founders of general systems theory. In this interdisciplinary conceptual framework, he described systems with interacting components. This general system approach was applicable to first-order cybernetics, biology and other fields. First-order cybernetics provided mathematical tools for the study of regulation and control. The basic epistemological and scientific approaches for the relationship analysis of cybernetics and general system theory are given in [21, 23]. The fundamental characteristic of general system theory is the inter-relationships and interactions between elements, which in unison form the entity [24]. This fundamental set-up, with the effort of many other scientists and philosophers, constructed a scientific platform for a general system theory with concrete and uniform mathematical descriptions, which was necessary for further theoretical developments of control system analysis and design.

2.3.4 Model-Based Design Paradigm

Mathematical models revolutionised all engineering disciplines and are abundantly used in engineering research and design [14]. The impact of model-based design in the control systems community has been tremendous. Both classical and modern design schools employ models in various forms. *The second paradigm shift* in control systems design is introducing general system theory and model-based design methodologies. The previously heuristic and designer-based techniques were replaced with fully model-dependent designs in a general system theoretical perspective. In this section, these design paradigms are critically reviewed. Model-based designs in this context include all the techniques that employ a mathematical model in any form in their design. The mathematical model may be analytically derived from the first principles, identified or estimated, or it may be of a soft computing nature, such as fuzzy or neural models.

2.3.4.1 Philosophical Discussions on Model Prevalence in Feedback Control

Induction, reductionism and *mechanism* are the main philosophical foundations and justifications for the substitution of a real physical phenomenon with various model forms in science and engineering, and in the control systems design and feedback context, models are predominantly mathematical models.

Following Bacon's description of scientific methodology, in the model-based control system design techniques, physical facts and laws governing the dynamical systems are derived along with input–output observations and other available measurements to formulate a mathematical model of the dynamical systems. The model is then validated through simulations and experimentally. This method of model derivation can be considered *induction-based*. However, several prominent philosophers have raised considerable objections to the use of induction in scientific methodologies that will be discussed later. There are philosophical questions regarding the ontology of model-based analysis and design approaches. From an ontological perspective of system dynamics modelling, three approaches are evident: Realism, Idealism and Moderation. In the realism approach, a model is assumed to exist for all natural phenomena. In the idealism approach, models cannot describe natural phenomena and are only intellectually formed as mental concepts. In the moderate approach, systems can, in some cases, be described by models, and sometimes the phenomenon is so complex that it is beyond our comprehension.

Modelling and system dynamical behaviour analysis are key steps in current scientific and engineering methodologies. The general modelling cycle is depicted in Figure 2.2. *Explanation* and *confirmation* are fundamental to the system modelling cycle. Where an explanation is the basic understanding of the physical

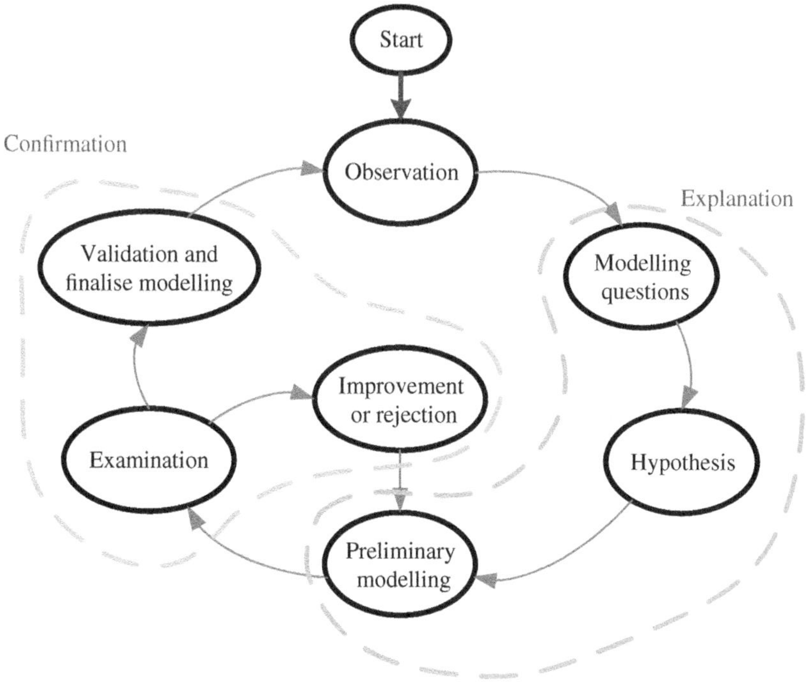

Figure 2.2 The modelling cycle.

phenomenon with *assumptions*, *hypotheses*, *theories* and *laws* as its core axis. In real-world applications, assumptions and hypotheses may vary with time, and this may enforce modifications in theories and laws. The explanation process involves *idealisation* and *unification*. Idealisation is the densification of empirical facts into a simple statement. In the densification procedure, some details are inevitably omitted. Idealisation may also involve isolating the phenomenon from the environment and other elements, as in reductionism. Also, the unification of apparently unrelated phenomena is the other procedure in the explanation process. In the philosophy of science, confirmation indicates data and events that approve and support a scientific theory. In this procedure, different tests and experiments are performed to confirm a theory or a law to validate the assumptions and hypotheses. The confirmations can be qualitative or quantitative.

In Plato's view, gaining a true understanding of what is constantly changing is impossible. The world of nature is constantly changing, and true cognition cannot be therefore achieved. Plato argued that everything has a potentially perfect form. Accordingly, Plato does not consider sensory perception to be true knowledge. Plato's famous definition of knowledge is: knowledge is a justified true belief. Based on this, Plato advocates deductive reasoning. Deductive reasoning is a type of logical thinking that begins with a general idea and reaches a specific conclusion and is sometimes referred to as a top-to-down thinking or moving from the general to the specific. A deductive approach is concerned with developing a hypothesis based on the existing theory and then designing a research strategy to test the hypothesis.

On the other hand, Aristotle believed that inductive reasoning was necessary to establish some basic assumptions prior to scientific tests. Inductive reasoning makes generalisations from specific observations. Inductive reasoning starts with data, and then conclusions are drawn from the data. In causal inference inductive reasoning, inductive logic is used to draw a causal link between a premise and a hypothesis. Aristotle used the term First principle (*Primum movens*) to prove his belief that states: knowledge gathering is the process of gaining experience from what we already know about the truth. He believed in the science of observations and measurements to create a general rule and construct a model.

Aristotle viewed scientific research as going from observations to general principles and returning to observations. He believed that the scientist should deduce the explanatory principles from observations and then deduce the theorems about phenomena deductively from the premises. In Figure 2.3, both inductive and deductive reasoning is briefly demonstrated.

Inductive reasoning is the philosophical justification of system modelling. To summarise, based on system data, a model is developed (inductive logic) utilising system identification techniques or modelling based on physical laws from the first principles. Then, the models are validated with the available information for

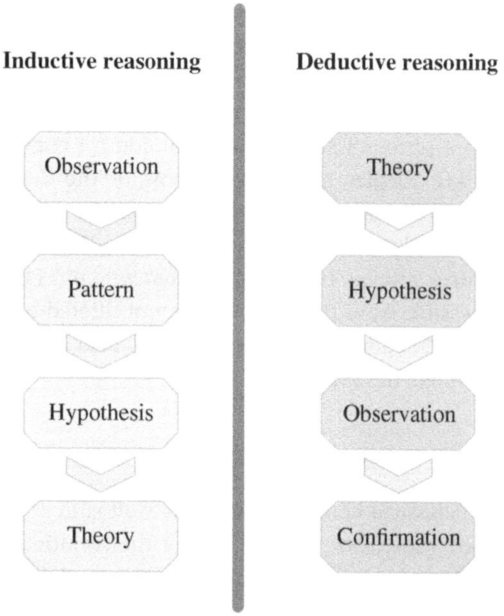

Figure 2.3 Inductive reasoning versus deductive reasoning.

confirmation (deductive logic), and finally, models are used for analysis, design and simulations (inductive logic).

2.3.5 Classical Control Design

The pre-classic control of 1930–1935, and particularly the classical control of 1935–1950, fundamentally transformed the control system design. In [18], the anomalies and incompetence of the settled control design paradigm before the classical control are highlighted. It is stated in [18] that: 'As control devices and systems began to be used in many different areas of engineering, two major problems became apparent: (1) there was a lack of theoretical understanding with no common language in which to discuss problems, and (2) there were no simple, easily applied analysis and design methods'. and 'As applications multiplied, engineers became puzzled and confused, controllers that worked satisfactorily for one application, or one set of conditions, were unsatisfactory when applied to different systems or different conditions: problems arose when a change in one part of the system (process, controller, measuring system or actuator) resulted in a change in the major time constant of that part. This frequently caused instability in what had previously been or seemed to have been a stable system'. These issues were resolved by the classical control methodologies and the newly established control design paradigm based on the novel system theory approach.

A significant advancement during the classical control era was the production of field-adjustable instruments with proportional-integral-derivative (PID) control. This allowed the broad application of PID control in different systems and under various working conditions. However, finding appropriate tuning parameters was a serious issue for practical control engineers. This problem was resolved by J.G. Ziegler and N.B. Nichols of the Taylor Instrument Companies, that proposed a data-driven technique called the Ziegler–Nichols tuning rules [25]. The Ziegler–Nichols tuning rules of PID controllers form the first data-driven control systems based on *pure system data*. Pure data is defined as data directly accessible from the plant that is not filtered or manipulated. The control design methodologies in this paradigm are also called the *error-based empirical design paradigm* in [13].

2.3.6 Modern Control Design

The classical control theory was well suited for simple and (relatively) complicated systems of the first half of the twentieth century. However, at the end of the 1940s and the second half of the twentieth century, many complicated and complex systems evolved that required automatic feedback control. The classical control techniques could not respond to the emerging engineering community's needs. Although the classical control techniques successfully solved the problems of simple and relatively complicated systems, as complications grew and complexities evolved, the weaknesses of the normal science in the established model-based control paradigm started to appear. It is worth mentioning that even today, in many practical controller designs for simple, and relatively complicated systems, the classical control approach is utilised, and the transfer function models are still widely employed. The quest for optimality and the expectations of an optimal design, and the stability issues of complicated and complex systems were the other bottlenecks of the classical control methods. In the case of high nonlinearities, time-varying parameters, multi-inputs and outputs, large plant orders and strong interactions, anomalous design results were building up, and design crises showed warning signals to control system designers. The time was ripe for an innovative and evolutionary design methodology called the *modern control theory*. The new approach changed the playground of analysis and design from frequency domain and *s*-plane of transfer function models to state-space time-domain models. This facilitated the analysis of time-varying and nonlinear systems. The introduction of matrix theory paved the way for the control of multi-input and output systems. The notion of optimality and estimation were also conveniently addressed in the new time-domain structure.

The advent of modern control theory with the central view of internal modelling and description of systems paved the way for the emergence of many new ideas in control theory analysis and design. The Lyapunov-based stability theory and optimal control design are two notable examples that we will discuss.

Stability ideas go back in history as far as Aristotle and Archimedes. Still, the Age of Reason (the seventeenth century) can be chosen as the starting point of the concrete mathematical and systematic approach to stability analysis of dynamical systems. Lyapunov's stability methodology, discovered in the West at the beginning of the twentieth century, marks the beginning of modern stability theory [26].

The Russian mathematician, mechanician and physicist Aleksandr Mikhailovich Lyapunov introduced his stability theory in [27]. His work was later translated into English, which was very well fitted to the structure of state-space and modern control theory. This is mentioned in [26] as: 'Unfortunately, though his work was applied and continued in Russia, the time was not ripe in the West for his elegant theory, and it remained unknown there until its French translation in 1907. Its importance was finally recognised in the 1960s with the emergence of control theory'. The Lyapunov stability theory soon became the dominant stability analysis tool in the control community.

According to Sussmann and Willems [28]: 'Optimal control was born in 1697-300 years ago-in Groningen, a university town in the north of The Netherlands, when Johann Bernoulli, professor of mathematics at the local university from 1695 to 1705, published his solution to the *brachystochrone problem*'. However, owing to the mathematical developments of model-based control, the notion of optimality became plausible in the modern control context in the first half of the twentieth century [29]. It is argued by Sussmann and Willems [28] that the conventional wisdom holds that optimal control theory was born in the former Soviet Union, with the work on the *Pontryagin maximum principle.* However, some mathematicians would prefer to imply that this new theory was no more than a minor addition to the classical calculus of variations, essentially involving the incorporation of inequality constraints. This would imply that optimal control could have been solved in the pre-modern control paradigm. He adds that this is highlighted in the unenthusiastic reaction at the 1958 International Congress of Mathematicians to the announcement of the maximum principle by the Soviet group. Sussmann and Williems [28] infers, 'it is likely that non-mathematical factors may have contributed to the negative reaction. Among these, two reasons stand out: first of all, Pontryagin's personality and, in particular, his notorious anti-Semitism, and second, the feeling that many held that the result was primarily intended for military applications'. Optimal control theory is an interesting example of extending a purely mathematical idea into an engineering design context. Calculus of variations as the primary optimisation tool provides the solutions of optimising a functional in terms of the space of all curves. At the same time, in the optimal design context, we are interested in the set of curves describing the system performance in terms of real dynamical constraints with some specific input. Furthermore, it is with optimal control that the *minimum time problems* imperative to some control engineering problems can be solved.

Although these ideas proved helpful in many essentially theoretical and scientific investigations [30]; however, reports of their malfunction were also reported in practical engineering design applications [31, 32]. The abstract of Ref. [32] is: 'Optimal control with a quadratic performance index has been suggested as a solution to the problem of regulating an industrial plant in the vicinity of a steady state. It is shown that such control is usually not feasible, and if feasible can have serious defects'. Uncertainty was the number one culprit. In [33], this is stated harshly: 'Lyapunov stability theory was the fad in the early 1960s. In that, it did not consider uncertainty. Optimal control was the huge fad of the 1960s in that it too almost completely neglected uncertainty'. These examples represent cases where normal science in the settled model-based control paradigm showed clear signs of incompetence in dealing with evolving real design problems.

The issue of *transparency* in control systems design is also notable. Transparency is the quality or condition of being transparent; perviousness to light; diaphaneity, and pellucidity, as stated in the Oxford English Dictionary. Transparency is a continual demand in global governance. The main expected accomplishments of transparency are visibility and controllability. If the power is visible, it could then become more controllable. Transparency has been raised in two control theory approaches.

The advocates of robust control theory and the scientists and engineers in the quantitative feedback theory (QFT) community have always insisted that the modern control theory and the state-space–based control system designs are non-transparent [34]. It is argued that the non-transparent nature of such design techniques widens the already existing theory–practice gap in scientific and control engineering methods. Transparency in the QFT literature is defined as the ability to visually relate the implementation of the design parameters to the real-world problem, from the onset of the design and throughout the individual design steps [34].

On the other hand, it can be argued that the control design methods based on transfer function or input–output models are black-box models that do not provide an insight into the internal structure of systems. While the state-space representation of systems provides an internal description of dynamical systems, and by selecting the state variable as the physical system variables, a transparent view of the internal interconnections and internal system behaviour is readily accessible. By defining the controllability and observability (reachability) concepts and providing appropriate evaluation tests, the designer is aware of the degree of the input influence on each part of the system and the overall impact of control on the system performance. Also, the observability property indicates the effectiveness of the output system measurements in representing a picture of the internal system variables or the system visibility. Hence, the controllability and observability concepts of internal dynamical system variables concerning a selected input–output set provided transparency in system representation absent in the input–output approaches. Such information overwhelmed the control scientist and boosted the state-space stand in the control community. This, to many, was seen as the end of the classical view of control, and a new design paradigm with advanced scientific and mathematical tools put an end to the pre-modern control theory era. This is described in [33]: 'In 1959, the prophet and father of Modern Control Theory publicly declared "Laplace Transform is dead and buried". Academia hastened to try to fulfil his prophecy and Frequency Response was neglected and derided'. However, the Achilles' heel was to be uncertainty.

The model-based paradigm has two distinctive eras: classical control and modern control. Classical control is distinguished by the novel theories of Bode, Nyquist, Nichols and Evans. Modern control theory laid its foundations by Kalman on the state-space platform. However, the modern control theory, with all its promises and attractive outlook and the solutions provided to the previously unsolved problems of the classical control regime, could not replace the classical control concepts. To date, classical control, even in its initial form, is still solving some of the control engineering routine problems or puzzles, as Kuhn calls them. It continued to participate in real routine problem-solving tasks. At the same time, control theorists tried to find remedies for parts of its shortcomings and hence made significant theoretical advancements in the settled normal science of that time and many new techniques evolved.

In the early 1970s, signs of the unforeseen failures of modern control theory started to appear. Many practising engineers were having trouble with modern control design implementations. The cause of these defects was identified as the model uncertainty. From the bitter experiences of such unexpected failures arose new crises in the control engineering community. The problem was sometimes referred to as the *theory–practice gap* in control engineering. In [19], the shift from practical control designs of the first decades of the twentieth century to the mathematically oriented designs of modern control is described as 'the parity that had been achieved between theory and practice after many decades was once again breached, this time in the reverse direction. Pure theory seized attention to a significant extent, and there emerged a perception among some that there was a *gap* and that the holistic view had been lost'. Control scientists tried to resolve the problem, and different schools were established based on previous experiences and preferences. Finally, after nearly a decade of debate, research and discussions, three areas of *robust control*, *(robust) adaptive control* and *soft computing-based* control techniques, were established as the new paradigms of control theory. The common motivating factor was the *uncertainty combat*. Hence, we have categorised them as one global paradigm shift from modernism (optimality) to robustness.

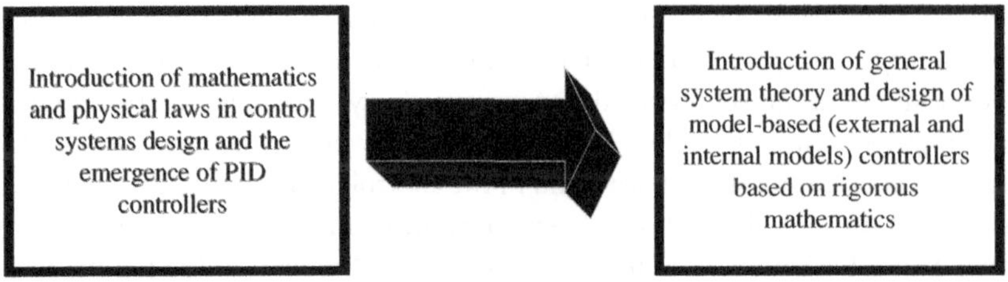

Figure 2.4 The second paradigm shift.

Many researchers followed a line of amalgamation of robustness, adaptation and intelligence in the design of control systems. Figure 2.4 depicts the second paradigm shift to model-based design techniques with the assumption of exact plant modelling by external (transfer function) or internal (state-space) models.

2.4 Uncertainty Combat Paradigm

2.4.1 Uncertainty and Performance Problem

There is no certainty, only less uncertainty [14]. Although this statement may not be valid for many events occurring in our everyday life, it is *certainly* valid for engineering designs and, in particular, control system designs. Uncertainties in engineering designs are called *practical uncertainty* by some researchers in the philosophy of science [14]. We often encounter control system design techniques that lack practicability. Such designs are founded on mathematical bases but with given assumptions and successful simulation validations that fail in practical implementations. The closed-loop responses are either unstable, have poor and unacceptable performances or performance degradations are observed over time. In most cases, uncertainties are to be blamed.

Two broad categories of uncertainties are defined as *aleatory* and *epistemic* uncertainties [14]. Aleatory means random or of stochastic nature. On the other hand, epistemic means related to knowledge. Both uncertainties are often present in systems, and uncertainty-type identification may be difficult in practice.

The sources of uncertainty in simple systems are *time, randomness, contingency* and *human error* [14]. In the case of complicated systems, in addition to the above sources of uncertainty, interaction can cause an uncertainty magnitude amplification. Finally, in complex and complex adaptive systems, emergence from the interactions and interconnectedness among the subsystems is another uncertainty source. The *butterfly effect* in chaotic systems is an example.

Engineers employ various engineering *heuristic* approaches to mitigate uncertainty in engineering designs. However, these approaches can introduce their uncertainties [14]. In the case of model-based control system design methodologies of simple and complicated systems, uncertainties are mathematically modelled. It is stated by Michelfelder and Doorn [14] that 'models are one of the primary ways to mitigate uncertainties from whatever source they arise. All models are imperfect representations of reality, so the use of models also introduces uncertainty, referred to as *model uncertainty*'. In a control engineering environment, control system designers closely work with system model developers and sometimes simplify the models to suit their design needs. Also, models are developed based on objective empirical phenomena and are required to represent them.

In most cases, modellers collaborate with control system designers to determine the model structure and objectives, and subsequently validate the models based on the predefined criteria. Model validation will be performed with the data from real experimentations under different working conditions that are finite and may not encompass all the possible working modes of a system. Hence, due to limitations in human rationality and cognitive capacity, information availability, time, economic and technological resources, the modelling process ends until an *accurate enough* model for control system design purposes is achieved. Therefore, it is clear that models that are to be employed mitigating uncertainties are potentially becoming new sources of uncertainty themselves. This could seriously question the use of modelling to control physical systems. However, since the second paradigm shift in control theory, system modelling has become a routine exercise for nearly all control engineers.

Notably, the control engineer's proposed solution for stability improvement and performance enhancement, including uncertainty mitigation, is the use of *feedback*. Feedback is essentially employed to change the *behaviour* of a system by its *behaviour*. That is, changing its input by considering its output. Serious design philosophical issues arise in this context. A key question would be whether to use feedback in a data-driven or model-based school of thought. Moreover, in the case of model-based techniques, one must be vigilant of the added model uncertainty. The *third paradigm shift* in control system theory is the *uncertainty combat paradigm*. Control engineers soon realised that certain accurate mathematical models are rarely available in real applications. On the other hand, the failed implementation of strict model-based techniques in many upcoming industrial and practical applications naturally shifted the control theorists to consider uncertainty inclusion and consideration of time-varying parameters in the models and to search for new methodologies to replace the strict model-based techniques. Many started to question the deep reliance of modern control theories on detailed mathematical models of physical systems and the deductive reasoning behind them. A crucial point was raised about whether the techniques of the normal science of the model-based paradigm are about controlling mathematical models instead of the actual physical plants [13]. Three distinctive schools were formed for the uncertainty combat task: The robust control approaches, the adaptive control approaches and the intelligent or soft computing-based approaches.

2.4.2 Uncertainty Combat: the Robust Control Approach

In the mid-1970s, control system scientists started to develop new tools to deal with uncertainty. The first reaction to the shortcoming of modern control theories in the face of uncertainty was the attempts of control theory modernists to find remedies within the settled modern paradigm. As the biography of one of the prominent control theory researchers stated: 'His research interests are to make modern control theoretical methods work' [35]. In 1963, the concept of the *small gain* principle, which later played a vital role in the robust stability criteria, was introduced [36]. Also, it was shown that under certain assumptions, optimal single input single output state-feedback control laws had some very strong robustness properties [29]. Extending these results to complicated multivariable systems, it was shown that the results remain valid for each input channel to the plant [37]. However, a pioneer in robust control, in a short but effective paper, studied the robustness of optimal controllers (linear quadratic Gaussian (LQG) regulators) under different conditions. The abstract has only three words, '*There are none*'. He finally remarks that 'It may, however, be possible to improve the robustness of a given design by relaxing the optimality' [38]. That is moving from optimality to robustness.

Uncertainty was included in the systems models as the first step. Introducing and establishing the notions of robust stability and robust performance was the second step. Then, the final step and top on the agenda was to reconsider the classical and modern methodologies to accommodate uncertainty in their analysis and design hence leading to optimised robustness of feedback control systems. Other innovative tools and methods within the modern control context were introduced for the new paradigm [31, 39].

In a parallel attempt, and at least from a decade earlier, the frequency techniques were revisited to encounter model uncertainty in a framework called the *Quantitative Feedback Theory* (QFT). The founder of QFT argues in [33] that: 'Ph.D. graduates from the most prestigious universities were convinced that the purpose of feedback was to achieve a desired input–output relation (if they belonged to the classical school) or eigenvalue realisation (if they were "modernists", which is even more restrictive than the classical objectives) for a fixed plant. Consequently, today, any feedback theory which considers uncertainty is adorned with the unnecessary label of "robustness"'. Hence, in the QFT design philosophy, uncertainty must be inherently embedded in any feedback design, and additions such as robustness are superfluous. It is as if one says 'live human begins' whenever human beings are discussed [33]. In the QFT philosophical approach to feedback, the two main elements of feedback design are *closed-loop specifications* and the *cost of feedback*. The consequence of ignoring or undermining these two elements is over-conservativeness in robust control design methodologies.

Also, in similar criticisms to the robust control paradigm after its establishment and declarations that uncertainty is conquered by robust control in [31], it is stated that 'despite the considerable progress of robust control theory, control designs are still not always sufficiently robust' and after stating several failures of robust control in practical implementations, continues that 'At design time the nature and size of model uncertainties were not, and perhaps even could not, be anticipated by engineers'. Moreover, it is correctly claimed that the mainstay robust control methods generally require accurate prior assumptions about uncertainty size and structure. However, such required online information is not available and not accessible by the control system, and attempts to derive the online data rely on further prior assumptions about the system.

The control engineering community, confronted with many practical failures of robust control approach due to erroneous uncertainty and probability assumptions, has looked into another paradigm shift. 'Useful answers to the problems now faced by robust control theory will require a somewhat revolutionary paradigm shift, abandoning the common belief that "one must assume something to know something"' [40].

2.4.3 Uncertainty Combat: the Adaptive Control Approach

The other leading philosophical school in control theory devoted to coping with uncertainty was the *adaptive control* school. The interest in the adaptive control approach goes back to the 1950s [41]. In a 1961 Ph.D. thesis, it is stated: 'In recent years, interest has been developing rapidly in designing control systems capable of adjusting to changing environmental conditions which affect the system's performance. In particular, this interest has centred on the design of systems that will sense the need for and carry out the adjustments automatically. Such systems have come to be called "self-adaptive" or more simply "adaptive"' [42].

The main elements of this approach are to identify the uncertain plan and then automatically update the control design using the latest identified (estimated) model. This reduced the conservativeness in the robust control design but had problems with stability and convergence. The theory underlying treating identified or estimated values as the actual values is the *certainty equivalence principle*. The idea was first raised in economics and decision-making under uncertainty [43]. The employment of the certainty equivalence principle is a common practice in most adaptive control techniques. There are certain assumptions in proving and employing the principle regarding the unconditional expected values or the joint probability distribution of the variables. It is claimed in [43] that 'Hence, the assumption made explicitly here is no more restrictive than the assumption made implicitly in servomechanism design when the range of admissible alternatives is limited to linear systems'.

Adaptive control theory tried to fill the theory–practice gap and provide practical designs in uncertain, time-varying complicated and complex systems. However, unfortunately, most adaptive methodologies, as in robust control, depend on prior assumptions about the system. It is pointed out by Safonov [40] that 'possibly disturbing consequence of this is that any theory for system identification or adaptive control that begins with prior assumptions about the structure or form of either the "true plant" (linear time invariant (LTI), parameters, order, etc.) or the "true noise" (Gaussian, unknown but bounded, etc.) may not be a scientific theory by Popper's definition'.

The adaptive methods that have tried to get free from assumptions have unfortunately not guaranteed acceptable closed-loop performance. Adaptive control methodologies have limitations and problems, both theoretical and with practical implementations, as is stated in [44]: 'Much theory was over-sold; experimentation or practical application did not always run according to plan; and practitioners are understandably cautious about the implementation of the technology'. It is mentioned in [41] that 'The role of simplified models and the robustness to neglected dynamics were other questions that have also arisen'. The model assumptions in adaptive control could be fatal in practical applications. This is very well demonstrated in Rohr's counterexample and its conclusion that 'The practical engineering consequences of the existence of the infinite-gain operator are disastrous' and 'none of the adaptive algorithms considered can be used with confidence in a practical control system design because instability will set in with high probability' [45]. The leading cause can be summarised in assumptions about the plant model and the un-modelled dynamics that are inevitable in practice.

The other assumption in many adaptive control techniques is the *persistent excitation condition* of inputs. This is a property of the input signal, which requires that the signal has a richness corresponding to the number of parameters in the controller or the number of parameters in an adjustable model of the plant. Persistent excitation is a strong condition to obtain a unique system model identification from the input–output data. Even in the direct adaptive control design methods, persistently exciting data must hold such that it is theoretically possible to identify the system model using the same data.

The adaptive control community soon tried to rectify the situation by modifying and building bridges to robust and intelligent control theories. Anderson and Dehghani [44] puts the situation as: 'Rohr's counterexample was a wake-up call for people to contemplate robustness issues in adaptive control'. There are other difficulties with the adaptive control approach that are referred to as 'Generic problems of adaptive control' in [44].

The search for practical solutions to implementation issues led to other adaptive control schemes, such as the *multiple models adaptive control*, with new assumptions in favour of some relaxed assumptions and applicability in nonlinear systems.

2.4.4 Uncertainty Combat: the Soft Computing-based Control Approach

The soft computing-based control design approaches include three main methodologies: artificial neural networks, random optimisation techniques and fuzzy logic. They were developed in parallel to other conventional control design methodologies with a different philosophical background and were in contrast to the conceptual, theory-intensive control sciences with a *rational* process, in which theorems and standard mathematical analysis granted them an indubitable legitimacy and the burden of mathematical proof attempted to distinguish truth from falsehood [46]. Although the philosophical backgrounds of the adaptive and robust control theories differ from the soft-computing–based theories, the common factor that places these three approaches in a unified normal science framework of the settled uncertainty combat paradigm is their model-based nature. These approaches, with different depths and types, partially depend on a mathematical model. They require comprehensive knowledge and certain assumptions of the plant under control to establish fuzzy rules or select the neural network structures. Hence, the control design process depends on fuzzy rules or a neural network model; in either case, the fundamental problems of model-based control still exist. The random optimisation techniques are essentially employed as parameter optimiser tools in a defined controller. Therefore, in this case, all of the issues associated with models and assumptions persist. It is argued in [46] that 'The pragmatic attitude, as manifested by much of the work in neural networks and fuzzy logic, attempts to solve the true problem directly. Few explicit simplifying assumptions are made, but of course, since the problem is intractable, the attempt is only partly successful. We end up with an approximate solution to the exact problem'. While in the case of conventional model-based control design methodologies, all attempts are exerted to derive an exact solution to an approximate problem, where the problem is made approximate through assumptions of linearity, affine model forms and convex optimisation criteria. A proposed solution for some of the control scientists working in the uncertainty combat paradigm would be a hybrid design methodology to integrate the approaches' analytic and pragmatic characteristics.

As a final remark before proceeding to the fourth paradigm shift towards data-driven control, it is worth mentioning that two notable surveys were carried out on the application of control theory in the industry. The first survey was carried out by the control technology committee in the Japanese Society of Instrument and Control Engineering (SICE) as a control technology survey of the Japanese industry, and the results were published in [47]. This followed an earlier investigation that was performed in 1989. The second survey was carried out about two decades later by some of the members of the International Federation of Automatic Control (IFAC) Industry Committee, formally established in 2017, and the results were published in [48]. To summarise some of the key points related to the present discussion, it is reported in [47] that based on the received responses, advanced PID type control has been widely applied, and about 30% of respondents have already used this type of control in their factories. Model predictive control (MPC) and fuzzy control are reported to be the most widely used modern control techniques. However, slightly less than 40% of the factories have applied these techniques. Application of modern control theory types such as LQG controller, observer, Kalman filter and H-infinity control/μ-analysis is 10% of responding factories, but the applications were shown to be increasing. Compared with the survey of 1989, it was shown that applications of certain advanced PID, dead time compensation, Kalman filter, MPC, H-infinity control/μ-analysis, rule-based control, fuzzy control and optimisation were increasing. It was also reported that satisfaction with the widely applied advanced PID was high, and satisfaction with modern control theory in its very limited applications was acceptable. MPC and fuzzy logic-based controllers had a high rate of satisfaction among the respondents. Generally, the respondents did not appreciate robust and adaptive control techniques. It is also interesting that the application issues for all the advanced considered control techniques were, ‘required engineering power is too high’, ‘there is no need to apply a new control in use’, and ‘benefit of a new technique is not clear’ obtained high points by the respondents, modelling is also considered as a hard issue in the report. The plants and industries that were subjects of the report can be categorised into the simple, complicated and complex plant classifications proposed in Section 2.2.5. It is noteworthy that the report was prepared during the period that marked the start of research and interest in data-driven control theory. The report clearly shows the success of control theories in certain applications, but presumably, the gap between theory and practice avoided a broad application of advanced control techniques based on analytical mathematical foundations. There is no report on the emerging advanced complex systems, and the application of model-based techniques in complex plants is not reported. The reported complex plant control systems are mainly MPC and fuzzy logic-based.

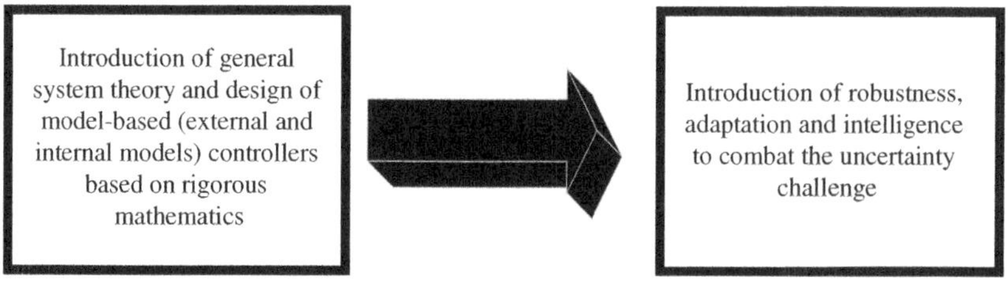

Figure 2.5 The third paradigm shift.

In the second report [48], it is shown that the survey respondents indicated a very high PID current impact in practice, and the next advanced control technique with a high impact is MPC. Other advanced model-based techniques of the second and third paradigm shifts are in a much lower impact status. This contradicts some of the predictions of [47] regarding the future demand for such techniques. MPC is expected to gain more impact over the next five years. A report on the complex adaptive systems, as defined in Section 2.2.5, is not available. Remarkably, MPC is a well-established control design methodology that has emerged in the control design paradigm shifts. MPC has its roots in industrial applications. Its power lies in the prediction ability, rather simple structure, and implementation issues. The robust, adaptive and intelligent versions of MPC are all reported. The data-driven predictive controllers are also appearing in the literature. However, in its present model-based form, MPC applications are limited to simple, complicated and complex plants. Figure 2.5 depicts the third paradigm shift.

2.5 The Paradigm Shift Towards Data-driven Control Methodologies

The advanced model-based techniques with enhanced robustness, adaptation and intelligence characteristics have successfully solved control problems of simple and complicated systems and some complex systems. In fact, as the complexity increases, the risks of control failure increase, and outcomes and reverberation of system faults also increase with system size and complexity.

The control of future complex and complex adaptive systems is studied by Lamnabhi-Lagarrigue et al. [49], and the *grand societal challenges* are reported in the areas of transportation, energy, water, healthcare and manufacturing. The attempts and beliefs of many control scientists in the settled paradigm of model-based control are to cultivate further and pursue the present techniques [19, 49, 50]. As an example, it is stated in [49] that model-based design and optimisation is the dominant paradigm in the present complex engineering systems and future engineering developments require addressing much more complex, interactive and autonomous dynamical systems. It is claimed that in all future developments, models will continue to serve as a basis of accumulated knowledge to facilitate optimal design and operational strategies. However, these models will need to be complemented by data-driven approaches. Although, it is also stated that variable dynamic characteristics, strong nonlinearities, heavy coupling, unclear mechanisms, mathematically un-modellable and online un-measurable key parameters constitute challenges to existing control theory and technology. Furthermore, in another section, it is stated that the complexity of the stochastic processes involved makes data-driven mechanisms an absolute necessity. A more philosophical standpoint is taken on the amalgamation of different methodologies to encounter new challenges in [46] based on *pluralism*. Quoting from [51], it is stated that promoting pluralism in science leads to a deliberate strategy of devising multiple competing theories to enhance testability. Although a pluralist philosophy will indeed enhance and further develop the existing control design methodologies in the settled paradigm of model-based control, it cannot provide a remedy for the inherently built limitation of models and assumptions of these techniques. The model-based advanced techniques are mature branches of control theory with rich and immense literature. However, as previously stated, these methodologies can encounter serious defects in the face of real unpredicted uncertainties and complexities. This is stated by Safonov [40] 'The possibility of mismatch between prior uncertainty modeling assumptions and reality is a problem both for robust control and for model-based adaptive control algorithms that aim to use real-time data to identify and correct such problems adaptively. Mismatches between prior model assumptions may fool adaptive algorithms intended to improve robustness into persistently preferring destabilising controllers over stabilising ones even when the instability is patently obvious to the eyes of the most casual observer'.

The common key feature in the proposed uncertainty combat paradigms of the last decades is that the actual system must conform to the modelling assumptions required in their theoretical developments and mathematical proofs. Indeed, 'Assumptions are the Achilles' heel of mathematical system theory. When modeling assumptions fail to hold, so do conclusions that rest on those assumptions' [40]. It is elegantly stated in [52]: 'control system design is a difficult area. If all engineering aspects are considered, the problem becomes very messy. If the problem is simplified to be solved elegantly, many aspects have to be neglected. This has caused much controversy. It will probably take a long time before we have good design tools that will deal with the real issues of control system design'.

The problems facing robust control and model-based adaptive control techniques are *safely* controlling real, uncertain, complicated and complex systems and the control of evolving complex adaptive systems in the present century. This will require another *revolutionary paradigm shift*. This time to get over all the fancy assumptions that come one after the other and a mindset change from dependency on bounds and other prior knowledge to reliance on actual field measurements and a truly data-driven control paradigm. In this new paradigm, control systems will automatically acquire the necessary available data and redesign or restructure the controller when necessary. The paradigm shift was predicted by Åström and Kumar [19] as 'It is of course difficult to have a good perspective on recent events but our opinion is that there are indications that yet another major development and spurt is now in progress'. The elements of the new data-driven control systems paradigm have flourished during the past two decades, and there are a number of truly data-driven control systems design strategies [53]. However, it appears that the state of normal science in the new paradigm has not yet been reached. Successful implementations and practical solutions to the puzzles of the new paradigm that are controlling complex and complex adaptive systems must be reported.

An essential question in the new paradigm concerns data characteristics. As in the adaptive control and the persistently exciting condition, the new paradigm would require some minimum data conditions. The objective is to obtain a controller from data that are not as rich and informative with required pre-assumptions as in the unique system identification approach. This has led to a new general framework to study data *informativity* conditions for data-driven analysis and control. Necessary and sufficient conditions must be derived under which the data are assuredly informative for certain properties of data-driven nature [54]. Figure 2.6 depicts the fourth paradigm shift.

The new paradigm should overcome *the three enemies of knowledge*, as acknowledged in [55]. The three enemies are *ignorance*, *uncertainty* and *complexity*.

Ignorance is not merely the absence of knowledge, but also the unawareness of one's lack of knowledge [55]. Ignorance occurs in the control engineering profession, where the control engineers must learn from previous failures. In the control system design, the statement that 'you have to assume something to know something' is a premise that the control system scientists have repeatedly experienced its falseness. It is nevertheless tacitly accepted by many researchers

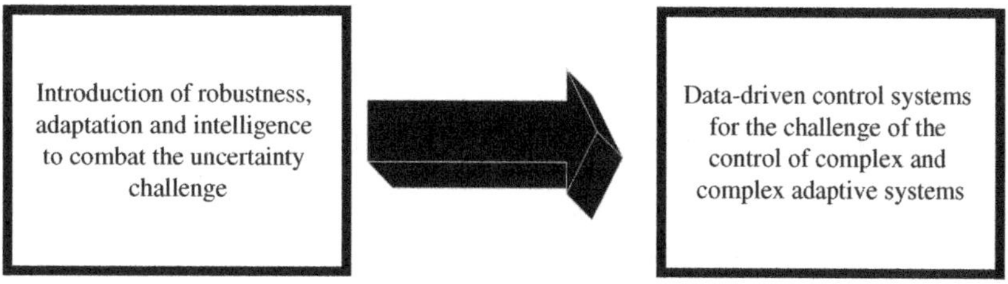

Figure 2.6 The fourth paradigm shift.

in learning theory, estimation and control. This is a *practical* ignorance, where designers attempt to design yet another learning system that fails to handle future experimental data inconsistent with prior assumptions.

The new paradigm must be able to tackle uncertainty, where essential knowledge is lacking in system models, and its absence is known to the designer. The techniques in the new paradigm must avoid uncertainty by relying on measured pure input–output data. In [19], the assumption or belief that the system to be controlled is given a priori is conceived as a *cardinal sin of automatic control.*

The main difficulty associated with complexity is unpredictability, and its main difference from an intricate or complicated system with many components and interconnections lies in unpredictability. Another point to note is that there is no effective tool to mathematically handle certain types of complexity, such as fast varying parameters or changing structures over time.

The present control methodologies successfully control simple systems and can, in some cases (although not guaranteed to) solve the design puzzles of complicated systems. However, they are showing incompetence in solving design problems of complex and complex adaptive systems. This must be resolved in the new paradigm. The control theories and methods must merely employ input–output data with no prior assumed knowledge of the system and no required explicit information or mathematical model, with guaranteed stability, convergence, resilience and performance specifications under minimum acceptable assumptions [53]. Many in the control community of the settled paradigm are working on solving the problems of the control methodologies within the paradigm. For an example of such attempts, as is indicated in [19], 'An important aspect anticipated of future control systems is the interaction between the physical world often modeled by differential equations and the logical dynamics of the computational world'. This refers to the current model-based techniques. The recommended solution is to establish the properties of the composite systems comprising both new models and assumptions in the emerging field of *hybrid systems*. Figure 2.7 depicts a two-dimensional picture of paradigm shifts versus plant classifications. Finally, Figure 2.8 provides a general overview of the developments and paradigm shifts in control theory based on Kuhn's perspective. Table 2.1 presents a summary of paradigm shifts, exemplars, puzzles, normal science, anomalies and non-competitiveness.

2.5.1 Unfalsified Philosophy in Control Systems Design

The new paradigm shift should eliminate prior assumptions in control systems design. Any prior assumption that requires knowledge of the system model would breach the purpose of data-driven philosophy. An assumption-free methodology

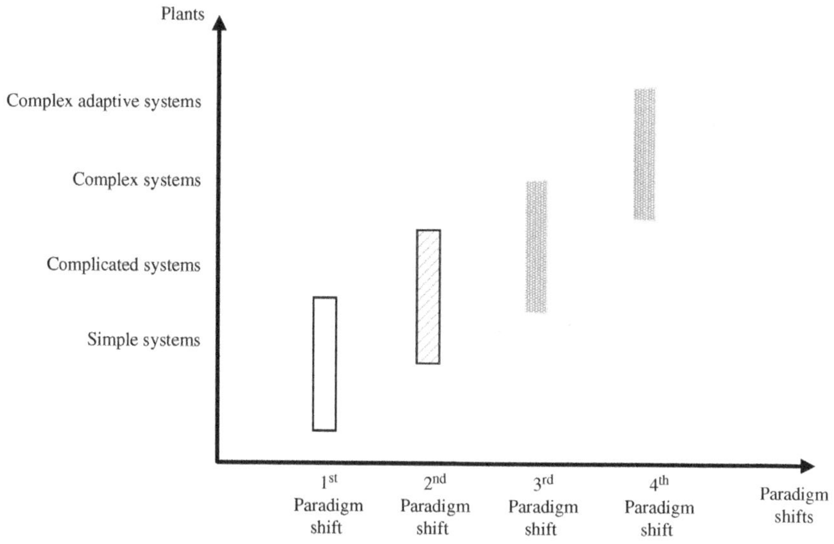

Figure 2.7 A two-dimensional picture of paradigm shifts versus plant classifications.

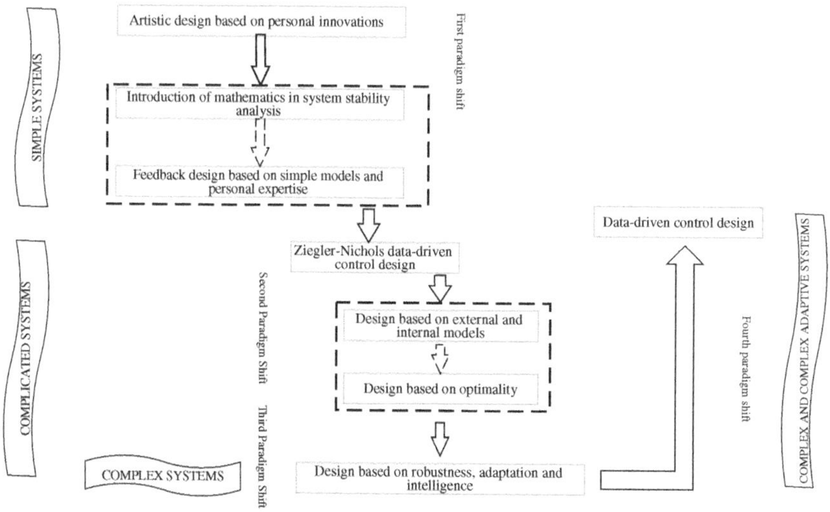

Figure 2.8 A general overview of the paradigm shifts in control system design.

Table 2.1 Summary of the main features of control systems design paradigm shifts modelled in Kuhn's theory.

Paradigm Shifts	Exemplars	Puzzles	Normal Science	Anomalies	Non-Competitiveness (the notions that are obsolete in the proceeding paradigm)
1st Paradigm Shift: Mathematical Emergence	Basic Mathematical Analysis Blended with Heuristics and Personal Expertise in Solving Case-Based Control Design Problems	CONTROL OF SIMPLE SYSTEMS: e.g. Early Ship and Aircraft Control, Process Applications, Servo and Communications Engineering	The Family of Three-Term Controllers	Emerging Complicated Systems	Artistic, Heuristic and Personal Case-Based Design Procedures
2nd Paradigm Shift: System Theory and Model-Based Designs	Rigorous Mathematical and Analytical Tools Based on Exact Input–Output or State-Space Models	CONTROL OF SIMPLE-COMPLICATED SYSTEMS: Systems with Optimality Necessities, Multivariable and Nonlinearity Characteristics	Classical and Modern Control Methodologies, Optimisation Techniques	Complex Uncertain and Time-varying Systems	Strict Model-Based Design Analysis Techniques with NO Uncertainty Considerations. Examples: Pole-Zero Assignment and Cancellation Designs, Exact Decoupling and Optimal control

3rd Paradigm Shift: Uncertainty Combat	Rigorous Mathematical and Analytical Tools in Robustness, Adaptation and Intelligence Frameworks	CONTROL OF COMPLICATED–COMPLEX SYSTEMS: Systems with Uncertainty and Time-varying Parameters: e.g. Advanced Aerospace Systems, Power Plants and Complex Petrochemical and Process Industries, as Sample Plants	Robust and Adaptive Control Methodologies with Soft Computing-Based Design Techniques-The Hybrid Approaches	Complex Adaptive Systems	The Notions of Structured, Non-Structured Uncertainty Analysis and Design, Classical Stable Adaptive Techniques, Conventional Intelligent Control Based on Soft Computing Models
4th Paradigm Shift: Data-Driven Control	Methodologies Based on Measured Input–Output Data	CONTROL OF COMPLEX–COMPLEX ADAPTIVE SYSTEMS: Systems with Complexity and Emergence such as Bio-Plants, Complex Networks	It is in its Formation Age	None Yet	None Yet

using pure, raw, noise-corrupted, past input–output data can overcome ignorance and uncertainty for the control of complex and complex adaptive systems. Hence, a performance criterion must be defined to select a candidate controller using only the available real data. However, the control system has a finite amount of past plant input–output data available and cannot predict the plant response to future signals unless assumptions are made. The future signals may or may not have been previously used. Corroboration or unfalsification of a controller is the ultimate achievement of the past data criterion up to that time. Based on Popper's falsification theory, a proper scientific data-driven control theory is limited to performance criterion checks using the available pure data. The formulae-driven mindset of closed-loop stability proofs is built on an assumption-based platform. While in the falsification approach, the inappropriate controllers are falsified and rejected with no prior assumption and only under mild data informativity and *cost detectability* conditions. Cost detectability is only a property of the performance criterion or cost function and the controller. This implies that it is free of any plant or signal assumptions. Cost detectability ensures that for each candidate controller, the cost function boundedness is a necessary and sufficient stabilisability condition or a corroboration permit.

2.6 Conclusions

It is shown in this chapter that philosophy has played an essential role in paradigm shifts in control systems design methodologies. Three main paradigm shifts have occurred in automatic control, and the fourth paradigm shift is emerging. The fourth paradigm shift solves the problems of assumptions, uncertainty and complexity, which are the three enemies of control system design in complicated, complex and complex adaptive systems. The control design puzzles within a settled paradigm have successfully been solved by the theories of that paradigm. However, it is shown that the introduction of design problems that have inherently evolved with technological advancements and greater expectations of safety, resilience and performance has given rise to anomalous results build up and led to crises and, finally, a new paradigm shift. The historical timeline of paradigm shifts in automatic control was divided into six main periods, and each period is separately discussed. The critical point noted is the re-emergence of data-driven methodologies requiring extensive research on the philosophical and theoretical issues within the coming paradigm shift. For more than two decades, parts of the control community have been involved in such developments; however, many unresolved theoretical questions remain before the data-driven approach can become the normal science of a settled data-based paradigm.

References

1 J. C. Maxwell, "I. On governors," *Proceedings of the Royal Society of London,* vol. 16, pp. 270–283, 1868.

2 S. Psillos, *Philosophy of Science AZ*. Edinburgh University Press, 2007.

3 S. Thornton, "Karl Popper," ed. Edward N. Zalta: *The Stanford Encyclopedia of Philosophy*, 2019.

4 T. Lewens, *The Meaning of Science: An Introduction to the Philosophy of Science*. Hachette UK, 2016.

5 B. J. Caldwell, "Clarifying popper," *Journal of Economic Literature,* vol. 29, no. 1, pp. 1–33, 1991.

6 A. Bird, "Thomas Kuhn," ed. Edward N. Zalta: *The Stanford Encyclopedia of Philosophy*, 2018.

7 T. Kuhn, *The Structure of Scientific Revolutions. (1962)* University of Chicago Press: Chicago, IL, 1996.

8 W. H. Newton-Smith, *The Rationality of Science*. Routledge, 2002.

9 F. L. Lewis and K. Liu, "Towards a paradigm for fuzzy logic control," *Automatica,* vol. 32, no. 2, pp. 167–181, 1996.

10 A. P. Engelbrecht, *Computational Intelligence: An Introduction*. John Wiley & Sons, 2007.

11 Z. Zang, R. R. Bitmead, and M. Gevers, "Iterative weighted least-squares identification and weighted LQG control design," *Automatica,* vol. 31, no. 11, pp. 1577–1594, 1995.

12 Z. Gao, Y. Huang, and J. Han, "An alternative paradigm for control system design," in *Proceedings of the 40th IEEE Conference on Decision and Control (Cat. No. 01CH37228)*, 2001, vol. 5: IEEE, pp. 4578–4585.

13 Z. Gao, "Active disturbance rejection control: a paradigm shift in feedback control system design," in *2006 American Control Conference*, 2006: IEEE, p. 7.

14 D. P. Michelfelder and N. Doorn, *The Routledge Handbook of the Philosophy of Engineering*. Routledge, 2020.

15 M. Bickley and K. White, "Control system design philosophy for effective operations and maintenance," *arXiv preprint physics/0111079,* 2001.

16 S. Bennett, *A History of Control Engineering, 1930–1955* (vol. 47). IET, 1993.

17 F. L. Lewis, *Applied Optimal Control and Estimation*. Prentice Hall PTR, 1992.

18 S. Bennett, "A brief history of automatic control," *IEEE Control Systems Magazine,* vol. 16, no. 3, pp. 17–25, 1996.

19 K. J. Åström and P. R. Kumar, "Control: a perspective," *Autom.,* vol. 50, no. 1, pp. 3–43, 2014.

20 B. Friedland, *Control System Design: An Introduction to State-space Methods*. Courier Corporation, 2012.

21 D. Pouvreau and M. Drack, "On the history of Ludwig von Bertalanffy's "General Systemology", and on its relationship to cybernetics: Part I: elements on the origins and genesis of Ludwig von Bertalanffy's "General Systemology"," *International Journal of General Systems,* vol. 36, no. 3, pp. 281–337, 2007.

22 A. N. Whitehead, *Science and the Modern World.* Macmillan, 1925.

23 M. Drack and D. Pouvreau, "On the history of Ludwig von Bertalanffy's "General Systemology", and on its relationship to cybernetics–part III: convergences and divergences," *International Journal of General Systems,* vol. 44, no. 5, pp. 523–571, 2015.

24 L. Von Bertalanffy, "General system theory, a new approach to unity of science. 5. Conclusion," *Human biology,* vol. 23, no. 4, pp. 337–345, 1951.

25 T. Hägglund and K. J. Åström, "Automatic tuning of PID controllers," *The Control Handbook.* CRC Press, 1996, pp. 817–826.

26 R. I. Leine, "The historical development of classical stability concepts: Lagrange, Poisson and Lyapunov stability," *Nonlinear Dynamics,* vol. 59, no. 1, pp. 173–182, 2010.

27 A. M. Lyapunov, "General problem of the stability of motion," Ph.D., Kharkov Mathematical Society, XI (in Russian). 1892.

28 H. J. Sussmann and J. C. Willems, "300 years of optimal control: from the brachystochrone to the maximum principle," *IEEE Control Systems Magazine,* vol. 17, no. 3, pp. 32–44, 1997.

29 R. E. Kalman, "When is a linear control system optimal?," 1964.

30 A. E. Bryson, "Optimal control – 1950 to 1985," *IEEE Control Systems Magazine,* vol. 16, no. 3, pp. 26–33, 1996.

31 M. G. Safonov, "Origins of robust control: early history and future speculations," *Annual Reviews in Control,* vol. 36, no. 2, pp. 173–181, 2012.

32 H. Rosenbrock and P. McMorran, "Good, bad, or optimal?," *IEEE Transactions on Automatic Control,* vol. 16, no. 6, pp. 552–554, 1971.

33 I. M. Horowitz, *Quantitative Feedback Design Theory (QFT).* QFT Publications, 1993.

34 C. Houpis, S. Rasmussen, and M. Garcia-Sanz, *Quantitative Feedback Theory: Fundamentals and Applications.* CRC Taylor & Francis, 2006.

35 C. Harvey and G. Stein, "Quadratic weights for asymptotic regulator properties," *IEEE Transactions on Automatic Control,* vol. 23, no. 3, pp. 378–387, 1978.

36 G. Zames, "Functional analysis applied to nonlinear feedback systems," *IEEE Transactions on Circuit Theory,* vol. 10, no. 3, pp. 392–404, 1963.

37 M. Safonov and M. Athans, "Gain and phase margin for multiloop LQG regulators," *IEEE Transactions on Automatic Control,* vol. 22, no. 2, pp. 173–179, 1977.

38 J. C. Doyle, "Guaranteed margins for LQG regulators," *IEEE Transactions on automatic Control*, vol. 23, no. 4, pp. 756–757, 1978.

39 P. Dorato, "A historical review of robust control," *IEEE Control Systems Magazine*, vol. 7, no. 2, pp. 44–47, 1987.

40 M. G. Safonov, "Robust control: fooled by assumptions," *International Journal of Robust and Nonlinear Control*, vol. 28, no. 12, pp. 3667–3677, 2018.

41 K. J. Astrom, "Adaptive control around 1960," *IEEE Control Systems Magazine*, vol. 16, no. 3, pp. 44–49, 1996.

42 M. L. Moe, "The use of time-moments in model-reference adaptive control systems," Ph.D., Northwestern University, 1961. [Online]. Available: https://www.proquest.com/openview/0d5ec29da8ab4d88e025df05e7e9e98e/1?pq-origsite=gscholar&cbl=18750&diss=y

43 H. A. Simon, "Dynamic programming under uncertainty with a quadratic criterion function," *Econometrica, Journal of the Econometric Society*, pp. 74–81, 1956.

44 B. D. Anderson and A. Dehghani, "Challenges of adaptive control–past, permanent and future," *Annual Reviews in Control*, vol. 32, no. 2, pp. 123–135, 2008.

45 C. E. Rohrs, "Adaptive control in the presence of unmodeled dynamics," Ph.D., Massachusetts Institute of Technology, 1982. [Online]. Available: http://hdl.handle.net/1721.1/15780

46 T. Samad, "Postmodernism and intelligent control," in *Proceedings of 12th IEEE International Symposium on Intelligent Control*, 1997: IEEE, pp. 37–42.

47 H. Takatsu and T. Itoh, "Future needs for control theory in industry-report of the control technology survey in Japanese industry," *IEEE transactions on control systems technology*, vol. 7, no. 3, pp. 298–305, 1999.

48 T. Samad *et al.*, "Industry engagement with control research: perspective and messages," *Annual Reviews in Control*, vol. 49, pp. 1–14, 2020.

49 F. Lamnabhi-Lagarrigue *et al.*, "Systems & control for the future of humanity, research agenda: current and future roles, impact and grand challenges," *Annual Reviews in Control*, vol. 43, pp. 1–64, 2017.

50 R. M. Murray, K. J. Astrom, S. P. Boyd, R. W. Brockett, and G. Stein, "Future directions in control in an information-rich world," *IEEE Control Systems Magazine*, vol. 23, no. 2, pp. 20–33, 2003.

51 D. M. Clarke, "Philosophical papers: Vol. 1: Realism, rationalism, and scientific method," *Philosophical Studies*, vol. 31, pp. 480–481, 1986.

52 H. Elmqvist, S. Mattsson, M. Otter, and K. Åström, "Modeling complex physical systems," in *Control of Complex Systems*: Springer, 2001, pp. 21–38.

53 Z.-S. Hou and Z. Wang, "From model-based control to data-driven control: survey, classification and perspective," *Information Sciences*, vol. 235, pp. 3–35, 2013.

54 H. J. Van Waarde, J. Eising, H. L. Trentelman, and M. K. Camlibel, "Data informativity: a new perspective on data-driven analysis and control," *IEEE Transactions on Automatic Control,* vol. 65, no. 11, pp. 4753–4768, 2020.

55 D. G. Elms, "Achieving structural safety: theoretical considerations," *Structural Safety,* vol. 21, no. 4, pp. 311–333, 1999.

3

Unfalsified Adaptive Switching Supervisory Control

3.1 Introduction

A fundamental change in the adaptive control approach prevailed in the late 1980s and early 1990s to handle the control of systems with rapid changes in structure or parameters. The classical adaptive control approach is incompetent in dealing with such system variations. Switching was introduced in the adaptive control scheme, while it was previously utilised in gain scheduling control and the control of systems with a switching nature. Another bottleneck of the conventional adaptive control schemes is the necessary assumptions on plant characteristics. It is well-known that:

- The plant order
- The plant's relative degree
- The sign of the plant's instantaneous or high-frequency gain

must be known a priori for classical adaptive control. However, it has been shown that the only necessary information to stabilise an unknown system is the order of the stabilising controller [1]. This is a key property for the design of data-driven switching control algorithms.

Adaptive switching supervisory control (ASSC) has attracted the attention of many researchers in recent years. The ASSC algorithms encompass a pre-designed control bank and a supervisor unit for controller selection. The task of the high-level control unit, or the supervisor unit, is to estimate each controller's performance based on the system input–output data and select the appropriate controller. A cost function estimates the performance of controllers. For each controller, the cost function in the supervisor unit is constantly calculated and based on the value of the current cost function, the controller with the superior performance is selected as the active controller. Adaptation is an instant in the adaptive switching control approaches, and this feature significantly improves the control performance and overcomes some of the limitations of classical

An Introduction to Data-Driven Control Systems, First Edition. Ali Khaki-Sedigh.

adaptive control. There are various methods proposed for controller selection in the supervisory control unit. The supervisory control unit strategies can be divided into three general categories:

- Pre-routed
- Estimator-based
- Performance-based

The pre-routed approach. In this case, the stabilising controller is searched through a controller set using a pre-determined procedure. The pre-routed switching is one of the first proposed methods in which switching to the next controller is performed based on a pre-determined sequence, and the goal is to achieve the desired performance asymptotically [2]. In the pre-routed approach, the main issue is determining the switching time to the next controller in the loop, and switching is terminated after an acceptable performance is achieved. However, this method is inefficient for a controller set with many members.

The estimator-based approach. In the case of estimators-based switching, also called the *indirect adaptive supervisory control*, the most appropriate model is selected utilising the system-measured data and the associated errors of the multiple estimators in the models' bank. In this approach, nominal models are considered to fully describe the system dynamics, and controllers are designed based on the nominal models. Closed-loop asymptotic stability is usually assured by the assumption that one of the nominal models accurately models the system. This assumption is called the *exact matching* condition. In the estimator-based context, the estimation error of the nominal model is often considered as the cost function. The model-controller set with the least error is introduced as the best option for the next controller. In the presence of extensive uncertainties, these methods usually require a large number of nominal models to satisfy the exact matching condition.

The performance-based approach. In the performance-based supervisory control method, which is also called the *direct adaptive supervisory control*, the performance of the controllers' set is estimated solely by utilising the input–output system data with minimum plant information and assumptions.

The supervisory switching control systems comprise the control set and the supervisor. The control set includes all the designed controllers or the *candidate* controllers. The controller employed in the control loop at time t is the *active* controller, and the rest of the controllers are *inactive* at t. Also, the supervisor includes the model set (usually multiple estimators), the monitoring signal generator and the switching logic.

In the switching procedure, utilising the nominal models, the system output is estimated. Then according to the norm of the weighted output errors, the monitoring signal is calculated. Finally, the switching logic is employed to select the active controller. The minimum necessary assumptions in the adaptive switching control approaches are as follows:

- The *matching property* that requires at least one of the estimation errors to be small.
- The *detectability property* that requires the closed-loop system to be detectable relative to the estimation error.
- The *small error gain property* that requires the switching signal to be selected based on the smallest bounded error.
- *The non-destabilising property* that enforces the detectability property to hold for the switching system.

The adaptive control strategies utilise the *certainty equivalence principle.* This heuristic principle permits the use of inaccurate current estimation of the plant's models. Although the estimation errors may not be small, the controllers are designed based on the estimated models. The detectability property justifies using a controller with a small estimation error. Intuitively, the smallest error provides the closest approximation of the real process, and thus the associated controller is considered to be the most appropriate controller in the set. It is important to note that a small monitoring signal does not necessarily lead to the closed-loop stabilisation property of the corresponding controller unless the persistent excitation requirement is satisfied [2].

The performance-based supervisory control approaches circumvent the need for model estimations. These approaches are data-driven and do not require the above general model-based assumptions. The prominent performance-based supervisory control structures are the unfalsified adaptive switching control (UASC), and the multiple models unfalsified adaptive switching control (MMUASC) structures. Closed-loop stability in the UASC methodology is investigated through the relative gain from the input to the output of the closed-loop system. This investigation is carried out with the minimum necessary assumption of the existence of a *stabilising* controller. The UASC is presented in this chapter, and the MMUASC that combines the features of the multiple model's structures and the unfalsified control philosophy will be discussed in Chapter 4.

3.2 A Philosophical Perspective

The UASC is based on Popper's *falsification theory* for demarcation in the philosophy of science. The falsification theory requires that a scientific theory should entail the following three features [3]:

(1) *Falsifiability*: A theory can be falsified if data from new observations (experiments) contradicts the theory's claims.

(2) *Parsimony*: The theory should be simple with no additional assumptions beyond the necessary assumption for the observed data.
(3) *Validation*: Theory must be thoroughly validated with rigorous experiments that seek a counterexample.

Popper rejected the classical scientific methodology of *observation–deduction*, also known as *positivism*. By advancing the method of experimental falsification, Popper rejected the idea that knowledge originated from analogy and classical reasoning (Aristotelian–Platonic), and he opted for critical rationalism.

In the falsification theory, the relationship between theories and facts is not affirmative but dissentient, abrogations and denials are the principles, and affirmations play a secondary role.

Popper points out that observation is always based on prejudices, and he thus neglected the share of observations' repetition to the share of thought. In other words, we are constantly trying to interpret our experiences rationally. In this view, experiments do not prove theories, but they can only refute them. A theory can only be temporarily reliable as long as it resists rigorous tests.

These philosophical concepts were introduced in the control system engineering context nearly three decades ago. The term *falsifying* was initially used by Jan Camiel Willems in the system identification framework [4]. He proposed that a model is *unfalsified* if it contains all the available information from the system at that moment. Also, the *most powerful unfalsified model* (MPUM) is the model that has the best data fit. Later, following the Willems approach, Michael George Safonov introduced falsification into the field of control system design. Based on the main features of the falsification theory for demarcation, the unfalsified adaptive control algorithms pursue stabilising controllers with the system input–output data and minimum assumptions. In the UASC methodology, instead of identifying a mathematical model for the system or using estimation models, a supervisor is trained in line with the input–output data. Figure 3.1 shows the general structure of the UASC methodology. In this structure, the supervisor selects a controller from the existing pre-designed control set or the controller bank using the observed system input–output data and places the selected controller as the active controller in the control loop.

Although no system model is used in the UASC strategy; however, a model may be used in the controller design procedures to produce the controller set. The data-driven nature of the UASC emanates from the fact that neither the cost function nor the switching algorithm rely on or make use of any a priori assumption about the plant to be controlled.

As depicted in Figure 3.1, it is assumed that the input–output system data are available, and the set of pre-designed controllers and the desired performance characteristics are given. A controller is called *unfalsified* if it does not

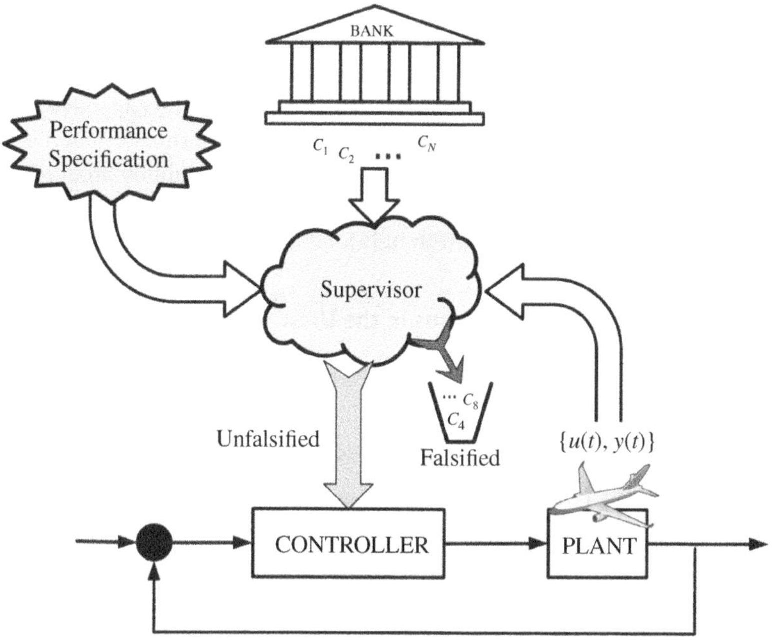

Figure 3.1 The concept of the controllers' falsification.

destabilise the system. *Falsified* or *refuted* controllers are destabilising controllers. Stabilisation is verified using the available input–output data and evaluating a performance criterion or cost function.

3.3 Principles of the Unfalsified Adaptive Switching Control

The falsification process of the candidate controllers is performed without placing the controllers in the control loop. An appropriate control set is assumed to be available and the adaptive control problem is called *feasible* if there is at least one stabilising controller in the control set. To evaluate the performance of inactive controllers, the supervisor uses the *virtual reference signal* concept. The virtual signals are derived from the system input–output data and the controllers' structure. The *performance function* or the *cost function* determines which controller stabilises the closed-loop system solely based on the input–output data.

Unfalsified adaptive control is one of the basic performance-based methods that ensure closed-loop system stability. Closed-loop stability is verified in the input–output sense, with a finite gain, and arbitrary initial conditions.

Interestingly, for stability assurance, knowledge of the system dynamics is not required, and the system is considered to be completely unknown.

Historically, there are papers prior to 1995 addressing the UASC approach, but Ref. [5] is the first paper to present an algorithm for the unfalsified adaptive control with mathematical formulations. In [5], based on a performance-based cost function, the controller is selected; however, the closed-loop stability proof is not provided. The input–output stability of the UASC using the Hysteresis algorithm, and an appropriate cost function is given in [6].

3.3.1 Basic Concepts and Definitions in the UASC Methodology

In discussions on unfalsified control, all the signals $x(t)$ are considered in the l_{2e} space defined as

$$l_{2e} = \{x(t) \colon \|x_t\| < \infty, \quad \text{For every bounded } t \in \mathbb{Z}^+\} \tag{3.1}$$

where $\mathbb{Z}^+$ denotes the positive integer numbers and $\|x_t\|$ denotes the truncated 2-norm of the signal $x(t)$ and is defined as follows[1]:

$$\|x_t\| = \sqrt{\sum_{\tau=0}^{t} |x(\tau)|^2} \tag{3.2}$$

and for some $t \in \mathbb{Z}^+$ the truncated signal x_t is defined as $x_t = \{x(0), \ldots, x(t)\}$.

Let a pre-designed control set with N members be denoted as $\mathbb{C} = \{C_1, C_2, \ldots, C_N\}$. The designed candidate controllers are linear time-invariant and can have different orders. The supervisor selects the active controller from $\mathbb{C}$ based on an appropriate cost function. However, there are $N-1$ inactive controllers for any given time t that their performance should be analysed as candidate controllers. For the assessment of the inactive controllers, the concept of *virtual* or *fictitious* reference is introduced and defined in the following definition. Fictitious reference signals allow the designer to experimentally evaluate the performance of the candidate controllers or even falsify them before any of these inactive controllers are inserted into the feedback loop.

Definition 3.1 Let the input–output data $z(t) = [y(t), u(t)]^T$ be collected up to a given time t. The *virtual reference signal* $\tilde{r}(C_i, z(t))$ or $\tilde{r}_i(t)$ for short, is a hypothetical reference input that would exactly reproduce the same input–output data $z(t)$ up to time t, had the controller C_i been in the control loop for the same period up to t.

This signal is used to evaluate the controllers in the controller set. The virtual reference signal and a so-called *potential loop* are shown in Figure 3.2. Note that

1 This can be equivalently denoted as $\|x\|_t$.

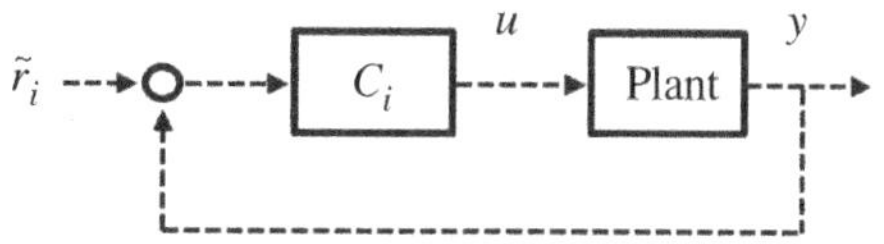

Figure 3.2 The virtual reference signal in the potential loop.

C_i represents active or inactive controllers in the controller set. The input–output data and the controller structure are used to calculate the virtual reference signal, and no system model is required.

Definition 3.2 The controller C_i is *stably causally left invertible* (SCLI) if the mapping $z \rightarrow r$ is causal and incrementally stable.

Incremental stability is an extension of the classical asymptotic stability concept of an equilibrium of a nonlinear system to consider the asymptotic behaviour of different solutions with respect to each other. That is, any two solutions will asymptotically converge to each other regardless of their initial conditions. A system $S: l_{2e} \rightarrow l_{2e}$ is said to be incrementally stable if for any two v, $w \in l_{2e}$, $v \neq w$, we have $\sup_{\tau \in (0,\infty)} \frac{\|(Sv-Sw)_\tau\|}{\|(v-w)_\tau\|} < \infty$.

If the controller C_i is LTI, necessary conditions for SCLI are that the controller transfer function be minimum-phase and biproper.[2] In this chapter, LTI controllers are considered. Hence, assuming that the controller C_i is SCLI, the virtual reference signal $\tilde{r}_i(t)$ that is the signal corresponding to C_i is obtained from the following equation:

$$e_i(t) = C_i^{-1}u(t), \quad \tilde{r}_i(t) = e_i(t) + y(t) \tag{3.3}$$

The cost function should evaluate the potential closed-loop system performance without invoking any system assumptions. The conventional UASC cost function is

$$V(C_i, z, t) = \max_{\tau \leq t} \frac{\|W_1(y - \tilde{r}_i)_\tau\|^2 + \|W_2 u_\tau\|^2}{\|(\tilde{r}_i)_\tau\|^2 + \varepsilon} \tag{3.4}$$

where W_1 and W_2 are the weighting coefficients, and ε is a small positive constant that prevents the cost function from becoming infinite when $\tilde{r}_i = 0$.

Definition 3.3 The adaptive control problem is *feasible* if at least one stabilising controller exists in the controller set at any time.

3.3.2 The Main Results

An essential issue in the practical application of switching adaptive control systems is to prevent the insertion of a destabilising controller in the control loop.

---

2 A biproper transfer function is a transfer function with the same number of finite poles and zeros, or zero relative degree.

Methodologies that focus on the prevention or reduction of the likelihood of an active destabilising controller in the control loop are called *safe adaptive control* techniques in the adaptive control literature. In the UASC, the activation of a destabilising controller is prevented by utilising a cost function mechanism. However, the detection of a future destabilising controller is not possible based on the previous system input–output data and no prior information. That is, with the input–output data up to the present time and no prior information, it is not possible to determine which controller is destabilising. Consequently, in the UASC, the strategy is to detect and immediately disconnect the destabilising controller from the closed-loop system. Hence, the concept of stability unfalsification is introduced. Consider the general closed-loop switching supervisory control Σ shown in Figure 3.3. The following definition is given for a controller's closed-loop stability unfalsification of Σ.

Definition 3.4 Closed-loop stability of a given system Σ is *unfalsified* if for the obtained system input–output data $z(t) = [y(u), u(t)]^T$ and the reference input $r(t)$, there exist parameters $\gamma_1, \gamma_2 > 0$ such that the following inequality is satisfied:

$$\|z_t\| \leq \gamma_1 \|r_t\| + \gamma_2 \quad \forall t \geq 0 \tag{3.5}$$

The *cost function V* is a casual map as

$$V: \mathbb{C} \times \mathbb{Z} \times \mathbb{T} \rightarrow \mathbb{R}_+ \cup \{\infty\} \tag{3.6}$$

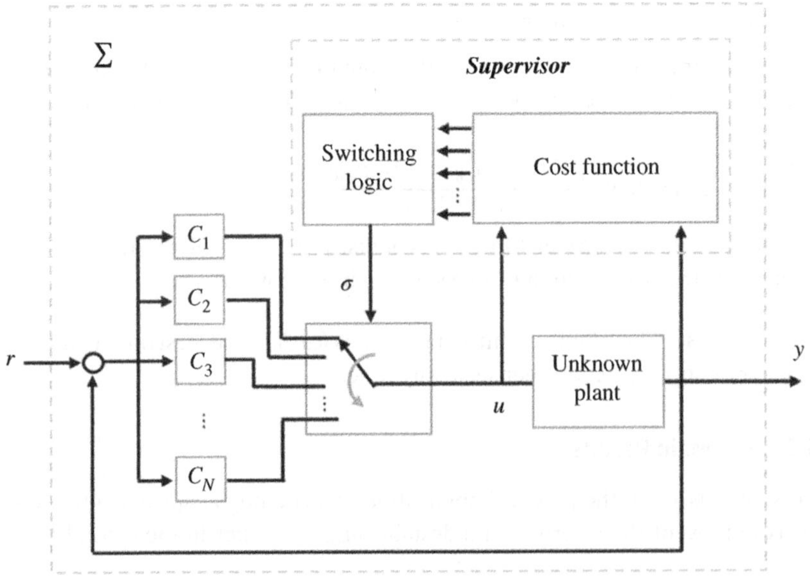

Figure 3.3 General structure of a switching supervisory control.

where $\mathbb{C}$ is the controller set, $\mathbb{Z}$ is the input–output system data, and $\mathbb{T}$ is time. The *active controller* is denoted as $C_{\sigma(t)}$ at time t and $\sigma(t)$ is the *switching signal*. The cost function for the active controller $C_{\sigma(t)}$ at time t is denoted by $V(C_{\sigma(t)}, z, t)$. For the closed-loop switching system Σ, the *true* cost denoted by V_{true} for the control set, $\mathbb{C}$ is defined as follows:

$$V_{true}(C) = \sup_{z \in \mathbb{Z},\, t \in \mathbb{T}} V(C, z, t) \tag{3.7}$$

The cost function V is used to select the appropriate controller. The active controller $C_{\sigma(t)} \in \mathbb{C}$ at time t with data z is falsified if:

$$V(C_{\sigma(t)}, z, t) > \gamma \tag{3.8}$$

The parameter γ is called the *falsifying level*. It can be argued that the controller $C_{\sigma(t)}$ is unfalsified with respect to cost level γ by plant data z. This process is called *self-falsification* since the controller is falsified if its cost function exceeds the given γ, and no comparisons are made with the other candidate controllers.

Another approach for controller falsification is to falsify the controller if its cost function value exceeds that of another controller. Hence, falsification is based on comparing the active controller with the inactive controllers in the controller set. This falsification process is called *interfalsification* and is formulated as follows:

$$V(C_{\sigma(t)}, z, t) > \min_{C \in \mathbb{C}} V(C, z, t) + h \tag{3.9}$$

where h is a small positive constant to avoid chattering.

Note that both the above falsification processes use only finite plant input–output data. However, Definition 3.4 is based on an infinite data verification, and the finite data used may result in erroneous falsification or unfalsification of a controller. It is, therefore, crucial to ensure closed-loop plant stability under the feasibility assumption for the given cost function and switching algorithm after a finite number of controller switching.

In what follows, definitions, assumptions and proof of closed-loop plant stability of the switching system given in Figure 3.3 with the cost function (3.4) and the Hysteresis switching algorithm are provided.

The *robust optimal controller* denoted by C_{RSP} is a controller that stabilises the system and also minimises the true cost function V_{true}, which is obtained from

$$C_{RSP} = \operatorname*{argmin}_{C \in \mathbb{C}} V_{true}(C) \tag{3.10}$$

Although C_{RSP} is not necessarily unique; there is at least one C_{RSP} in the controllers set under the feasibility assumption, it will be shown that its cost function is bounded.

Consider the ASSC shown in Figure 3.3. The notion of *cost detectability* is introduced to establish a relation between a cost function and the stability of the closed-loop system Σ.

Definition 3.5 The cost function V and the controller set $\mathbb{C}$ are said to be a *cost detectable* pair if for every $C_{\sigma(t)} \in \mathbb{C}$ with a finite number of switching, the following conditions are equivalent:

(1) $V(C_f, z, t)$ remains bounded when $t \to \infty$, and C_f is the final selected controller.
(2) The closed-loop system stability with the controller $C_{\sigma(t)}$ is unfalsified by the system input–output data.

With the cost detectability condition, closed-loop instabilities are detected with no prior system assumptions.

Definition 3.6 The cost function V with the controller set $\mathbb{C}$ is l_{2e} *gain related* if for each $C_i \in \mathbb{C}$ and $z \in l_{2e}$, the following statements hold:

- $V(C_i, z, t)$ is uniformly non-decreasing in time.
- The virtual reference input signal exits and $\tilde{r}(C_i, z) \in l_{2e}$.
- For every $C_i \in \mathbb{C}$ and $z \in l_{2e}$, the cost function $V(C_i, z, t)$ remains bounded when $t \to \infty$, if and only if closed-loop system stability is unfalsified with respect to $\tilde{r}(C_i, z)$ and z.

It is clear from Definitions 3.5 and 3.6 that cost detectability implicitly expresses the property of being l_{2e} gain related. In fact, in a special case where the controller is fixed and non-switching, l_{2e} gain related is the same as cost detectability.

The following theorem provides a sufficient condition for cost detectability; in the case of LTI systems, this will be a necessary and sufficient condition.

Theorem 3.1 If the cost function is l_{2e} gain related, a sufficient condition for detectability of $(V, \mathbb{C})$ is that all controllers in the control set be SCLI. This condition is necessary and sufficient for linear time-invariant controllers.

Proof: Refer to [6] for details.

The following lemma is necessary for the stability proof of Σ in Theorem 3.2 that subsequently follows.

Lemma 3.1 If controller switching is finally terminated, there will be a time $t_f > 0$ such that:

$$C_{\sigma(t)} = C_f, \quad C_f \in \mathbb{C}, \quad \forall t > t_f \tag{3.11}$$

If the controller C_f is SCLI and the corresponding virtual reference signal is $\tilde{r}_{(C_f)}$, the mapping $r \rightarrow \tilde{r}_{(C_f)}$ is stable with a gain of 1. Then, there exists a constant $\alpha > 0$ such that

$$\|(\tilde{r}_{C_f})_t\| \leq \|r_t\| + \alpha, \quad \forall t \in \mathbb{Z}_+ \tag{3.12}$$

Proof: If the controller is SCLI, the controller inverse should be incrementally stable. According to the incrementally stable definition, there exist positive constants $\overline{\alpha}$ and $\overline{\beta}$ such that

$$\|(r - \tilde{r}_{C_f})_t\| \leq \overline{\alpha}\|(z - z_f)_t\| + \overline{\beta} \tag{3.13}$$

where z_f is the data after the final controller is activated. By using the triangular inequality, it can be shown that

$$\|(\tilde{r}_{C_f})_t\| \leq \|r_t\| + \overline{\alpha}\|(z - z_f)_t\| + \overline{\beta} \tag{3.14}$$

Consequently, the proof is completed by considering $\alpha = \overline{\alpha}\|(z - z_f)_t\| + \overline{\beta}$. ■

Unfalsification of the controllers is done based on the system input–output data z, according to (3.5). Consequently, the controller is selected as follows:

$$C_{\sigma(t)} = \underset{C_i \in C}{\text{argmin}} \left(V(C_i, z, t) - h\delta_{i,\sigma(t-1)}\right) \tag{3.15}$$

where $h > 0$ is the Hysteresis constant. The symbol arg denotes the argument of the minimisation problem, and $\delta_{i,j}$ is the Kronecker δ defined as:

$$\delta_{i,j} = \begin{cases} 0 & \text{if } i \neq j \\ 1 & \text{if } i = j \end{cases} \tag{3.16}$$

The Hysteresis switching logic is used in the UASC methodology to select the stabilising controller. The Hysteresis algorithm is presented in Algorithm 3.1.

Algorithm 3.1 Hysteresis Switching Algorithm

Input $t = 0$, select an arbitrary $C_{\sigma(0)}$ and $\varepsilon > 0$ (used in V) and $h > 0$

Step 1 Let $t \leftarrow t + 1$

If $V(C_{\sigma(t-1)}, z, t) > \min_{C_i \in C} V(C_i, z, t) + h$, *then*

$$C_{\sigma(t)} \leftarrow \underset{C_i \in C}{\text{argmin}}\ V(C_i, z, t)$$

Else, Then

$$C_{\sigma(t)} \leftarrow C_{\sigma(t-1)}$$

END

Step 2 Go to step 1.

According to Algorithm 3.1, the active controller (the controller in the loop) remains in the control loop in the next step time if its cost function is less than the minimum cost function of all the controllers plus h, where h is the Hysteresis parameter that prevents chattering. Considering non-decreasing cost functions in time, as well as the fact that at least one of the cost functions is always bounded, there is a final time t_f so that there is no switching afterwards. In addition, the cost function of the last controller is bounded. If the cost function is uniformly non-decreasing in time for all $C \in \mathbb{C}$, then for each falsifying level $\gamma \in \mathbb{R}$, the set of unfalsified controllers at time t is represented as $\mathbb{C}_{unf}(\gamma, t)$. The set of unfalsified controllers decreases over time. Hence, for $\tau_1 \leq \tau_2$, the control set satisfies $\mathbb{C}_{unf}(\gamma, \tau_1) \geq \mathbb{C}_{unf}(\gamma, \tau_2)$.

Also, the upper number of switching with the cost function (3.4) and the Hysteresis algorithm is $(N+1)\frac{V_{true}(C_{RSP})}{h}$, where V_{true} and C_{RSP} are introduced in Eqs. (3.7) and (3.10), respectively. If the input is rich enough, the following relation is satisfied for each controller [7]:

$$\lim_{\tau \to \infty} \max_{z \in \Sigma} V(C, z, \tau) \geq \{V_{true}(C_{RSP}) = \min_{C \in \mathbb{C}} V_{true}(C)\} \tag{3.17}$$

Sufficiently rich input contains frequencies that excite all dynamics of the system, especially the unstable dynamics, and this excitation increases the cost function. If the cost function is cost detectable, uniformly non-decreasing in time, and uniformly bounded from above for a stabilising controller, then the proposed algorithm with a limited number of switching guarantees stabilisation of the closed-loop system. Also, if the input signal is rich enough, the optimal cost function tends to $V_{true}(C_{RSP})$.

$$V(C_{\sigma(t)}, z, t) \to V_{true}(C_{RSP}) \pm h \tag{3.18}$$

Theorem 3.2 If the adaptive control problem is feasible, the cost functions are l_{2e} gain related, and all the controllers are SCLI in the control set, the closed-loop system of Figure 3.3 with the switching Algorithm 3.1 is input–output stable.

Proof: According to the feasibility condition, there exists a stabilising controller in the controller set. Let C_j be the stabilising controller. Considering Figure 3.2, there exist positive constants γ_1 and γ_2 such that the following inequality is satisfied:

$$\|z_t\| \leq \gamma_1 \|(\tilde{r}_j)_t\| + \gamma_2 \tag{3.19}$$

Squaring both sides give

$$\|z_t\|^2 \leq \gamma_3 \|(\tilde{r}_j)_t\|^2 + \gamma_4 \tag{3.20}$$

where $\gamma_3 = {\gamma_1}^2$ and $\gamma_4 = {\gamma_2}^2 + 2\gamma_1\gamma_2\|(\tilde{r}_j)_t\|$.

By defining $\varepsilon = \gamma_4/\gamma_3$, the following inequality is derived from (3.20)

$$\frac{\|z_t\|^2}{\|(\tilde{r}_j)_t\|^2 + \varepsilon} \leq \gamma_3 \tag{3.21}$$

In this relation $\|z_t\|$ is

$$\|z_t\|^2 = \left\| \begin{matrix} y_t \\ u_t \end{matrix} \right\|^2 = \|y_t\|^2 + \|u_t\|^2 \tag{3.22}$$

By substituting (3.22) in (3.21) and some mathematical manipulations, the following inequality is achieved:

$$\frac{\|W_1\, y_t\|^2 + \|W_2\, u_t\|^2}{\|(\tilde{r}_j)_t\|^2 + \varepsilon} \leq \gamma_5 \tag{3.23}$$

where $\gamma_5 = \max\{W_1{}^2, W_2{}^2\} \times \gamma_3$. This shows that the relative cost function of the stabilising controller C_j remains bounded for all time.

In the next step, let switching terminate after a finite number of switches, and the final controller be C_f. Therefore, its cost function should be bounded, which implies that there exists a positive constant γ_6 such that

$$\frac{\|W_1\, y_t\|^2 + \|W_2\, u_t\|^2}{\|(\tilde{r}_f)_t\|^2 + \varepsilon} \leq \gamma_6 \tag{3.24}$$

This inequality can be rewritten as follows:

$$\left\| \begin{matrix} y_t \\ u_t \end{matrix} \right\|^2 \leq \frac{\gamma_6}{\min\{W_1, W_2\}} \left(\|(\tilde{r}_f)_t\|^2 + \varepsilon\right) \tag{3.25}$$

Also, according to Lemma 3.1, the virtual reference of the final controller satisfies:

$$\|(\tilde{r}_f)_t\| \leq \|r_t\| + \gamma_7 \tag{3.26}$$

where γ_7 is a positive constant. By substituting (3.26) in (3.25), it can be shown that

$$\|z_t\|^2 \leq \gamma_8 \|r_t\|^2 + \gamma_9 \tag{3.27}$$

where

$$\gamma_8 = \frac{\gamma_6}{\min\{W_1, W_2\}} \tag{3.28}$$

$$\gamma_9 = \frac{\gamma_6}{\min\{W_1, W_2\}} \left(2\|r_t\|\gamma_7 + \gamma_7{}^2 + \varepsilon\right) \tag{3.29}$$

Consequently, the proof is completed. ■

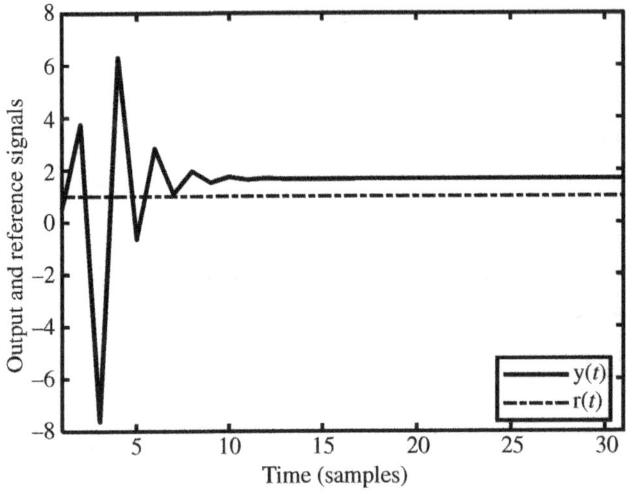

Figure 3.4 The output and reference input.

Example 3.1 Consider the following system:

$$G(d) = \frac{d}{1 - 2d} \tag{3.30}$$

and d is the delay operator.[3] A controller bank is designed as

$$\mathbb{C} = \{2.5{,}5.5\} \tag{3.31}$$

The stabilising controller in the controller set is $C = 2.5$, and therefore the feasibility condition is satisfied. Cost function (3.4) is used with the following parameters:

$$h = 5, \quad \varepsilon = 0.01, \quad W_1 = W_2 = 1 \tag{3.32}$$

Consider the closed-loop system shown in Figure 3.3 with $y(0) = 0.5$. The initial controller is $C_{\sigma(0)} = 5.5$, which is a destabilising controller.

The system output is depicted in Figure 3.4 for a unit step reference input. After three sample times, the supervisor selects the stabilising controller. The cost functions and switching signal are shown in Figure 3.5.

3 That is $d = z^{-1}$, where z is the forward shift operator.

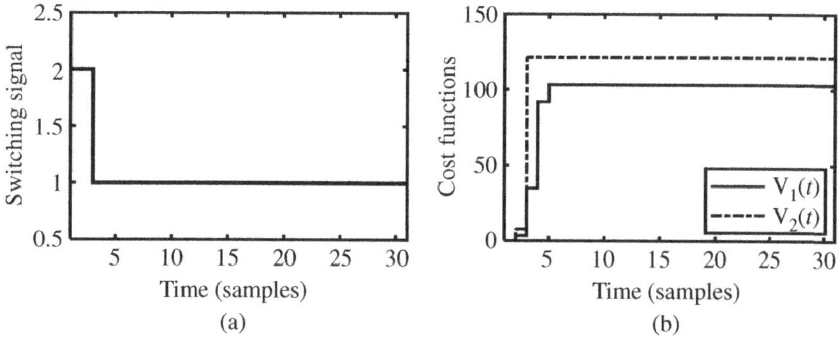

Figure 3.5 (a) The switching signal and (b) the cost functions.

3.4 The Non-Minimum Phase Controller

So far, the controllers in the controller set are assumed SCLI. In this section, the virtual reference and the cost function are modified such that non-SCLI controllers can be used. To calculate the modified virtual reference signal for non-minimum phase controllers, the coprime factorisation decomposition is used for the controllers.

Definition 3.7 Any two stable, proper rational functions $A(d)$ and $B(d)$ are coprime if there exists stable, proper rational functions $X(d)$ and $Y(d)$ such that the Bezout's identity given below is satisfied.

$$X(d)A(d) - Y(d)B(d) = I \tag{3.33}$$

Note that $A(d)$ and $B(d)$ are coprime if and only if they have no common zeros, and in the case of proper rational matrices, left and right coprime factorisations are defined.

Let the coprime factorisation of the controller $C_i(d)$ be

$$C_i(d) = D_i(d)^{-1}N_i(d) \tag{3.34}$$

where $N_i(d)$ and $D_i(d)$ are stable, proper rational functions. The modified virtual reference signal is calculated as follows:

$$\tilde{v}_i(t) = N_i(d)y(t) + D_i(d)u(t) \tag{3.35}$$

By this definition, the potential loop is modified, as shown in Figure 3.6, and the new cost function is represented as follows [7]:

$$V(C_i, z, t) = \max_{\tau \le t} \frac{\|y_\tau\|^2 + \|u_\tau\|^2}{\|(\tilde{v}_i)_\tau\|^2 + \varepsilon} \tag{3.36}$$

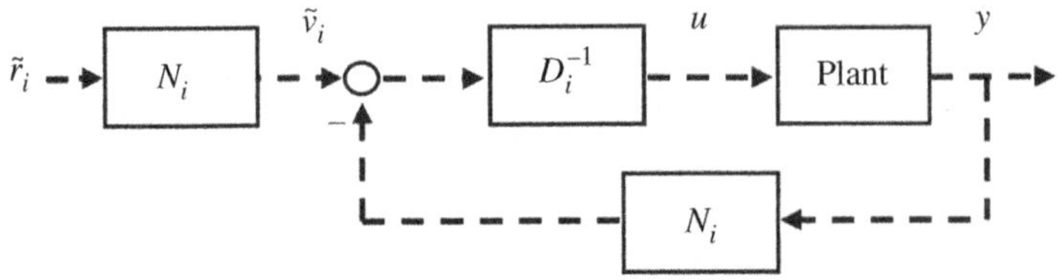

Figure 3.6 The new potential loop for a non-minimum phase controller.

In the cost function (3.36), the controller no longer needs to be SCLI. The stability analysis of the cost function (3.36) is similar to the cost function (3.4).

3.5 The DAL Phenomena

Adaptive control systems, including the UASC methodology, can exhibit poor transient time performance with large input and output signals, which is called the *DAL*[4] *phenomena* [8]. The DAL phenomenon occurs when a destabilising controller is inserted into the loop. The magnitudes of the control and output signals increase to unacceptable levels before the plant is finally stabilised. The following example shows the DAL phenomenon:

Example 3.2 Consider the following system and controllers:

$$G(s) = \frac{1}{s-1} \quad C_1 = 2, \quad C_2 = 0.5 \tag{3.37}$$

The controllers C_1 and C_2 are stabilising and destabilising controllers, respectively. For a sine reference input, the closed-loop system responses are shown in Figure 3.7. The sampling time is $T_s = 0.1$ and $\varepsilon = 0.1$. The closed-loop responses for the stabilising controller in the loop, and three different values of the Hysteresis parameter $h = 0.01, 0.05, 0.2$, are presented. As is shown in Figure 3.7, decreasing the Hysteresis parameter intensifies the DAL phenomenon, and very large transient responses are observed.

An important point to note is that the definition of norm plays a key role in the UASC stability study. This is discussed in [9]. Utilising the cost function (3.4), it will remain bounded in the presence of large overshoots. This comes from the fact that the l_2 norm is employed in (3.4). That is, although the system may not be stable in the sense of the l_∞ norm, it will be stable in the sense of the l_2 norm. Let

$$V_p(C_i, z, t) = \max_{\tau \in [0,t]} \frac{\|u_\tau\|_p^{\,2} + \|(\tilde{r}_i - y)_\tau\|_p^{\,2}}{\|(\tilde{r}_i)_\tau\|_p^{\,2} + \varepsilon}, \quad p = 2, \infty \tag{3.38}$$

4 Dehghani-Anderson-Lanzon.

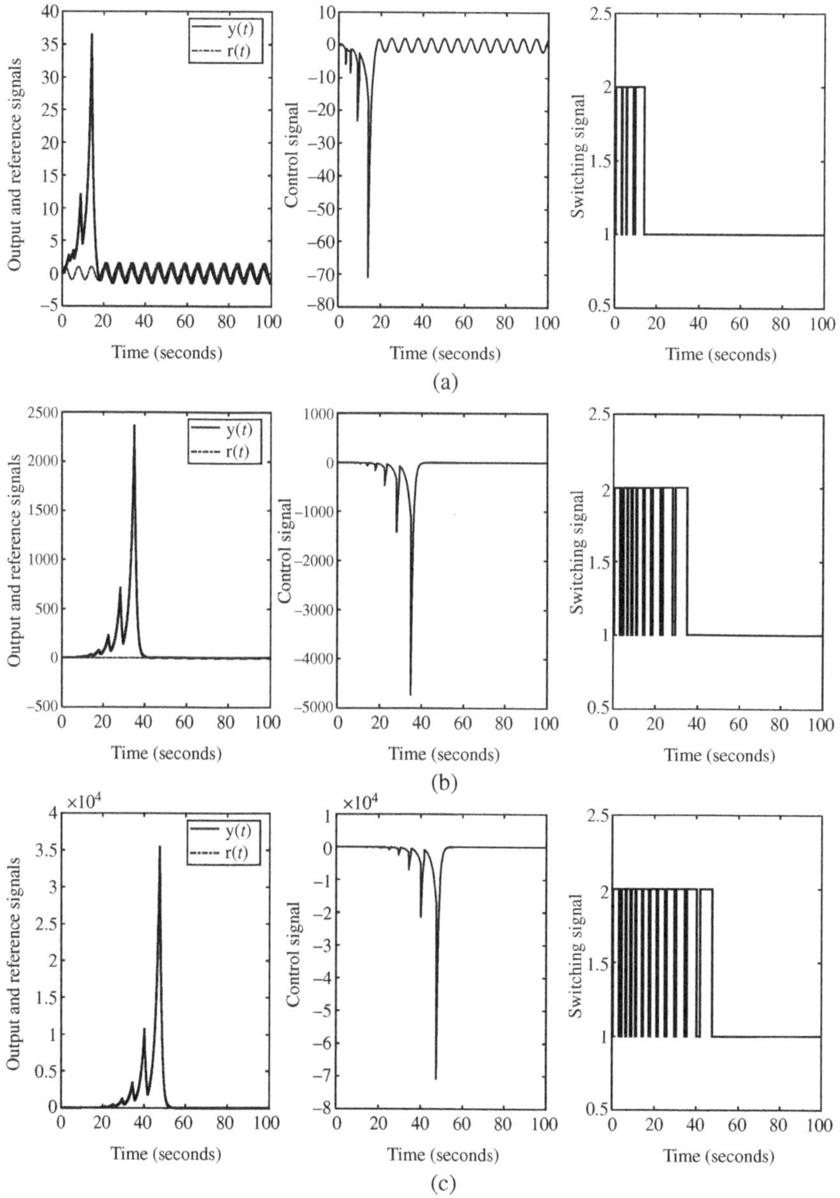

Figure 3.7 The switching signal, control input signal and output for different values of h: (*a*) $h = 0.2$, (*b*) $h = 0.05$, (*c*) $h = 0.01$.

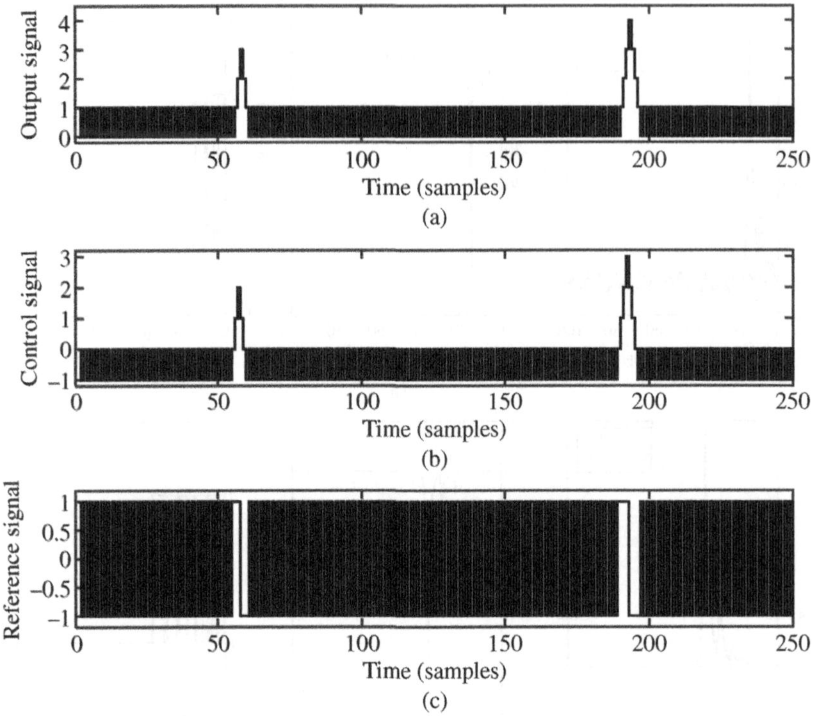

Figure 3.8 (a) Output signal, (b) control input signal and (c) reference input.

where the l_2 norm $\|\cdot\|_2$ is defined in (3.2) and the l_∞ norm $\|\cdot\|_\infty$ is defined as:

$$\|s_t\|_\infty = \max_{\tau \le t} |s(\tau)| \tag{3.39}$$

It is shown via an example that although the amplitudes of u and y may be increasing, the cost function based on the l_2 norm ($\mathcal{V}_2$) can remain bounded, and in this case, the use of the l_∞ norm in the cost function is proposed [9].

The simulation results of the example in [9] are shown in Figures 3.8 and 3.9. The cost function calculated with the l_∞ norm ($\mathcal{V}_\infty$) is increasing. However, the cost function with the infinity norm is not cost detectable, and if the cost function with the infinity norm is finite, it is not possible to guarantee input–output stability.

To solve the detectability issue and reduce the DAL phenomenon, a combination of cost functions is proposed. The combined cost function is

$$V(C_i, z, t) = \mathcal{V}_2(C_i, z, t) + \varrho\mathcal{V}_\infty(C_i, z, t) \tag{3.40}$$

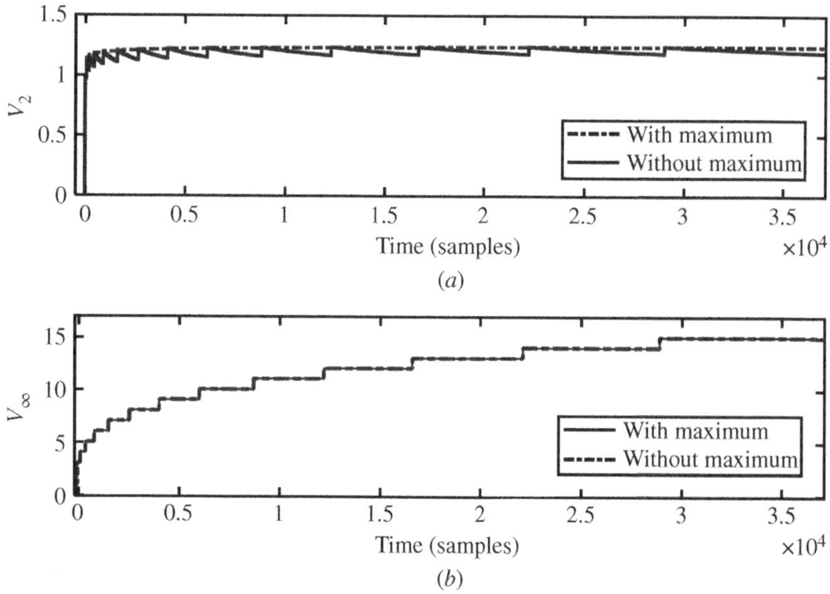

Figure 3.9 Cost function: (a) norm 2 versus (b) norm ∞.

where ϱ is a positive weighting constant. This function is cost detectable, and thus the input–output stability of the closed-loop system is assured. Also, instead of using a linear combination in (3.40), the maximum value of the cost functions can be used. In this case, the final cost function (V) for the controller C_i is:

$$V(C_i, z, t) = \max_p \mathcal{V}_p(C_i, z, t), \quad p = 2, \infty \tag{3.41}$$

In the case of a combination of cost functions, it can be shown that for cost detectability assurance, at least one of the costs $\mathcal{V}_j$ must be cost detectable [10].

3.6 Performance Improvement Techniques

In this section, four basic modifications in the UASC literature are introduced to enhance the closed-loop performance of the UASC approach.

3.6.1 Filtered Cost Function

A performance-enhancement approach is to filter the signals using N designed filters and provide additional degrees of freedom. The filters are placed in a certain bandwidth determined by the designer, and the cost functions for different controllers are evaluated as:

$$W(C_i, z, t) = \max_j \frac{\|(F_j u)_t\|^2 + \|F_j(\tilde{r}_i - y)_t\|^2}{\|(F_j \tilde{r}_i)_t\|^2 + \varepsilon} \tag{3.42}$$

and the final cost function is selected as

$$V(C_i, z, t) = \max_{\tau \in [0,t]} W(C_i, z, \tau) \tag{3.43}$$

Simulation results show the superior performance of the proposed approach to the conventional UASC. However, this cannot be analytically proved. Also, the computational load is increased in comparison with the conventional UASC [11]. Finally, the filter location selection can be difficult, and the number of switching is generally unknown.

3.6.2 Threshold Hysteresis Algorithm

To improve the performance of the UASC, a threshold can be considered for the cost function. If the cost function value is higher than a specified threshold level M, then switching is performed according to the Hysteresis algorithm; otherwise, there is no switching. The *threshold hysteresis algorithm (THSA)* is proposed in [12] and is implemented in Algorithm 3.2. The closed-loop stability under the assumptions of feasibility and SCLI controllers is proved in [12].

Algorithm 3.2 Threshold Hysteresis Algorithm

Input Let $t = 0$, select an arbitrary $C_{\sigma(0)}$ and $\varepsilon > 0$ (used in V) and $h > 0$

Step 1 Let $t \leftarrow t + 1$ collect r, u, y, update $\tilde{r}_i, V$

Step 2 *If* $V(C_{\sigma(t-1)}, z, t) > \min_{C_i \in \mathbb{C}} V(C_i, z, t) + h$ *and* $V(C_{\sigma(t-1)}, z, t) > M$, *Then*

$$C_{\sigma(t)} \leftarrow \operatorname*{argmin}_{C_i \in \mathbb{C}} V(C_i, z, t)$$

Else, Then

$$C_{\sigma(t)} \leftarrow C_{\sigma(t-1)}$$

END

Step 3 Go to step 1.

Example 3.3 Consider the system in Example 3.2. The THSA parameter is $M = 50$, and the Hysteresis constant is $h = 0.01$. The output, control input and switching signals of THSA and UASC are depicted in Figures 3.10 and 3.11, respectively. Also, the cost functions of THSA and UASC are shown in Figures 3.12 and 3.13. As shown in Figure 3.10, the stabilising controller is selected faster than the conventional UASC and the DAL phenomenon is attenuated. The cost functions of the THSA indicate the rapid selection of the first controller.

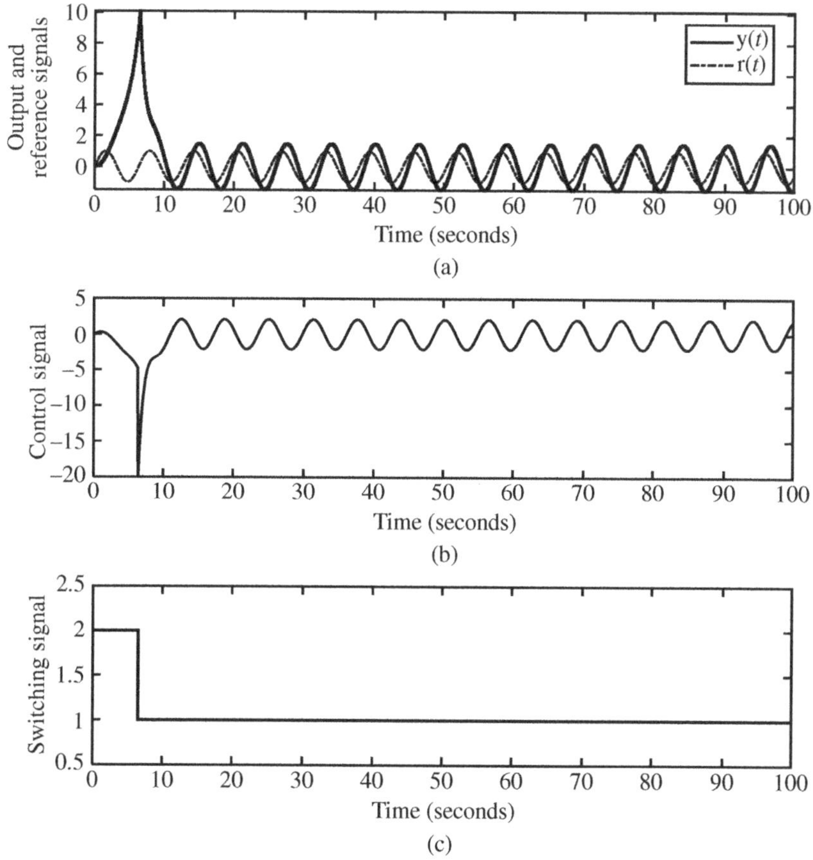

Figure 3.10 THSA algorithm: (a) output and reference input, (b) control input and (c) switching signal.

3.6.3 Scale-Independent Hysteresis Algorithm

The multiplicative Hysteresis constant is proposed in [12]. As the cost function value is not known in advance, the choice of the Hysteresis constant parameter has a significant effect on the system performance. Using the multiplicative Hysteresis approach, this dependency is removed. The multiplicative Hysteresis is called the *scale-independent hysteresis algorithm (SIHSA),* and its closed-loop stability under the assumptions of feasibility and SCLI controllers is proved in [12]. The SIHSA is summarised in Algorithm 3.3. In the SIHSA, step 2 in the Hysteresis Algorithm 3.1 is modified according to the multiplicative Hysteresis constant policy.

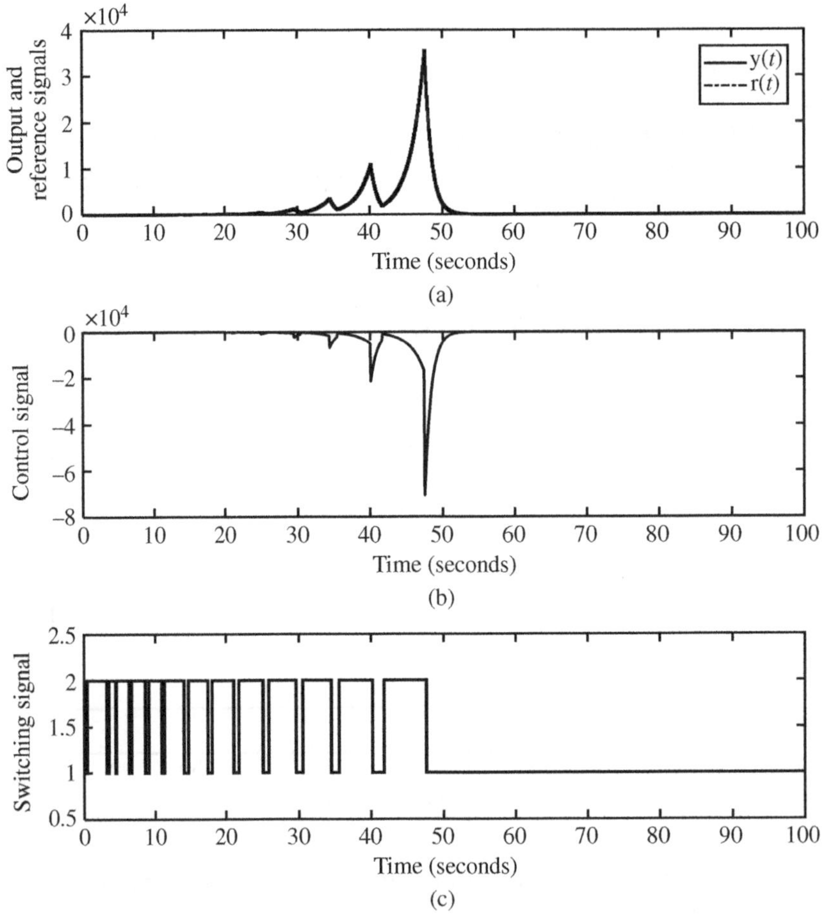

Figure 3.11 UASC algorithm: (a) output and reference input, (b) control input and (c) switching signal.

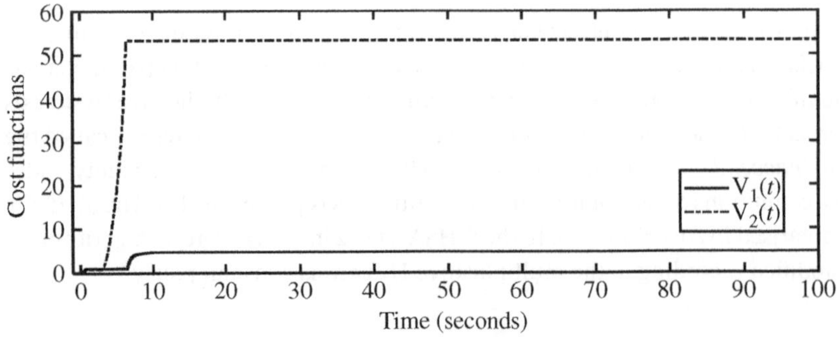

Figure 3.12 The cost functions for the THSA algorithm.

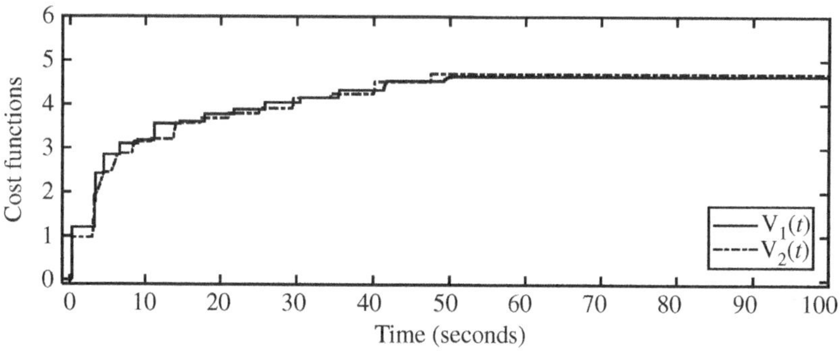

Figure 3.13 The cost functions for the UASC algorithm.

As the DAL phenomenon is directly related to the Hysteresis value h, simulation results have shown that the DAL phenomenon is significantly reduced by using the SIHSA algorithm.

Algorithm 3.3 Scale-Independent Hysteresis Algorithm

Input Let $t = 0$, select an arbitrary $C_{\sigma(0)}$ and $\varepsilon > 0$ (used in V) and $h > 0$

Step 1 $t \leftarrow t + 1$ collect r, u, y, update $\tilde{r}_i$, V

Step 2 *If* $V(C_{\sigma(t-1)}, z, t) > \min\limits_{C_i \in \mathbb{C}} V(C_i, z, t)(1 + h)$, *Then*

$$C_{\sigma(t)} \leftarrow \underset{C_i \in \mathbb{C}}{\operatorname{argmin}}\, V(C_i, z, t)$$

Else, Then

$$C_{\sigma(t)} \leftarrow C_{\sigma(t-1)}$$

END

Step 3 Go to step 1

Example 3.4 Consider the system in Example 3.2. The Hysteresis parameter in the SIHSA is $h = 0.1$. The output, control input, and switching signals are depicted in Figure 3.14, and the cost functions are shown in Figure 3.15. As shown in Figure 3.14, the stabilising controller is selected faster than the conventional UASC and the DAL phenomenon is attenuated.

3.7 Increasing Cost Level Algorithms in UASC

The previous interfalsification Hysteresis-based algorithms suffer from two limitations. First, the cost function must be uniformly non-decreasing, which limits the closed-loop system performance. In [13], by applying the concept of *self-falsification,* this limitation is circumvented, and with the application of

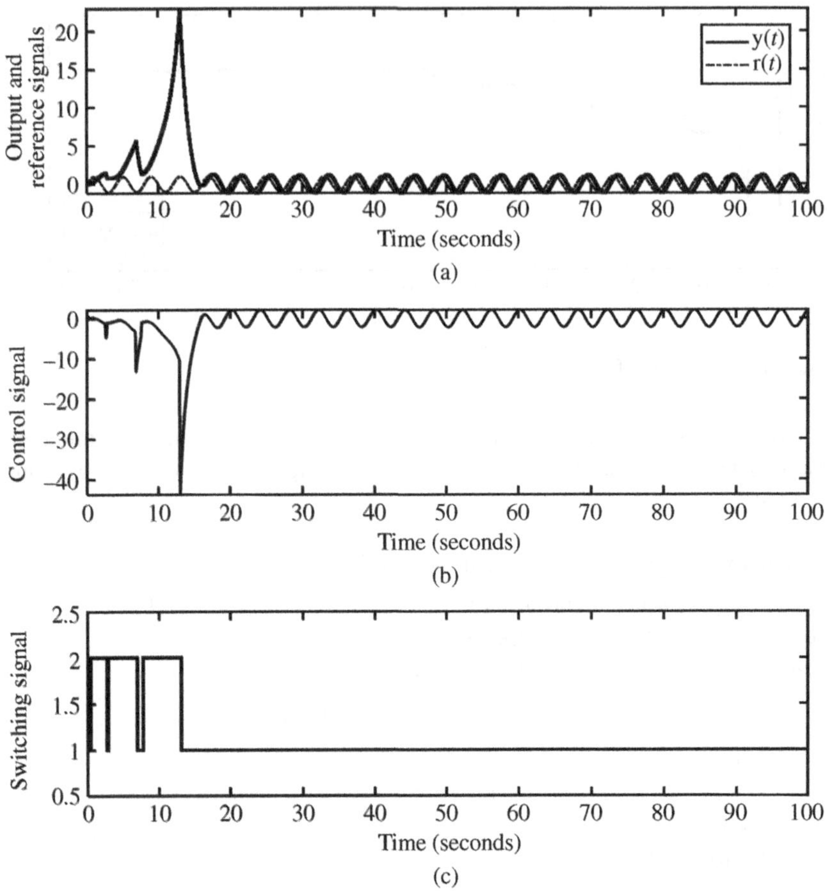

Figure 3.14 SIHSA algorithm: (a) Output and reference input, (b) control input and (c) switching signal.

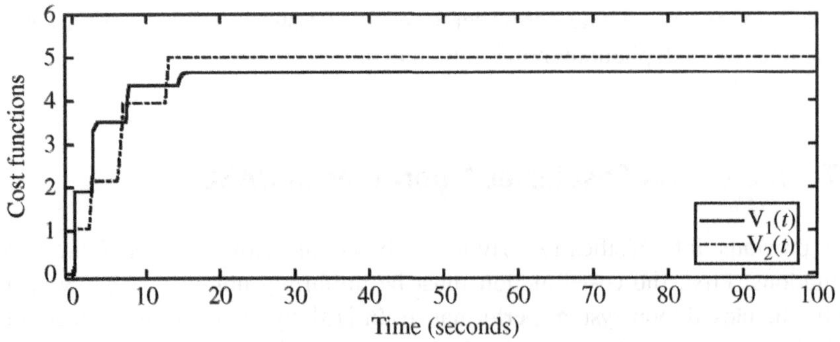

Figure 3.15 The cost functions for the SIHSA algorithm.

nonmonotone[5] cost functions, the employment of fading memory cost functions becomes feasible. The cost function is modified as

$$V(C_i, z, t) = \frac{\|W_1(y - \tilde{r}_i)_t\|^2 + \|W_2 u_t\|^2}{\|(\tilde{r}_i)_t\|^2 + \varepsilon} \tag{3.44}$$

Second, it was shown in Section 3.6 that even with cost detectable cost functions, a supervisor using the ε-Hysteresis algorithms might insert a destabilising controller in the feedback loop before finally stabilising the closed-loop system and thus deteriorate the transient response with very large control signals and outputs. In the inter-falsification, the process of falsification is done by comparing the active controller cost function with the minimum of the other cost functions. As previously described, the inter-falsification Hysteresis algorithm is used to prevent intermittent switching, and for stability, the cost function should be uniformly non-decreasing in time.

In the self-falsification, the cost function of the active controller is compared to a predetermined sequence γ. The Hysteresis switching logic is no longer used here, and the closed-loop stability proof must be reinvestigated. The Increasing Cost Level Algorithm (ICLA) and Linearly Increasing Cost Level Algorithm (LICLA) are the two algorithms that use the self-falsification concept. These algorithms proposed in [13] aim to resolve the uniformly non-decreasing requirement of cost functions. Also, although the self-falsification algorithms cannot prevent the insertion of destabilising controllers in the feedback loop, they can attenuate the DAL phenomenon. In these algorithms, the cost level sequence to falsify the active controllers is increasing, and with LICLA algorithms, the cost level sequence *linearly* increases at each switching time.

3.7.1 Increasing Cost Level Algorithm

The *Increasing Cost Level Algorithm (ICLA)* is presented in Algorithm 3.4, where $\eta_j > 0$ is the sequence that the cost function is compared with. This sequence should be increasing in time and has the following characteristics.

$$\lim_{j\to\infty} \eta_j = \infty \tag{3.45}$$

Algorithm 3.4 Increasing Cost Level Algorithm

Input Let $1 \leftarrow j$, $t \leftarrow 0$, select an arbitrary $C_{\sigma(0)}$ and choose the sequence $\{\eta_j\}$ *and* $\gamma \leftarrow \eta_1$

Step 1 Let $t \leftarrow t+1$ collect r, u, y, update $\tilde{r}_i, V$

5 Nonmonotone functions are functions whoes first derivatives change sign and can therefore be increasing or decreasing in time.

Step 2 *If* $V(C_{\sigma(t-1)}, z, t) > \gamma$, *Then*

$$\mathbb{C} \leftarrow \mathbb{C} \backslash C_{\sigma(t-1)}$$

If $\mathbb{C} = \emptyset$, *Then*

$$\mathbb{C} \leftarrow \{C_1, .. C_N\}$$

END

$$C_{\sigma(t)} \leftarrow C_{next}$$

$$\gamma \leftarrow \eta_{j+1}$$

$$j \leftarrow j+1$$

Else, Then

$$C_{\sigma(t)} \leftarrow C_{\sigma(t-1)}$$

END

Step 3 Go to Step 1

Algorithm 3.4 never terminates, and the supervisor tests the active controller for possible falsification at time instants, 0, 1, 2, ... The data is collected and updated at each sampling time, and the supervisor then calculates the value of the cost function for each candidate controller. The third step is the inter-falsification phase. If the inequality $V(C_{\sigma(t-1)}, z, t) > \gamma$ holds the currently active controller is self-falsified by the current cost level γ and is removed from the control loop. The removed controller is now inactive and is replaced by the next controller from the control set. The next cost level to falsify will be set to η_{j+1}. Note that if the inequality $V(C_{\sigma(t-1)}, z, t) > \gamma$ doesn't hold; the active controller would be unfalsified and will remain in the control loop for the next sampling time.

The main problem of the ICLA algorithm is to determine the sequence η_j especially when there is no available system information, and the cost functions' values are unknown.

The following theorem gives the stability conditions of the ICLA algorithm.

Theorem 3.3 Consider the closed-loop switching adaptive control system Σ in Figure 3.3. Suppose the pair $(V, \mathbb{C})$ is cost detectable. Then, with Algorithm 3.4, the closed-loop system will be stable if the following two conditions hold:

(1) The cost-level sequence $\{\eta_j\}$ satisfies $\eta_j < \infty, \forall j \in \mathbb{N}$ and $\lim_{j \to \infty} \eta_j = \infty$.

(2) The controller selector procedure is such that for each input r, switching stops after at most finitely many switches.

Proof: For proof, refer to [13].

3.7.2 Linear Increasing Cost Level Algorithm

In the ICLA algorithm, the controller selector procedure and the cost level sequences are not specified. The *Linearly Increasing Cost Level Algorithm (LICLA)* is more practical as it uses the minimum of the remaining cost functions to determine the next active controller and specifies the cost level sequences. Algorithm 3.5 presents the LICLA procedure.

Algorithm 3.5 Linearly Increasing Cost Level Algorithm

Input Let $\eta_0 > 0$, $\Delta\eta > 0$, $\gamma \leftarrow \eta_0$, $t \leftarrow 0$, $\varepsilon > 0$ and an arbitrary $C_{\sigma(0)}$

Step 1 Let $t \leftarrow t+1$ collect r, u, y, update $\tilde{r}_i, V$

Step 2 *If* $V(C_{\sigma(t-1)}, z, t) > \gamma$, *Then*

$$\mathbb{C} \leftarrow \mathbb{C} \backslash C_{\sigma(t-1)}$$

If $\mathbb{C} = \emptyset$, *Then*

$$\mathbb{C} \leftarrow \{C_1, \ldots C_N\}$$

END

$$C_{\sigma(t)} \leftarrow \underset{C_i \in \mathbb{C}}{\operatorname{argmin}} V(C_i, z, t)$$

$$\gamma \leftarrow \gamma + \Delta\eta$$

Else, Then

$$C_{\sigma(t)} \leftarrow C_{\sigma(t-1)}$$

END

Step 3 Go to Step 1

A comparison of Algorithms 3.4 and 3.5 indicates the LICLA is a special version of the ICLA with the following provisions:

$$\eta_j = \eta_0 + (j-1)\Delta\eta \tag{3.46}$$

and

$$C_{\sigma(t)} = \underset{C_i \in \mathbb{C}}{\operatorname{argmin}} V(C_i, z, t) \tag{3.47}$$

The following theorem presented and proved in [13] provides the convergence results of Algorithm 3.5.

Theorem 3.4 Suppose that the closed-loop switching adaptive control system Σ in Figure 3.3 is feasible, all candidate controllers in set $\mathbb{C}$ are SCLI, and the pair $(V, \mathbb{C})$ is l_{2e} gain related. Then switching will stop with Algorithm 3.5, and the system Σ is stable.

Proof: For proof, refer to [13].

Simulations have shown that self-falsification reduces the probability and severity of the DAL phenomenon [13].

Example 3.5 Consider the system in Example 3.2. The LICLA parameters are

$$\eta_0 = 0, \quad \Delta\eta = 1 \tag{3.48}$$

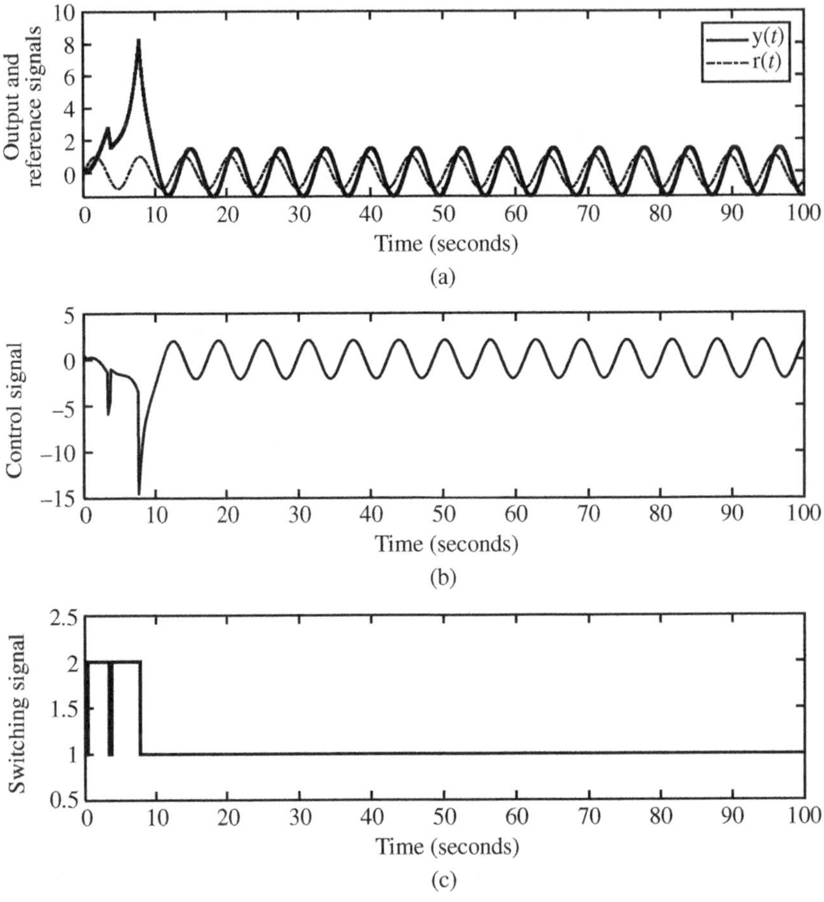

Figure 3.16 LICLA algorithm: (a) Output and reference input, (b) control input and (c) switching signal.

The output, control input and switching signal are depicted in Figure 3.16, and the cost functions are depicted in Figure 3.17. As shown in Figure 3.16, the stabilising controller is selected faster than the conventional UASC and the DAL phenomenon is attenuated.

In the self-falsification methods, the cost function is compared with the level of falsification γ. If the active controller is falsified in the LICLA algorithm, the γ parameter is linearly increased and is replaced by $\gamma + \Delta\eta$, where $\Delta\eta$ is a positive constant. It is clear that as long as all the cost functions are larger than γ, the controllers are constantly falsified. To modify this intermittent switching, the γ sequence is updated differently in [10] and is called the *Level Set Algorithm*. In this

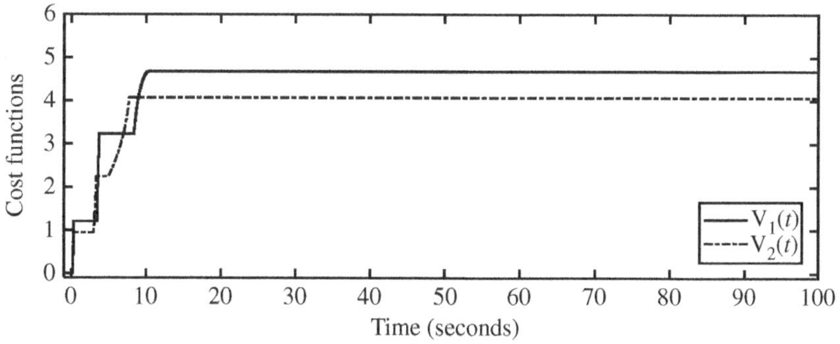

Figure 3.17 The cost functions for the LICLA algorithm.

modification, the minimum cost functions are used to update the γ sequence in the case of switching as follows

$$\gamma \leftarrow \max\left\{\gamma + \Delta\eta, \min_{C_i \in \mathbb{C}} V(C_i, z, t)\right\} \tag{3.49}$$

3.8 Time-varying Systems in the UASC

In the UASC, the maximum cost function $V(C_i, z, t)$ on the time window is calculated from zero to t. However, in the case of time-varying systems using the unfalsified adaptive control approach, a *time-varying window* for the UASC cost function is introduced to consider parameter variations. Also, a *forgetting factor* is presented in the norm calculations. Windowing the cost function is based on the *reset time* concept. This algorithm is called UASC with reset-time or UASC-R for short. The cost function $J(C_i, z, \tau)$ is defined as

$$J(C_i, z, t) = \frac{\|(y - \tilde{r}_i)_t\|_\lambda^{\ 2} + \|u_t\|_\lambda^{\ 2}}{\|(\tilde{r}_i)_t\|_\lambda^{\ 2} + \varepsilon} \tag{3.50}$$

where $\|\cdot\|_\lambda$ denotes the λ-exponentially weighted l_2 norm with the forgetting factor λ, and is defined as

$$\|s_t\|_\lambda = \sqrt{\sum_{\tau=0}^{t} \lambda^{2(t-\tau)}|s(\tau)|^2} \tag{3.51}$$

If the kth reset-time occurs in time t_k, the cost function is calculated as follows

$$V(C_i, z, t) = \max_{t_k \le \tau \le t} J(C_i, z, \tau), \quad t \in \mathbb{Z}_+, i \in \{1, \dots, N\} \tag{3.52}$$

The reset time specifies the time window $t_k \le \tau \le t$, where the maximum value of the cost function is calculated. The classical UASC approach uses a *persistent time window* $\tau \in (0, t)$, that equally treats the old and new data and disables the possible forgetting characteristics of the cost function. Hence, in the case of time-varying systems, where recent input-output data are more relevant, the windowed cost function has a superior performance. The reset time condition is selected as

$$t_{k+1} = 1 + \min\{t : t \geq t_k, | J_\star(t) - J(C_{\sigma(t+1)}, z, t) | \leq \varepsilon_1\} \tag{3.53}$$

where ε_1 is a positive arbitrary constant, and $J_\star$ is defined as

$$J_\star(t) = \frac{\|(y-r)_t\|_\lambda{}^2 + \|u_t\|_\lambda{}^2}{\|r_t\|_\lambda{}^2 + \varepsilon} \tag{3.54}$$

In the classical UASC approach, the only reset time occurs in $t = 0$ at the startup. The reset time abolishes the monotonicity of the cost functions over time, and therefore, using the Hysteresis algorithm cannot guarantee the switching termination. More details can be found in [14].

Example 3.6 Consider the following uncertain system

$$G(s) = \frac{K}{s - 0.4} \tag{3.55}$$

where $K \in [0.1,1]$ and the sampling time is $T_s = 0.2$ seconds. The controller set is designed as

$$C_1 = \frac{3.485 - 3.265d}{1 - d}$$
$$C_2 = \frac{27.36 - 25.63d}{1 - d} \tag{3.56}$$

For a square reference input, the basic UASC and the UASC-R closed-loop system performances are compared. For the basic UASC, the cost function (3.4) and the following parameters are considered.

$$h = 1, \quad \varepsilon = 1, \quad W_1 = W_2 = 1 \tag{3.57}$$

For the UASC-R, the cost function (3.50) is utilised with the parameters

$$\lambda = 0.98, \quad \varepsilon_1 = 1 \tag{3.58}$$

The system's initial condition is $y(0) = 0$. The uncertain parameter is $K = 0.9$ in the first time interval $t \in [0,142]$, and is $K = 0.16$ in the second time interval $t \in [142,180]$. The controller C_1 is stabilising in the first time interval, and both C_1 and C_2 are stabilising in the second time interval. The closed-loop outputs of UASC and UASC-R, the control selection signal $\sigma(t)$ and the reset time signal, and the cost functions are shown in Figures 3.18–3.20, respectively. As depicted in Figure 3.18, the UASC-R has a superior performance following the plant parameter change. Although the UASC selects a stabilising controller, it is incapable of selecting a controller with superior performance.

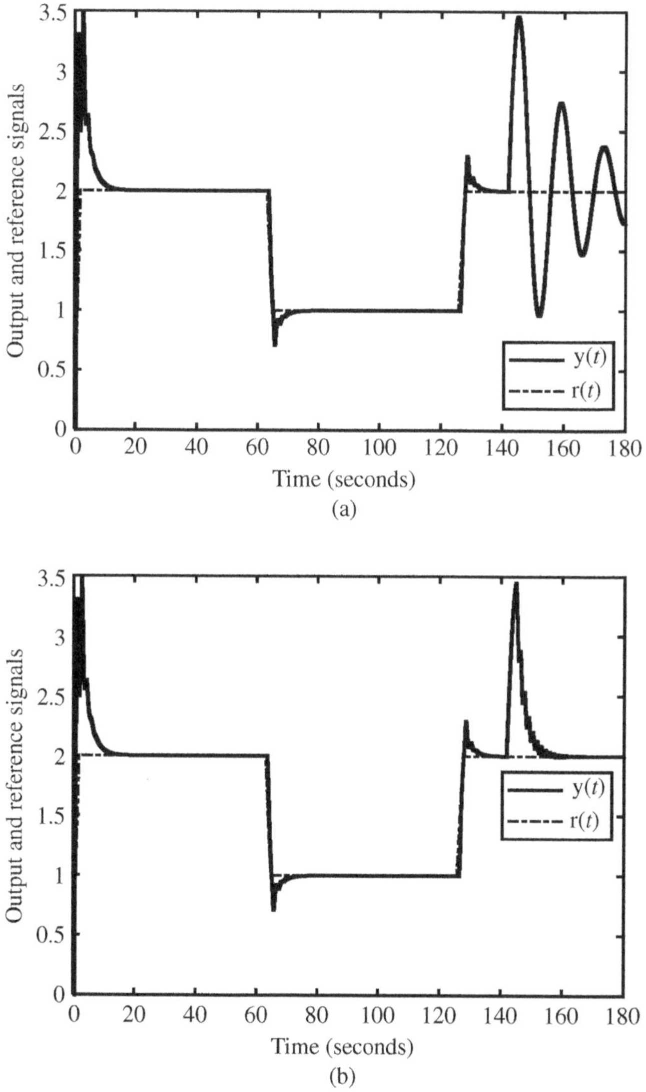

Figure 3.18 The closed-loop output: (a) UASC, (b) UASC-R.

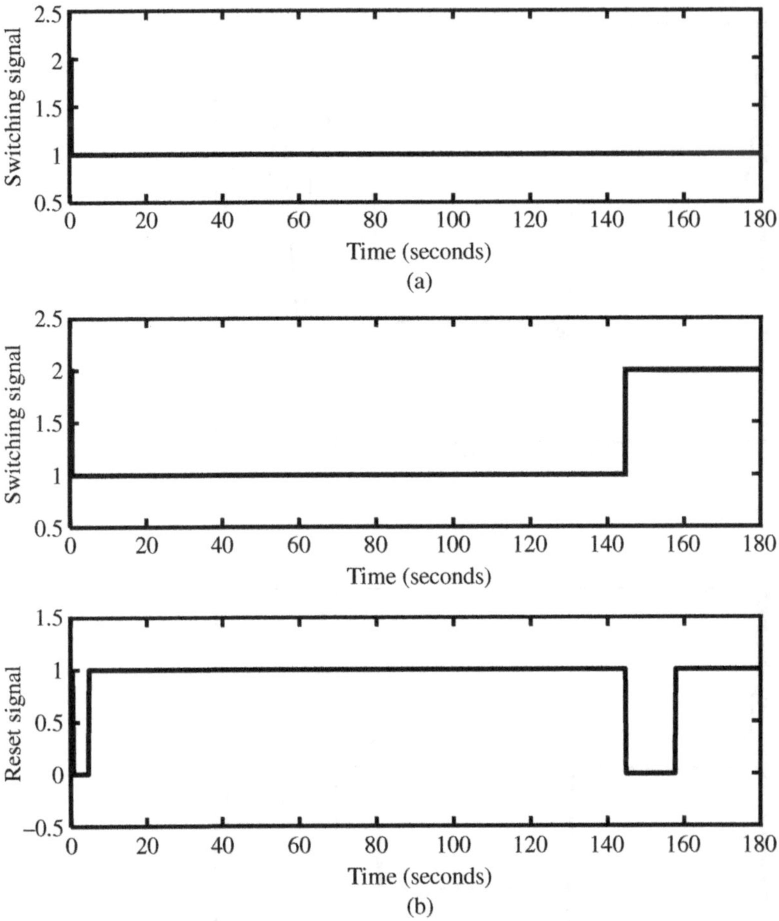

Figure 3.19 The control selection signal $\sigma(t)$ and the reset time signal: (a) UASC, (b) UASC-R.

3.9 Conclusion

In this chapter, the ASSC scheme is addressed, and the challenges of the model-based switching control systems are discussed. It is shown that a performance-based scheme is a practical approach for the switching control of real-world problems. Based on the philosophical theory of falsification, the unfalsified adaptive performance-based switching control scheme is presented. Unfalsified adaptive control is a well-established methodology in the data-driven control paradigm.

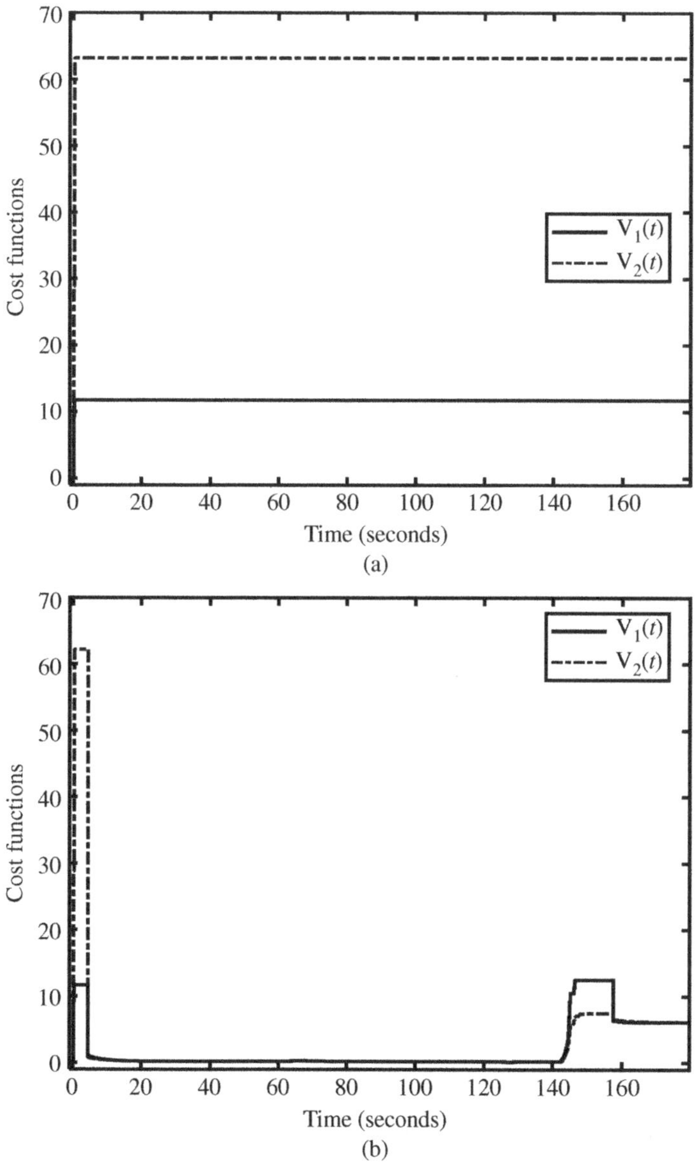

Figure 3.20 The cost functions: (a) UASC, (b) UASC-R.

The principles of UASC are presented, closed-loop stability theorems are given, and discussions on closed-loop performance are provided. The DAL phenomenon is introduced, and the UASC modifications to tackle the performance degradations are presented. Simulation results are provided for further insight into the theoretical results.

Problems

3.1 Consider the cost function given by

$$V(C_i, z, t) = \max_{\tau \leq t} \frac{\|(y - \tilde{r}_i)_\tau\|^2 + \| u_\tau \|^2}{\|(\tilde{r}_i)_\tau\|^2 + \varepsilon}$$

Let the reference signal in the adaptive switching control shown in Figure 3.3 be bounded for all time, that is, $|r(t)| \leq R\,\forall t$, where R is a positive real number. Also, the final selected controller is $C_f = D_f(d)^{-1} N_f(d)$, where $N_f(d)$ and $D_f(d)$ are stable, proper rational functions and their corresponding virtual reference signal is $\tilde{r}_{(C_f)}$.

a) Show that $N_f(d)\left(\tilde{r}_{(C_f)}(t) - r(t)\right) = 0$. Comment on this result.
b) Show that for the final selected controller C_f that unfalsified the closed-loop stability for the data $r(t)$ and $z(t)$, the cost function is bounded.
c) Show that if $V(C_f, z, t)$ is bounded, then the final selected controller C_f has unfalsified the closed-loop stability for the data $r(t)$ and $z(t)$.

3.2 Consider the following transfer function:

$$G(s) = \frac{1}{s^2 - 1}$$

The controller bank includes the following three PD controllers:

$$C(s) = \{C_1(s) = 0.5s + 1.1, C_2(s) = 2s + 0.5, C_3(s) = 0.2s + 0.7\}$$

where $C_1(s)$ is the only stabilising controller in the controller bank.

a) Employing Algorithm 3.1, show the DAL effect for the adaptive switching control system for the hysteresis constants $h \in \{0.1,1,5\}$, and compare the closed-loop output, control input and switching signal results.
b) Employing Algorithm 3.2, show the effect of the threshold level M on the DAL effect in the adaptive switching control system for $M \in \{30,100,200\}$, and compare the closed-loop output, control input and switching signal results. Select h from part a).

c) Employing Algorithm 3.3, show the effect of the hysteresis value h on the DAL effect in the adaptive switching control system for $h \in \{0.1, 1, 5\}$, and compare the closed-loop output, control input and switching signal results.

d) Compare the performance of Algorithms 3.1–3.3 in dealing with the DAL phenomenon.

3.3 Consider the following transfer function:

$$G(s) = \frac{1}{(s+1)^3}$$

The controller bank includes the following three PID controllers:

$$C(s) = \left\{ \begin{array}{c} C_1(s) = 2.112s + 4.8 + \dfrac{2.65}{s}, C_2(s) = 15.8s + 0.75 + \dfrac{0.095}{s}, \\ C_3(s) = 2s + 0.7 + \dfrac{6}{s} \end{array} \right\}$$

a) Analyse the step response of the closed-loop system for the controller C_1, C_2, and C_3.

b) Employing Algorithm 3.1, show the cost function, closed-loop output, control input and switching signal results. Comment on the results.

3.4 Consider the following plant and its corresponding controller bank:

$$G(s) = \frac{1}{s-1}, \quad C = \{C_1 = 2, C_2 = 0.5\}$$

Assume zero initial condition and let $r(t) = \sin(t)$ for the reference input. Also, the initial active controller in the loop is selected as C_2.

a) *Dwell time* is a restriction on the elapsed time between switching controllers and is a well-known approach for the stability guarantee of closed-loop switching control systems. Run Algorithms 3.1–3.3 with the following dwell times: 0.1, 0.2, 0.5, 1 and 5 seconds. Summarise the number of switchings between C_1 and C_2 in a table and discuss the results.

b) Fix the dwell time at 5 seconds. Run the Algorithms 3.1–3.3 with the following initial conditions: 0, 3, 9, 15, 30, 100, 150. Summarise the results in terms of the time durations which the destabilising controller C_2 is active, and the corresponding maximum control signal, in a table and discuss the results.

3.5 Consider the following multivariable plant:

$$G(z) = \begin{bmatrix} \dfrac{0.084\ 03z + 0.063\ 52}{z^2 - 1.343z + 0.4317} & \dfrac{-0.1994}{z - 0.9003} \\ \dfrac{0.1558}{z - 0.5326} & \dfrac{0.4583z - 0.2695}{z^2 - 1.711z + 0.7298} \end{bmatrix}$$

The multivariable controller bank includes the following controller transfer function matrices:

$$C(z) = \begin{cases} C_1(z) = \begin{bmatrix} \dfrac{3.48z - 3.18}{z - 0.9833} & \dfrac{0.696z - 0.6361}{z - 0.9833} \\ \dfrac{-0.12z + 0.1097}{z - 0.9833} & \dfrac{0.576z - 0.5264}{z - 0.9833} \end{bmatrix} \\ C_2(z) = \begin{bmatrix} \dfrac{0.9641z - 0.7341}{z - 1} & \dfrac{-0.5212z + 0.357}{z - 1} \\ \dfrac{3.127z + 2.787}{z - 1} & \dfrac{0.1597z - 0.1359}{z - 1} \end{bmatrix} \\ C_3(z) = \begin{bmatrix} \dfrac{2.9z - 1.543}{z - 0.9389} & \dfrac{0.58z - 0.3085}{z - 0.9389} \\ \dfrac{-0.1z + 1.053\,19}{z - 0.9389} & \dfrac{0.48z - 0.2553}{z - 0.9389} \end{bmatrix} \end{cases}$$

a) Show that the controllers $C_1(z)$ and $C_3(z)$ are stabilising controllers, and $C_2(z)$ destabilises the closed-loop plant.
b) Evaluate the interaction of the open-loop plant, and also compare interaction in the closed-loop plant with the controllers $C_1(z)$ and $C_3(z)$.
c) Assume that the multivariable plant is unknown and design a UASC system with the candidate controllers using Algorithm 3.1. Discuss the results.
d) Discuss the results of part c) from an interaction performance perspective and analyse the performance of the conventional UASC in effectively handling interaction to select the controller with a lower closed-loop interaction performance. How would you modify the cost function in the presence of strong open-loop plant interactions?

References

1 B. Mårtensson, "The order of any stabilizing regulator is sufficient a priori information for adaptive stabilization," *Systems & Control Letters,* vol. 6, no. 2, pp. 87–91, 1985.
2 B. D. Anderson, T. Brinsmead, D. Liberzon, and A. Stephen Morse, "Multiple model adaptive control with safe switching," *International Journal of Adaptive Control and Signal Processing,* vol. 15, no. 5, pp. 445–470, 2001.
3 M. G. Safonov, "Origins of robust control: early history and future speculations," *Annual Reviews in Control,* vol. 36, no. 2, pp. 173–181, 2012.
4 J. C. Willems, "Paradigms and puzzles in the theory of dynamical systems," *IEEE Transactions on Automatic Control,* vol. 36, no. 3, pp. 259–294, 1991.

5 M. G. Safonov and T.-C. Tsao, "The unfalsified control concept: a direct path from experiment to controller," in *Feedback Control, Nonlinear Systems, and Complexity*: Springer, 1995, pp. 196–214.

6 R. Wang, A. Paul, M. Stefanovic, and M. Safonov, "Cost detectability and stability of adaptive control systems," *International Journal of Robust and Nonlinear Control: IFAC-Affiliated Journal*, vol. 17, no. 5–6, pp. 549–561, 2007.

7 M. Stefanovic and M. G. Safonov, "Safe adaptive switching control: stability and convergence," *IEEE Transactions on Automatic Control*, vol. 53, no. 9, pp. 2012–2021, 2008.

8 B. D. Anderson and A. Dehghani, "Challenges of adaptive control–past, permanent and future," *Annual Reviews in Control*, vol. 32, no. 2, pp. 123–135, 2008.

9 G. Battistelli, E. Mosca, M. G. Safonov, and P. Tesi, "Stability of unfalsified adaptive switching control in noisy environments," *IEEE Transactions on Automatic Control*, vol. 55, no. 10, pp. 2424–2429, 2010.

10 K. S. Sajjanshetty and M. G. Safonov, "Multi-objective cost-detectability in unfalsified adaptive control," *Asian Journal of Control*, vol. 18, no. 6, pp. 1959–1968, 2016.

11 M. Chang and M. Safonov, "Unfalsified adaptive control: the benefit of bandpass filters," in *AIAA Guidance, Navigation and Control Conference and Exhibit*, 2008, p. 6783.

12 H. Jin, H. B. Siahaan, M. W. Chang, and M. G. Safonov, "Improving the transient performance of unfalsified adaptive control with modified hysteresis algorithms," *IFAC Proceedings Volumes*, vol. 47, no. 3, pp. 1489–1494, 2014.

13 H. Jin and M. Safonov, "Unfalsified adaptive control: controller switching algorithms for nonmonotone cost functions," *International Journal of Adaptive Control and Signal Processing*, vol. 26, no. 8, pp. 692–704, 2012.

14 G. Battistelli, J. P. Hespanha, E. Mosca, and P. Tesi, "Model-free adaptive switching control of time-varying plants," *IEEE Transactions on Automatic Control*, vol. 58, no. 5, pp. 1208–1220, 2013.

4

Multi-Model Unfalsified Adaptive Switching Supervisory Control

4.1 Introduction

Adaptive switching supervisory control (ASSC) was introduced in Chapter 3. ASSC strategies were developed to overcome the inherent limitations of classical adaptive control. For example, complications in the design of continuously parametrised controllers of classical adaptive control appear if unknown parameters enter the process model in complex ways, or in the case of highly uncertain systems, where the sign of the high-frequency gain changes, a key assumption of the classical adaptive control is violated, and the adaptive control design would fail. Also, the sensor-actuator limitations can make the implementation of continuous control impractical. The two main current approaches to ASSC design in a data-driven context are as follows:

- Unfalsified ASSC (UASSC)
- Multi-model ASSC (MMASSC)

The unfalsified ASSC was presented in Chapter 3. The UASSC selects a controller in finite time under the minimal assumption of the existence of a stabilising candidate controller with guaranteed closed-loop stability. No assumptions on the linearity or structural properties of the plant are required, which ensures closed-loop robustness properties. However, as shown in Chapter 3, the presented UASSC algorithms can experience significant initial transients and temporary trends of divergence before the final controller is selected and employed in the control system. Although several remedies for closed-loop performance improvements were proposed in Section 3.6, this is still a challenging issue in the UASSC approaches.

The basic idea in the MMASSC approach is to divide large plant uncertainty sets into smaller subsets. This results in a finite number of identified models M_1, M_2, …, M_N that describe the overall system behaviour, and each model has a substantially reduced uncertainty. Corresponding to each model M_i, a controller

An Introduction to Data-Driven Control Systems, First Edition. Ali Khaki-Sedigh.

C_i is designed to ensure closed-loop stability and desired performance. Then, based on a switching criterion, at each instant of time, a model-controller set $\{M_i, C_i\}$ is selected, and the output of C_i is utilised for control. The design problem is, therefore, the selection of $\{M_i, C_i\}$ and switching rules between controllers for closed-loop stability and desired performance. Switching between controllers is performed by the *supervisor*, that processes the input–output data and produces the switching sequence σ to specify the switching or *active controller* C_σ.

In the MMASSC algorithms, the norms of the sequences of estimation errors based on the models-controllers $\{M_i, C_i\}$ are compared, and the prediction error norm of minimum magnitude is selected as the most appropriate model-controller set.

The multi-model control methods commonly assume the existence of a *matching model* in the model set. This is also referred to as the *exact match condition* [1]. The exact matching condition assumes that the set of plant models M_i, $i = 1, 2, \ldots, N$ matches or equals the actual process model. The exact or near-exact match condition results in dense nominal models' distribution in the parameter space. The exact match condition is challenging when the system has large uncertainties. In this case, numerous models should be considered in the model set to satisfy the exact match condition. Nevertheless, if the exact match condition is not ensured, stability, signal boundedness and convergence to the final controller are not guaranteed. The proven asymptotic stability properties of the MMASSC algorithms are only guaranteed if the exact match condition holds and the actual unknown plant is closely approximated by at least one nominal model M_i. Note that if the difficult exact match condition holds, or in the case of very dense nominal model distribution, the transient performance of the MMASSC algorithms can be significantly improved in comparison to the UASSC algorithms. This chapter presents the multi-model unfalsified adaptive switching control (MMUASC) methodology to exploit the positive features of both the MMASSC and UASSC approaches and moderate their deficiencies, specifically to circumvent the need for the exact match condition and improve the closed-loop transient response. The MMUASC was initially introduced in [2]. In this method, a model set is utilised to select the appropriate controller based on the *falsifying strategy*. It is shown that the MMUASC has a superior performance in comparison with the basic unfalsified adaptive switching control (ASC).

The notions of the *virtual reference signal* and *potential closed-loop* introduced in Chapter 3 are central to the falsifying strategy of MMUASC. These tools are used in online stability inference of a potential closed-loop, also called a *potential control loop*. The MMUASC approach consists of embedding a family of nominal models associated with the given candidate controllers $\{M_i, C_i\}$ in the unfalsified ASC schemes of Chapter 3. To implement an inferring procedure of stability properties of the potential feedback loop, appropriate *cost functions*, such as those defined in (3.4), also called *test functionals*, must be introduced.

To improve performance, different cost functions are introduced in [3], and closed-loop performance and stability of these cost functions are discussed.

The MMUASC approach is extended in several directions. The MMUASC approach is developed for multivariable systems in [4]. Adaptive memory and forgetting factor are the two tools utilised in [5] for the control of time-varying systems. The reset concept is applied to detect the cost function memory length, and by using the reset strategy, data irrelevant to stability is omitted. The classic MMUASC uses the hysteresis switching logic, though the cost function should be monotone, non-decreasing in time for closed-loop stability. The self-falsification concept is used in the MMUASC [6] to enable the usage of non-monotone cost functions that results in performance improvement. The supervisory switching control with the aim of disturbance attenuating is proposed in [7] based on the MMUASC structure. The disturbance attenuation feature of each controller in the controller set is evaluated without activation, and also a selection of the best controller is guaranteed.

Constraints are inevitable features of real systems that imperil the stability of the control mechanism. Input constraints are considered in the MMUASC structure in [8]. A new virtual reference and a suitable cost function are introduced for stability assurance. Although the MMUASC supervisor finally selects a stabilising controller, when there exists more than one stabilising controller, selecting the controller with superior performance is not guaranteed. By using fuzzy logic, the cost function is improved in [9] to select the controller with superior performance. The MMUASC with a constrained generalised predictive controller (GPC) set is proposed in [10]. In the proposed scheme, the virtual reference signal is derived by solving an inverse optimisation problem.

In this chapter, the MMUASC strategy is presented. The MMUASC benefits from both the multi-model adaptive control approach and the unfalsified control strategy. First, the MMASSC control structure is presented briefly, and then the MMUASC algorithm is elaborated.

4.2 The Multi-Model Adaptive Control

In the multi-model adaptive control strategy, there are N designed *candidate controllers* $\{C_i, i = 1, \dots, N\}$, and an *active controller* must be selected and inserted in the control loop based on the performance assessment of all the candidate controllers. The common approach is to first identify the model that best fits the actual system in the model set $\{M_i, i = 1, \dots, N\}$, and then select the corresponding designed controller in the candidate controllers set. For the plant model selection,

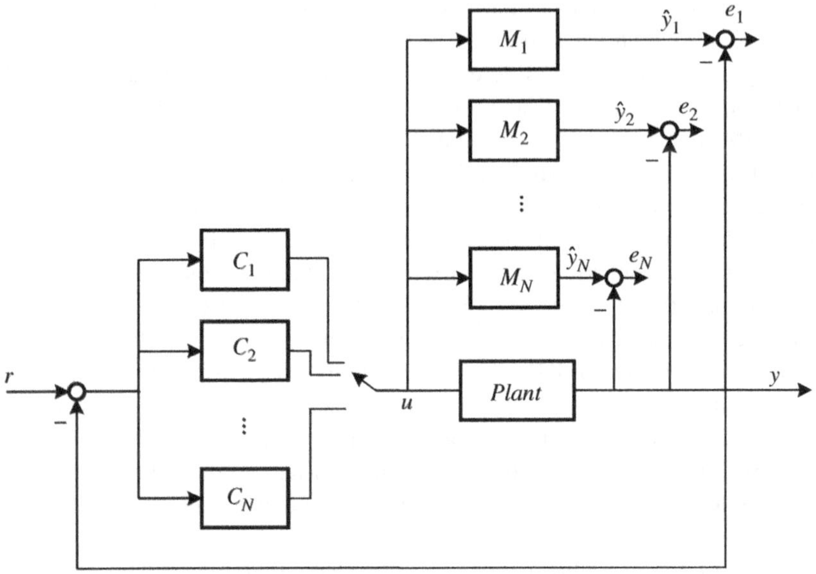

Figure 4.1 Multi-model supervisory control structure.

the system input is simultaneously applied to all the available models in the model set, and the errors between the actual plant output and the models' outputs are evaluated. Then, the model M_i with the least output error is selected. Finally, the corresponding controller C_i is *activated*, as shown in Figure 4.1.

Consider a model set $\mathbb{M} = \{M_1, \dots, M_N\}$, and the corresponding controller set $\mathbb{C} = \{C_1, \dots, C_N\}$. The performance index for model selection is

$$J_i(t) = \alpha e_i^{\,2}(t) + \beta \sum_{k=1}^{t} \lambda^k e_i^{\,2}(t-k), \quad i \in \{1, \dots, N\} \tag{4.1}$$

where $t \in \{1, 2, \dots\}$, α and β are weighting positive coefficients, and λ is the forgetting factor. The model set errors e_i, $i \in \{1, \dots, N\}$ are calculated online as

$$e_i(t) = \hat{y}_i(t) - y(t), \quad i \in \{1, \dots, N\} \tag{4.2}$$

where y and $\hat{y}_i$ are the system output and the model output, respectively as depicted in Figure 4.1. More details can be found in [11].

The general structure of the MMASSC approach is depicted in Figure 4.2. The supervisor consists of the multiple estimators $\mathcal{E}$, the monitoring signal generator $\mathcal{M}$ and the switching logic $\mathcal{S}$. The monitoring signal generator calculates the performance index J_i based on the estimation errors. The switching logic decides which controller and when it should be activated based on the cost function values.

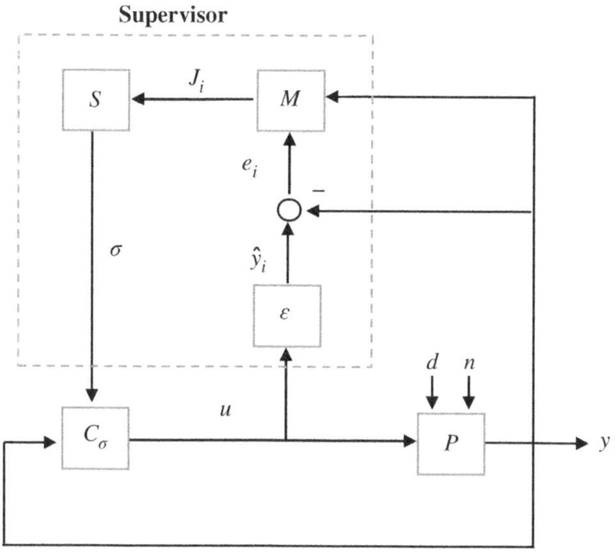

Figure 4.2 General structure of the multi-model adaptive switching supervisory control.

In case one of the models exactly fits the system and the exact match condition is satisfied, one of the performance indexes tends to zero under the disturbance and noise-free conditions. By using the hysteresis switching logic, the switching is terminated in a finite time and closed-loop stability is guaranteed.

However, even if the exact match condition is assumed, in the presence of uncertainty, bounded noise and disturbance, typically, the switching will not stop in a finite time. To prove closed-loop stability, the slow switching condition should hold, i.e. the hysteresis parameter h should be large enough to ensure the *average dwell time* condition.

Definition 4.1 The switching signal σ has an *average dwell time* $\tau > 0$ if there exists a positive integer N_0 such that the number of switching σ on the interval $(0, t)$ denoted by $N_\sigma(0, t)$ satisfies the following condition:

$$N_\sigma(0, t) \leq N_0 + \frac{t}{\tau} \tag{4.3}$$

To ensure closed-loop system stability in the presence of uncertainty, one of the models should describe the system approximately. The vulnerability of most of the MMASSC strategies is the assumption that plants under control must be *tightly approximated* by the nominal models in the model set and closed-loop stability in the face of un-modelled dynamics at large is dependent on this assumption. Indeed, the closed-loop stability of the MMASSC strategies can be ensured if at least one nominal model tightly approximates the plant. The analysis is based on the assumption that one of the models set errors e_i should be small, and as a result, the corresponding model is closest to the system. This assumption is referred to as the *small error property*. The model's stabilising controller is expected to stabilise the actual system. This intuition is referred to as the *detectability property* so that if the error is small, the states of the system remain bounded. Detectability property is required to prove closed-loop stability. To satisfy the small error property, a dense model set is required, which for a highly uncertain plant would require numerous model members [12]. The dense nominal model distributions assumption can be considered the Achilles' heel of the MMASSC strategies in the presence of large uncertainties.

In the MMUASC, the small error property is relaxed and is substituted with the problem *feasibility*. Feasibility requires the *existence* of a stabilising controller in the candidate controller set, and this is the only condition imposed on the model-controller set to ensure closed-loop stability in the MMUASC. ∎

4.3 Principles of the Multi-Model Unfalsified Adaptive Switching Control

In the MMUASC approach, several nominal models are considered for an indeterminate system. For the nominal models, linear controllers are designed to ensure closed-loop stability and desired performance. Considering the closed-loop structure depicted in Figure 4.3, the ASC structure is described as

$$y(t) = P(u + d_i)(t) + d_o \tag{4.4}$$

$$u(t) = C_{\sigma(t)}(r - y)(t) \tag{4.5}$$

where $P : u \rightarrow y$ is an indeterminate system, $y(t)$, $u(t)$, $r(t)$, d_i and d_o are the system output, the control signal, the reference input and unknown additive input and

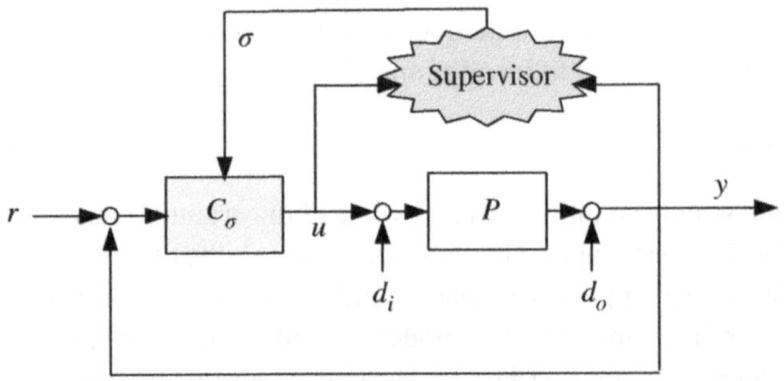

Figure 4.3 Typical switching control structure.

output disturbances, respectively. The subscript $\sigma(t)$ specifies the active controller in the closed-loop system at time t. Assume that the system dynamics are linear and described by the following equation:

$$P: A(d)y(t) = B(d)u(t) \tag{4.6}$$

where $A(d)$ and $B(d)$ are relatively prime polynomials, and d is the delay operator. Also, $A(d)$ is monic, i.e. the coefficient of the highest power in A is unity. Let the model set be denoted as $\mathbb{M} = \{M_1, \dots, M_N\}$, and the corresponding designed controller set be $\mathbb{C} = \{C_1, \dots, C_N\}$. The plant models and the respective designed controllers are described as follows:

$$M_i(d) = \frac{B_i(d)}{A_i(d)} \tag{4.7}$$

$$C_i(d) = \frac{S_i(d)}{R_i(d)} \tag{4.8}$$

where $i \in \{1, \dots, N\}$, $A_i(d)$, $B_i(d)$, $S_i(d)$ and $R_i(d)$ are polynomials, $R_i(d)$ and $A_i(d)$ are monic and $A_i(d)$ and $B_i(d)$ are relatively prime.[1]

In the classical adaptive control based on the certainty equivalence principle, adaptation mechanisms are used to update the controller parameters. In the MASSC algorithms based on the exact matching condition, a comparison of errors between the actual plant output and the model outputs leads to the control selection. In the MMUASC, with the feasibility assumption, a controller is selected based on the falsification strategy. That is, the active controller is either unfalsified or falsified via a comparison based on the virtual reference input, as defined in Definition 3.1.

The time-varying switching system is depicted in Figure 4.4(a). Virtual reference is the reference input to the *potential loop*, as shown in Figure 4.4(b). Hence, from the potential loop, $z = [u, y]$ can be obtained as

$$z = (P/C_i)\tilde{r}_i \tag{4.9}$$

and the virtual reference signal is calculated from

$$e_i(t) = (R_i(d)/S_i(d))\, u(t), \qquad \tilde{r}_i(t) = e_i(t) + y(t) \tag{4.10}$$

Definition 4.2 As is shown in Figure 4.4(c), the hypothetical feedback closed-loop system with the controller C_i and the system M_i is referred to as the *control candidate loop*. This loop is denoted by (M_i/C_i) and the corresponding reference input signal is the ith virtual reference signal $\tilde{r}_i$.

---

1 These polynomials can be equivalently written as A_i, B_i, S_i and R_i, for simplicity.

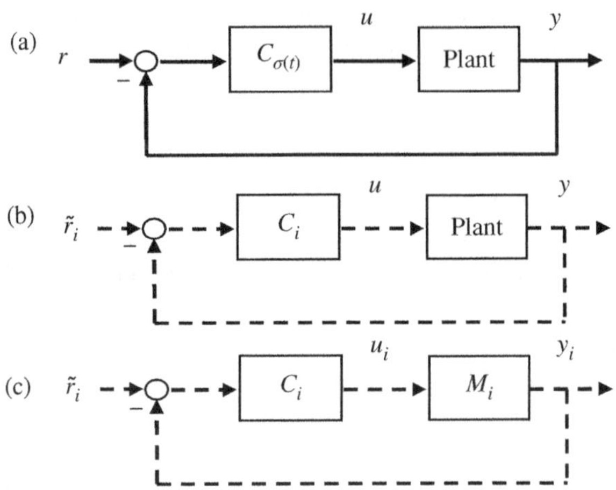

Figure 4.4 (a) Time-varying closed-loop switching system, (b) the potential loop, (c) the control candidate loop.

The controllers C_i, $i \in \{1, \ldots, N\}$ are designed to achieve a stable control candidate loop with desired closed-loop performance. According to the control candidate loop, $z_i = [u_i, y_i]$ is calculated as follows:

$$z_i = (M_i/C_i)\,\tilde{r}_i \tag{4.11}$$

Remark 4.1 Closed-loop stability requires the designed controllers to be SCLI (see Theorem 3.2). Hence, the polynomials $S_i(d)$ should be stable for $i \in \{1, \ldots, N\}$. For a non-SCLI controller, the coprime factorisation decomposition of the controller is used to modify the virtual reference signal. See Section 3.4 for more details.

Assumption 4.1 If σ includes all the possible switching signals, and $V(C_\sigma, z, t)$ is the cost function corresponding to the controller C_σ, let the following assumptions hold for the system (P/C_σ):

- For every σ and $i \in \{1, \ldots, N\}$, $V(C_i, z, t)$ admits a limit (even infinite) as $t \to \infty$.
- There always exists a $\mu \in [1, \ldots, N]$, such that $V(C_\mu, z, t)$ is finite.

The MMUASC problem is activating the controller C_σ such that the behaviour of (P/C_σ) resembles one of the control candidate loops (M_i/C_i). That is, the supervisor selects a switching signal σ for the uncertain system such that:

- The closed-loop system (P/C_σ) is stable.
- The behaviour of (P/C_σ) is similar to one of the control candidate loops (M_i/C_i).

To this extent, the cost function corresponding to C_i is calculated as

$$V(C_i, z, t) = \max_{\tau} \frac{\|(P/C_i)\,\tilde{r}_i - (M_i/C_i)\,\tilde{r}_i\|_\tau}{\|(M_i/C_i)\,\tilde{r}_i\|_\tau + \varepsilon} \tag{4.12}$$

where $\|\cdot\|_\tau$ denotes the truncated 2-norm, and ε is a small positive constant. In a disturbance-free case, the numerator in (4.12) decreases as the discrepancy between P and M_i reduces. As M_i gets closer to P, the cost function (4.12) tends to zero under the same initial condition.

For each controller, the following three steps are performed:

- The virtual reference signal is calculated according to (4.10).
- The signal z_i is calculated by (4.11).
- The cost function $V(C_i, z, t)$ is calculated by (4.12).

Note that the closed-loop outputs of the system (P/C_σ) with the reference input r, and the hypothetical system (P/C_i) with the reference input $\tilde{r}_i$, are the same. That is,

$$z = (P/C_\sigma)r = (P/C_i)\,\tilde{r}_i \tag{4.13}$$

As shown in Figure 4.5, the MMUASC cost function should demonstrate the distance between the two closed loops $(P/C_i)\tilde{r}_i$ and $(M_i/C_i)\tilde{r}_i$. This cost function is introduced based on the measured signal z and the calculated signal z_i as

$$J(C_i, z, t) = \frac{\|z - z_i\|_t}{\|z_i\|_t + \varepsilon}, \quad t \in \mathbb{Z}_+, \quad i \in \{1, \dots, N\} \tag{4.14}$$

and

$$V(C_i, z, t) = \max_{\tau \le t} J(C_i, z, \tau) \tag{4.15}$$

The hysteresis switching logic is used in the MMUASC approach to select the optimal controller. The controller switching signal is based on the hysteresis switching logic

$$\sigma(t) = \arg\min_i \,(V(C_i, z, t) - h\delta_{i,\sigma(t-1)}) \tag{4.16}$$

where $h > 0$ is the hysteresis constant, the symbol arg denotes the argument of the minimisation problem and $\delta_{i,j}$ is the Kronecker δ defined as

$$\delta_{i,j} = \begin{cases} 0 & \text{if } i \neq j \\ 1 & \text{if } i = j \end{cases} \tag{4.17}$$

The following lemma is proved in [13].

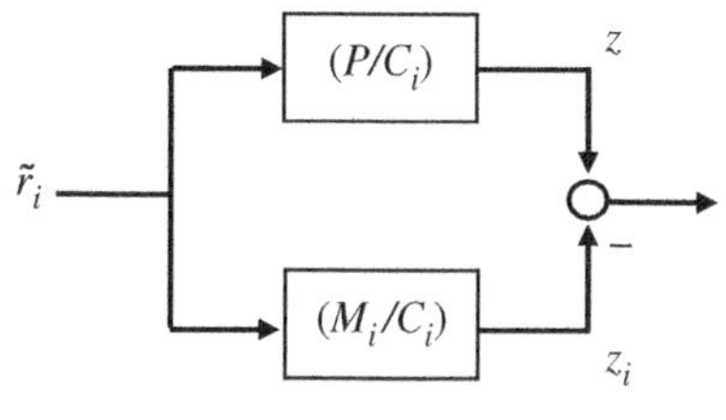

Figure 4.5 Error of the closed-loop system and the control candidate loop.

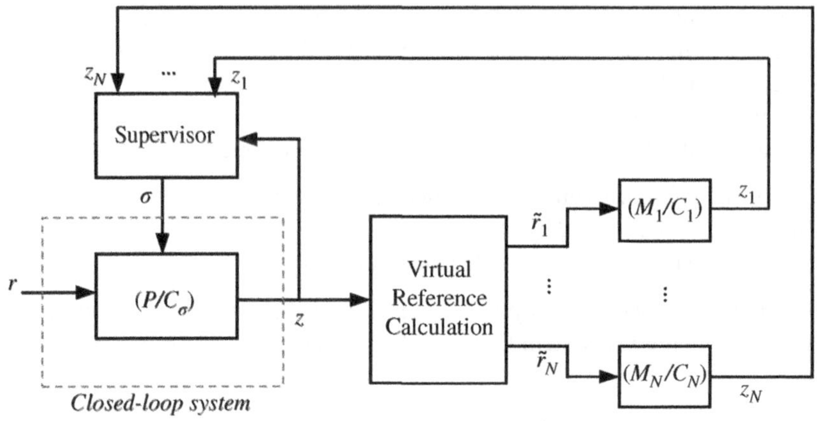

Figure 4.6 The MMUASC block diagram structure.

Lemma 4.1 Let z be the input–output system data, the switching logic be given by (4.16) and Assumption 4.1 holds. Then, for any reference input r, and any initial condition, switching is terminated in a finite time, and if C_f is the final controller, the cost function $V(C_f)$ is bounded.

The MMUASC overall algorithm is presented in Algorithm 4.1 and is summarised in the block diagram shown in Figure 4.6.

Algorithm 4.1 Multi-Model Unfalsified Adaptive Switching Control Algorithm

Input Let $t = 0$ and select an arbitrary $C_{\sigma(0)}$, $\varepsilon > 0$, and $h > 0$

Step 1 Let $t \leftarrow t + 1$ collect r, u, y

Step 2 Calculate the virtual reference: $\tilde{r}_i\,(t),\ \ i \in \{1, \ldots, N\}$

Step 3 Calculate the control candidate loops output: z_i, $i \in \{1, \ldots, N\}$

Step 4 Calculate the cost functions $V(C_i, z, t)$, $i \in \{1, \ldots, N\}$

Step 5 If $V(C_{\sigma(t-1)}, z, t) > \min_{C_i \in C} V(C_i, z, t) + h$

$$C_{\sigma(t)} \leftarrow \operatorname*{argmin}_{C_i \in C} V(C_i, z, t)$$

Else, then

$$C_{\sigma(t)} \leftarrow C_{\sigma(t-1)}$$

End

Step 6 Go to step 1

Theorem 4.1 If the adaptive control problem is feasible, and all the controllers in the control set are SCLI, then the closed-loop system of Figure 4.3 with the switching Algorithm 4.1 is input–output stable.

Proof: The proof is presented in two steps. First, it is proved that switching is terminated in a bounded time. Then, it is shown that the final controller is a stabilising controller.

Step 1. The feasibility condition ensures that there exists a stabilising controller in the controller set. Let C_j be the stabilising controller. According to the virtual reference input definition, there exist positive constants γ_1 and γ_2 such that the following inequality is satisfied:

$$\|z\|_t \leq \gamma_1 \|\tilde{r}_j\|_t + \gamma_2 \tag{4.18}$$

where $\tilde{r}_j$ is the virtual reference input corresponding to the stabilising controller C_j. Also, it is assumed that all of the control candidate loops are stable, including the jth loop. So, there exist positive constants γ_3 and γ_4 such that

$$\|z_j\|_t \leq \gamma_3 \|\tilde{r}_j\|_t + \gamma_4 \tag{4.19}$$

The following inequality holds from the triangular inequality property of norms:

$$\frac{\|z - z_j\|_t}{\|z_j\|_t} \leq \frac{\|z\|_t + \|z_j\|_t}{\|z_j\|_t} = \frac{\|z\|_t}{\|z_j\|_t} + 1 \tag{4.20}$$

Hence, to show that the cost function is bounded, it is sufficient to show that $\frac{\|z\|_t}{\|z_j\|_t}$ is bounded. Dividing (4.18) by (4.19), the following inequality is achieved:

$$\frac{\|z\|_t}{\|z_j\|_t} \leq \frac{\gamma_1 \|\tilde{r}_j\|_t + \gamma_2}{\gamma_3 \|\tilde{r}_j\|_t + \gamma_4} \tag{4.21}$$

The right-hand side of (4.21) is always bounded even when $t \to \infty$. This ensures that the cost function of the stabilising controller is always bounded, and based on Lemma 4.1, switching will be terminated in a finite time.

Step 2. Let switching terminate, and the final controller be C_f. Therefore, its cost function should be bounded, which implies that

$$\frac{\|z - z_f\|_t}{\|z_f\|_t} \leq \gamma_5 \tag{4.22}$$

where γ_5 is a finite constant. We have,

$$\|z\|_t \leq (\gamma_5 + 1)\|z_f\|_t \tag{4.23}$$

Also, the control candidate loop of the final controller is stable. So, there exist positive constants γ_6 and γ_7 such that

$$\|z_f\|_t \leq \gamma_6 \|\tilde{r}_f\|_t + \gamma_7 \tag{4.24}$$

Substituting (4.24) in (4.23) gives

$$\|z\|_t \leq \gamma_6(\gamma_5 + 1)\|\tilde{r}_f\|_t + \gamma_7(\gamma_5 + 1) \tag{4.25}$$

Also, the virtual reference of the final controller satisfies

$$\|\tilde{r}_f\|_t \leq \|r\|_t + \gamma_8 \tag{4.26}$$

where γ_8 is a positive constant. Hence, substituting (4.26) in (4.25) gives

$$\|z\|_t \leq \gamma_6(\gamma_5 + 1)\|r\|_t + \gamma_7(\gamma_5 + 1) + \gamma_6\gamma_8(\gamma_5 + 1) \tag{4.27}$$

Consequently, the proof is completed. Another stability proof is presented in [2]. ■

Note that the virtual reference is not directly used in the cost function (4.12), and it is only utilised to calculate the signal z_i. The following corollary presents a method for calculating z_i.

Corollary 4.1 The control candidate loop output z_i and the plant input–output data z satisfy the following equation:

$$(A_i(d)R_i(d) + B_i(d)S_i(d))\, z_i(t) = \begin{bmatrix} A_i(d)R_i(d) & A_i(d)S_i(d) \\ B_i(d)R_i(d) & B_i(d)S_i(d) \end{bmatrix} z(t), \quad i \in \{1, \dots, N\} \tag{4.28}$$

Proof: The control candidate loop relations are as follows:

$$y_i(t) = \frac{B_i(d)}{A_i(d)} u_i(t) \tag{4.29}$$

$$u_i(t) = \frac{S_i(d)}{R_i(d)} \tilde{r}_i(t) - \frac{S_i(d)}{R_i(d)} y_i(t) \tag{4.30}$$

Substituting (4.29) in (4.30), the virtual reference input is rewritten as

$$\tilde{r}_i(t) = \frac{A_i(d)R_i(d) + B_i(d)S_i(d)}{A_i(d)S_i(d)} u_i(t) \tag{4.31}$$

By substituting (4.31) in (4.10) and some mathematical manipulations, it can be shown that the signal z_i can be directly calculated as

$$(A_i(d)R_i(d) + B_i(d)S_i(d))u_i(t) = A_i(d)R_i(d)\, u(t) + A_i(d)S_i(d)\, y(t)$$

$$(A_i(d)R_i(d) + B_i(d)S_i(d))y_i(t) = B_i(d)R_i(d)\, u(t) + B_i(d)S_i(d)\, y(t) \tag{4.32}$$

■

Corollary 4.1 reduces computational load and enhances the practical implementation capability of the MMUASC. The MMUASC algorithm without virtual reference calculation is presented in Algorithm 4.2, and simulation results are provided thereafter.

Algorithm 4.2 MMUASC Algorithm Without Virtual Reference Calculation

Input Let $t = 0$, select an arbitrary $C_{\sigma(0)}$, and $\varepsilon > 0$ and $h > 0$
Step 1 Let $t \leftarrow t+1$ collect r, u, y
Step 2 Calculate the control candidate loops output: z_i, $i \in \{1, \dots, N\}$
Step 3 Calculate the cost functions $V(C_i, z, t)$, $i \in \{1, \dots, N\}$
Step 4 If $V(C_{\sigma(t-1)}, z, t) > \min_{C_i \in \mathbb{C}} V(C_i, z, t) + h$

$$C_{\sigma(t)} \leftarrow \arg\min_{C_i \in \mathbb{C}} V(C_i, z, t)$$

Else, then

$$C_{\sigma(t)} \leftarrow C_{\sigma(t-1)}$$

End
Step 5 Go to step 1

Example 4.1 Consider a two-carts plant connected mechanically by a spring shown in Figure 4.7. The plant state-space model is

$$\dot{\mathbf{x}}(t) = \begin{bmatrix} 0 & 1 & 0 & 0 \\ -\dfrac{\gamma}{m_1} & 0 & \dfrac{\gamma}{m_1} & 0 \\ 0 & 0 & 0 & 1 \\ \dfrac{\gamma}{m_2} & 0 & -\dfrac{\gamma}{m_2} & 0 \end{bmatrix} \mathbf{x}(t) + \begin{bmatrix} 0 \\ \dfrac{1}{m_1} \\ 0 \\ 0 \end{bmatrix} (u(t) + n_u(t))$$

$$y(t) = \begin{bmatrix} 0 & 0 & 1 & 0 \end{bmatrix} \mathbf{x}(t)$$

where the masses of the carts are $m_1 = m_2 = 1$, and the uncertain spring stiffness is $\gamma \in [0.25, 1.5]$ N/m. Also, $n_u(t)$ denotes the disturbance. The nominal models M_i correspond to the stiffness values:

$$\gamma_1 = 0.3, \quad \gamma_2 = 0.5, \gamma_3 = 1.0$$

Based on the H_∞ mixed sensitivity technique, the control set is designed as [2].

$$C_1 = \frac{43.442 - 128.997d + 127.871d^2 - 42.311d^3}{1 - 2.387d + 1.925d^2 - 0.520d^3}$$

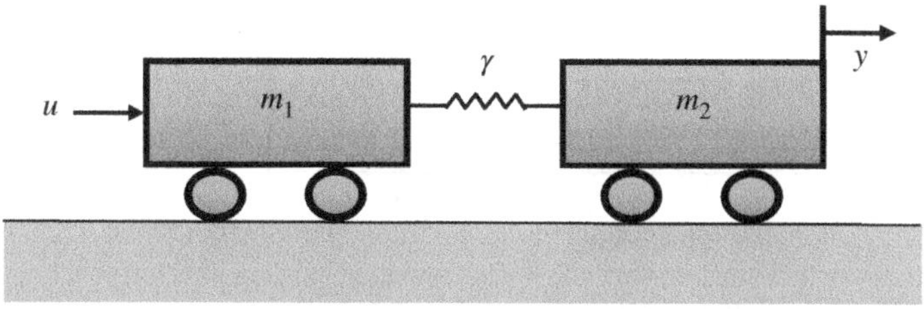

Figure 4.7 The connected two carts with a spring.

$$C_2 = \frac{26.509 - 79.066d + 78.797d^2 - 26.236d^3}{1 - 2.372d + 1.902d^2 - 0.511d^3}$$

$$C_3 = \frac{13.283 - 40.393d + 41.140d^2 - 14.026d^3}{1 - 2.322d + 1.829d^2 - 0.484d^3}$$

The controllers' stabilising intervals are

$$C_1 : \gamma \in [0.25, 0.499)$$

$$C_2 : \gamma \in (0.361, 0.830)$$

$$C_3 : \gamma \in (0.631, 1.5]$$

Thus, the feasibility condition is satisfied. For the simulation, the sampling time is 0.1 seconds, and the other parameters are

$$h = 0.01, \varepsilon = 0$$

The system's initial condition is $y(0) = 0$, and the system spring stiffness is $\gamma = 0.35$. The initial controller is C_3, which is a destabilising controller.

The system output is depicted in Figure 4.8 for a square reference input, and $n_u = 0$. After 11 samples, the supervisor selects the stabilising controller. The switching signals and the cost function are shown in Figures 4.9 and 4.10.

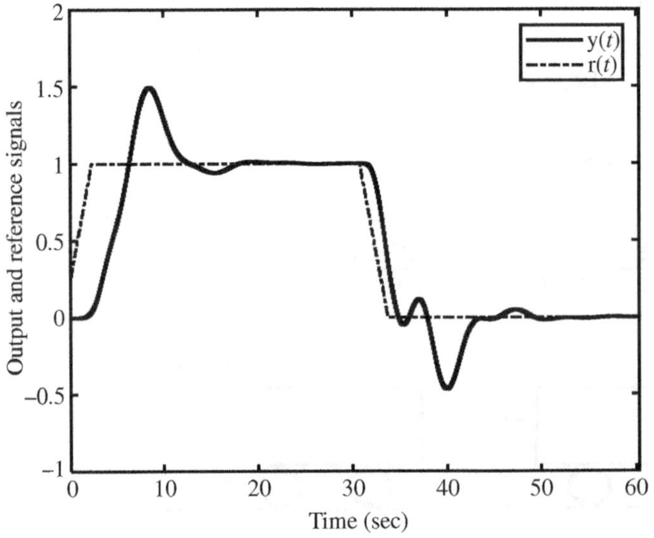

Figure 4.8 The MMUASC reference input and closed-loop output.

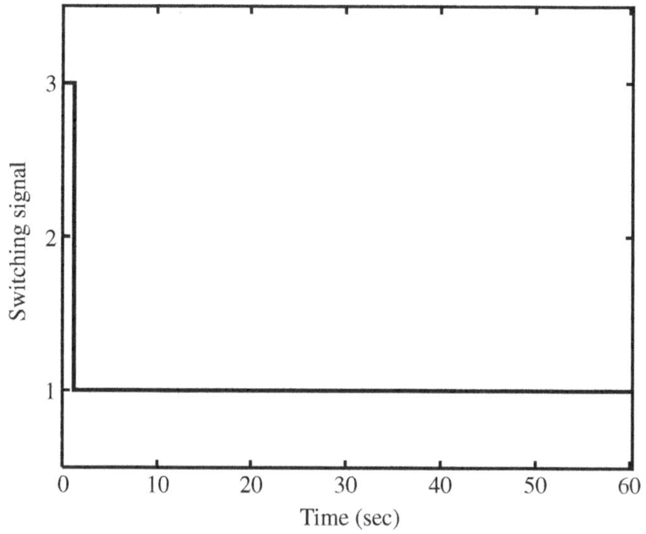

Figure 4.9 The MMUASC switching signal.

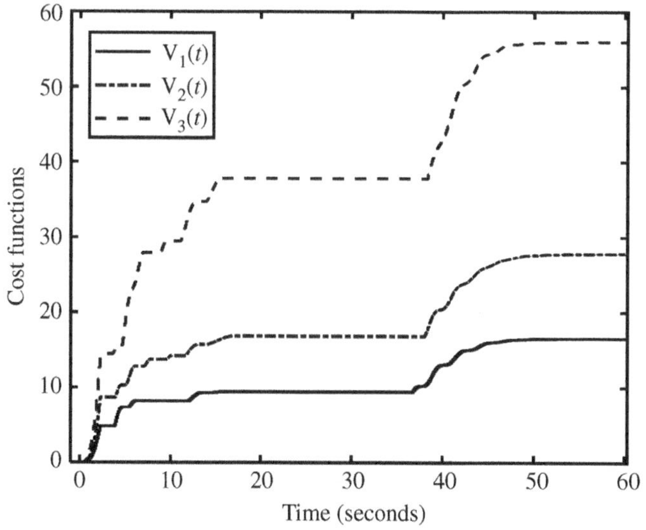

Figure 4.10 The MMUASC cost functions.

4.4 Performance Enhancement Techniques in the MMUASC

Performance enhancement in MMUASC is achieved through cost function modifications. In this section, five different MMUASC cost functions and their key features are briefly introduced. Then, the time windowing concept in the cost function is presented, and its performance enhancement consequences are shown.

4.4.1 Different MMUASC Cost Functions

Various cost functions for the MMUASC approach are reviewed in [3]. Comparison of various MMUASC systems, based on the employment of various cost functions, can be studied in terms of their generality of applicability, simplicity of implementation and performance.

The cost functions can be described in the following general structure:

$$V(C_i, z, t) = \max_{\tau \le t} \frac{\|(P/C_i)\,\tilde{r}_i - (M_i/C_i)\,\tilde{r}_i\|_\tau}{\|F_i[\tilde{r}_i]\|_\tau + \varepsilon} \tag{4.33}$$

With different F_i choices, different cost functions are formed. Five different types are considered in Table 4.1: Φ, [PA], [RA], [PC] and [RC]. Type Φ is the basic mode in which normalisation is performed by the virtual reference signal. In the [PA] type, the output error of the system and the virtual reference signal are used, and in the [RA] type, the output error of the control candidate loop and the virtual reference signal are used for normalisation. Normalisation in the [PC] type is with the control signal and the system output, and in the [RC] type, normalisation is with the control signal and the output of the control candidate loop. These are summarised in Table 4.1.

Closed-loop stability considerations. The feasibility condition (Definition 3.3) is necessary for stability guarantees in all cases. In Φ, [RA] and [RC] types, the feasibility condition will result in stability assurance. In other

Table 4.1 Different choices of F_i.

	F_i Filter	$F_i[\tilde{r}_i]$
Φ	$F_i = I$	$F_i[\tilde{r}_i](t) = \tilde{r}_i(t)$
[PA]	$F_i = (P/C_i) - [0, I]$	$F_i[\tilde{r}_i](t) = [u(t), y(t) - \tilde{r}_i(t)]^T$
[RA]	$F_i = (M_i/C_i) - [0, I]$	$F_i[\tilde{r}_i](t) = [u_i(t), y_i(t) - \tilde{r}_i(t)]^T$
[PC]	$F_i = (P/C_i)$	$F_i[\tilde{r}_i](t) = z(t)$
[RC]	$F_i = (M_i/C_i)$	$F_i[\tilde{r}_i](t) = z_i(t)$

types, in addition to the feasibility condition, the closed-loop system is stable only when the models have dense nominal model distributions. In other words, for stability assurance in [PA] and [PC] types, one of the models in the model set should describe the system well. In [RA] and [PA], the plant's unstable modes should be known a priori.

Closed-loop performance considerations. The virtual reference should be calculated in Φ, [PA] and [RA], and in [PC] and [RC], this calculation is not necessary. Hence, [PC] and [RC] have a less computational load, and this feature enhances the practical implementation capability. Any switching performance via [RC] can be achieved via [PC] and vice versa, although [RC] is preferred due to stability considerations. The performance of different cost functions is compared in Table 4.2.

Altogether, [RC] has noticeable advantages: does not require virtual reference calculation, only requires prediction error, and the stability guarantee is solely hinged upon problem feasibility. More details can be found in [3].

4.4.2 Adaptive Window in the MMUASC

The reset time concept in the cost function is elaborated in Section 3.8. The idea can be developed for the MMUASC structure called the *adaptive memory*, and the algorithm is denoted by MMUASC-R [14]. Adaptive memory reduces the dependence of the cost function on past data and can increase system performance.

The adaptive memory implies that the time window of the cost function changes in different conditions. The windowed cost function is calculated as

$$V(C_i, z, t) = \max_{\tau \in \{t-m(t),\dots,t\}} J(C_i, z, \tau), \quad t \in \mathbb{Z}_+, \quad i \in \{1, \dots, N\} \tag{4.34}$$

where $m(t)$ is a non-negative integer that specifies the time window length. The cost function $V(C_i, z, t)$ gives the cost associated with the controller C_i that has

Table 4.2 Performance comparison of different cost functions.

	Virtual reference calculation	**Stability condition**
Φ	Necessary	Feasibility condition
[PA]	Necessary	Feasibility condition Dense Model set
[RA]	Necessary	Feasibility condition
[PC]	Unnecessary	Feasibility condition Dense Model set
[RC]	Unnecessary	Feasibility condition

resulted in the worst performance on the time window $\{t - m(t), \ldots, t\}$. The presence of a maximum in the cost function causes the cost function V to be uniformly non-decreasing over the time period $\{t - m(t), \ldots, t\}$. According to the hysteresis switching logic, the switching termination in a finite time is ensured if the cost function is uniformly increasing over time [13]. Therefore, to guarantee a finite-time convergence of switching, the usual choice for the time window length is

$$m(t) = t \; \forall t \tag{4.35}$$

which is called a *persistent memory*. However, persistent memory prevents the devaluation of old data over time. Thus, the control based on persistent memory is not a good option for time-varying systems. On the other hand, if the cost function memory is fixed, $m(t) = M_{\star}$, and the hysteresis switching logic is utilised, the switching will not terminate in a finite time. With the hysteresis algorithm and the fixed memory cost function, the basic feasibility assumption is insufficient for ensuring an uncertain system's closed-loop stability, and further assumptions would be necessary.

The memory length $m(t)$ is a design parameter, and its selection is closely related to the closed-loop system stability assurance.

The process of $m(t)$ determination is such that the previous cost function data is utilised if it entails information relevant to stability. As shown in Figure 4.11, a new hypothetical loop is considered (M_i/C_i) with the reference input r.

The closed-loop output data are represented as $\bar{z}_i = (M_i/C_i)$. Therefore, a new cost function is introduced as

$$\bar{J}(C_i, z, t) = \frac{\|(z - \bar{z}_i)_t\|_\lambda}{\|(\bar{z}_i)_t\|_\lambda + \varepsilon}, \quad t \in \mathbb{Z}_+, \quad i \in \{1, \ldots, N\} \tag{4.36}$$

where $\|(\cdot)_t\|_\lambda$ denotes the truncated l_2 norm with the forgetting factor λ defined in (3.51). Note that the cost function $\bar{J}$ is similar to $J_\star$ in Section 3.8. The reset occurs when the behaviour of the system (P/C_σ) is similar to the behaviour (M_σ/C_σ) with the reference input r. Thus, the reset time is activated when $\bar{J}(C_{\sigma(t+1)}, z, t)$ is similar to $J(C_{\sigma(t+1)}, z, t)$.

$$t_{k+1} = 1 + \min\{t \geq t_k : |\bar{J}(C_{\sigma(t+1)}, z, t) - J(C_{\sigma(t+1)}, z, t)| \leq \varepsilon_1\} \tag{4.37}$$

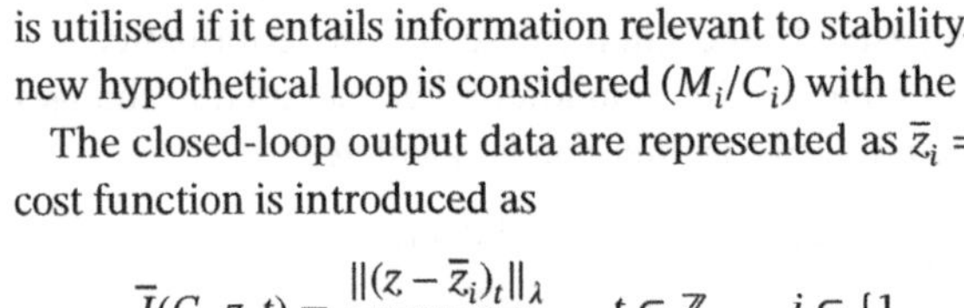

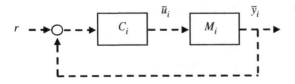

Figure 4.11 New control candidate loop.

where ε_1 is an arbitrarily positive constant, and t_k is the kth reset time. The adaptive time window is selected as

$$m(t) = \begin{cases} \min(t, M_\star) \text{ if } t = t_k \\ m(t-1) + 1 \text{ o.w.} \end{cases} \tag{4.38}$$

where $M_\star$ is an integer constant. According to (4.38), it is concluded that if the reset conditions are met, the time windows $m(t)$ will have a maximum length of $M_\star$. Otherwise, the time window length is increased linearly. If the plant variation is occasional, closed-loop stability can be achieved, and it is shown that the time window $m(t)$ is always bounded. The stability guarantees of a general time-varying system are still an open problem.

Example 4.2 Consider the system in Example 4.1. Performances of the MMUASC-R and the MMUASC algorithms are compared with the following parameters:

$$h = 0.1, \quad M_\star = 10, \varepsilon = 1, \lambda = 0.95, \quad \varepsilon_1 = 0.01$$

The reference input is zero, and n_u is uniformly distributed on $[-0.01, 0.01]$. An abrupt change in γ occurs at $t = 700$ from $\gamma = 1.5$ to $\gamma = 0.25$. The initial controller is C_3. The outputs of the MMUASC and the MMUASC-R are shown in Figure 4.12, the control selection signal $\sigma(t)$ and the spring stiffness $\gamma(t)$ and the cost functions are shown in Figures 4.13 and 4.14, respectively. Also, the reset time t_k and the adaptive memory $m(t)$ for the MMUASC-R are demonstrated in Figure 4.15. As depicted in Figure 4.12, the MMUASC-R has a superior performance in the sense of output deviation when γ abruptly changes at 700 [10]. The maximum output deviation of MMUASC-R and MMUASC are 0.1 and 0.33, respectively.

4.5 Input-constrained Multi-Model Unfalsified Switching Control Design

The combination of UASSC and MMASSC resulted in the proposed MMUASC methodology with enhanced closed-loop transient performance and guaranteed closed-loop stability with required minimum plant assumptions. However, actuator constraints are inevitable in practical controller implementations that can significantly affect closed-loop behaviour. Despite the deteriorating consequences of the presence of constraints in real applications, the input constraints are not generally considered in the structures of unfalsified control systems. In this section, the input-constrained systems are considered using a *general anti-windup structure*. Also, the constrained *generalised predictive control* (GPC) is presented in a *multi-model unfalsified constrained generalised predictive control* (MMUCGPC) framework. The results of this section are from [8, 10].

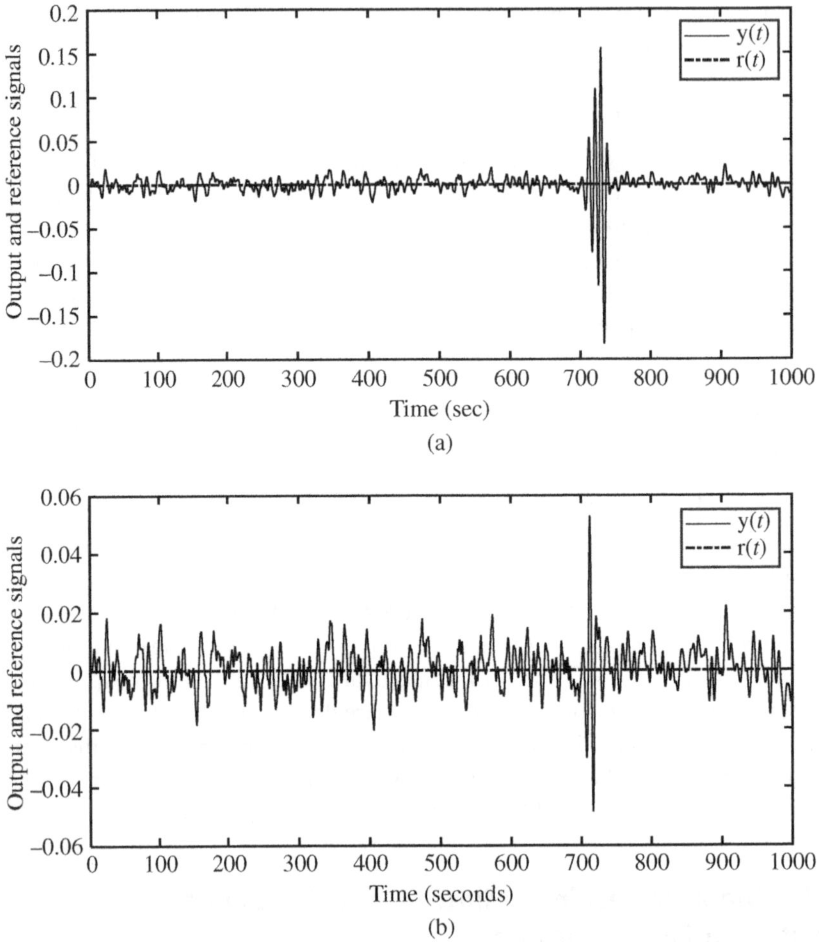

Figure 4.12 The output and reference: (a) MMUASC, (b) MMUASC-R.

4.5.1 Multi-Model Unfalsified Constrained Anti-Windup Control

Consider the following unknown LTI plant:

$$\begin{cases} \mathbf{x}(t+1) = \mathbf{A}\mathbf{x}(t) + \mathbf{B}\mathbf{v}(t) \\ \mathbf{y}(t) = \mathbf{G}\mathbf{x}(t) \\ \mathbf{v}(t) = \varphi(\mathbf{u}(t)) \end{cases} \tag{4.39}$$

where $\mathbf{v}(t) \in \mathbb{R}^m$ is the plant input, $\mathbf{y}(t) \in \mathbb{R}^p$ is the plant output and $\mathbf{x}(t) \in \mathbb{R}^n$ is the state. $\varphi(\cdot)$ denotes that input nonlinearity and $\mathbf{u}(t) \in \mathbb{R}^m$ is the controller's

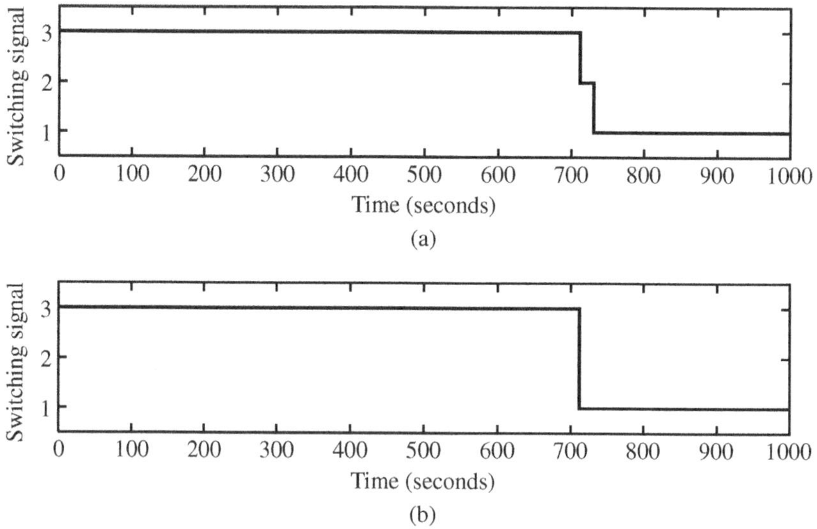

Figure 4.13 The switching signal: (a) MMUASC, (b) MMUASC-R.

output. The uncertain matrices (**A**, **B**, **G**) belong to an uncertainty set S. The considered input non-linearity $\varphi(\cdot): \mathbb{R}^m \rightarrow \mathbb{R}^m$ is of saturation type in the sense that $\varphi(\mathbf{u}) = \mathbf{u}$ for any $\mathbf{u} \in \mathbb{U}$, and $\varphi(\mathbf{u}) \in \mathbb{U}$ for any $\mathbf{u} \in \mathbb{R}^m$ for some compact set $\mathbb{U}$, with 0 belonging to the interior of $\mathbb{U}$. Hereafter, for ease of presentation, $\mathbb{U}$ is taken as a hyper-rectangle $\mathbb{U} = [\underline{u}^{(1)}, \overline{u}^{(1)}] \times \cdots \times [\underline{u}^{(m)}, \overline{u}^{(m)}]$, where $\underline{u}^{(\cdot)}$ and $\overline{u}^{(\cdot)}$ denote the corresponding minimum and maximum values.

Let a finite family of two-degrees-of-freedom dynamical output feedback LTI controllers $\{C_1, \ldots, C_N\}$ be available. For $i = 1, \ldots, N$, each controller is governed by $R_i(d)\mathbf{u}_i(t) = S_i(d)\mathbf{y}(t) + T_i(d)\mathbf{r}_i(t)$, and d is the delay operator, $R_i(d)$, $S_i(d)$ and $T_i(d)$ are stable transfer function matrices, and $R_i(d)$ is assumed monic. Also, $\mathbf{r}_i(t)$ is the corresponding virtual signal. There are several different strategies for implementing anti-windup control structures. In the anti-windup compensation based on the actual plant input, the past values of the control output are utilised. The switching controller C_σ is assumed to have the following generalised anti-windup structure realised:

$$\begin{aligned} \mathbf{u}(t) &= \hat{\varphi}(\mathbf{v}(t)) \\ \mathbf{v}(t) &= (I - R_{\sigma(t)}(d))\mathbf{u}(t) + T_{\sigma(t)}(d)\mathbf{r}(t) + S_{\sigma(t)}(d)\mathbf{y}(t) \end{aligned} \tag{4.40}$$

where $\mathbf{r}(t)$ is the reference input and the function $\hat{\varphi} : \mathbb{R}^m \rightarrow \mathbb{R}^m$ is an artificial nonlinearity that preserves the direction of input as proposed in [15]: $\hat{\varphi}(\mathbf{v}(t))$ takes the values $\mathbf{v}(t)$, if $\mathbf{v}(t) \in \hat{\mathbb{U}}$ and $\mathbf{v}(t)\min_j \{\varphi(v^{(j)}(t))/v^{(j)}(t)\}$ otherwise, where $v^{(j)}(t)$ denotes the jth component of $\mathbf{v}(t)$. Note that the concept of *directionality* is related

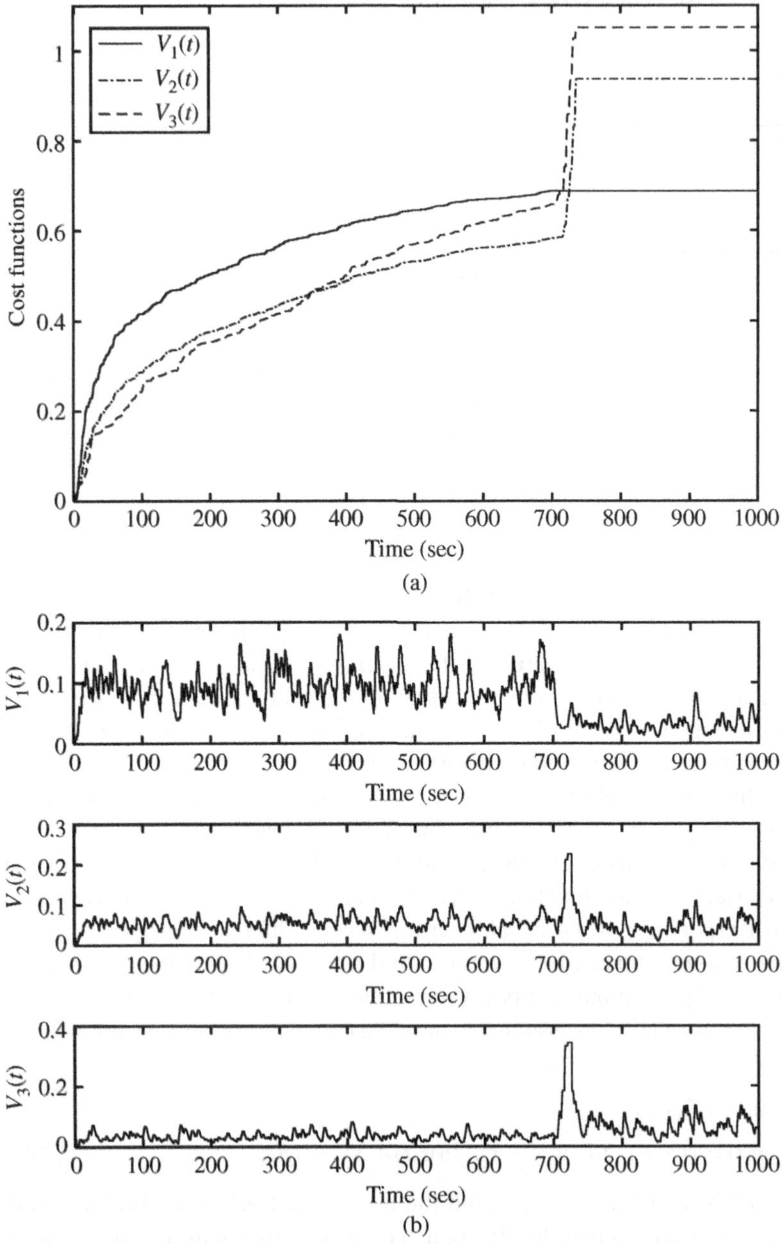

Figure 4.14 The cost functions: (a) MMUASC, (b) MMUASC-R.

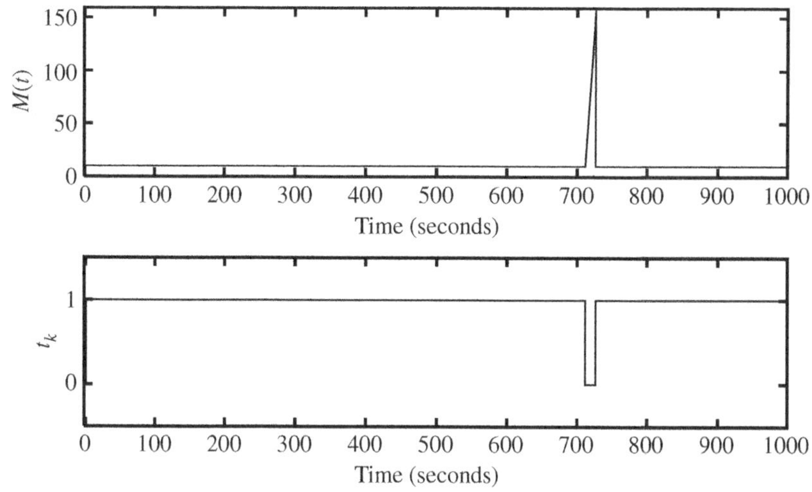

Figure 4.15 The adaptive memory $M(t)$ and reset time t_k for the MMUASC-R.

to the performance degradation of a controller in the face of actuator saturations (nonlinearities) that often occur in practical applications. It can be argued that directionality or performance degradation occurs when the direction of an unconstrained controller output differs from the actual and feasible plant input. To prevent this performance degradation from occurring, the direction of the unconstrained controller output must be preserved. For more details on this subject, refer to [16] and the references therein. To form the artificial nonlinearity, the linear range of the nonlinearity $\varphi(\cdot)$ should be known, although $\varphi(\cdot)$ is not completely known. The compact set $\hat{\mathbb{U}}$ limiting the control signal of the artificial nonlinearity is supposed to satisfy the condition $\hat{\mathbb{U}} \subseteq \mathbb{U}$ so that the function φ always remains in the linear zone, i.e. $\varphi(\hat{\varphi}(\mathbf{v}(t))) = \hat{\varphi}(\mathbf{v}(t))$. So, the nonlinearity $\varphi(\cdot)$ is never activated.

Following the unfalsified adaptive control philosophy, it is desired to evaluate the behaviour of the potential loop (P/C_i) in response to a properly defined *virtual reference*. The virtual reference $\mathbf{r}_i(t)$ is chosen to satisfy the following relationship:

$$
\begin{aligned}
&\mathbf{u}(t) = \hat{\varphi}(T_i(d)\mathbf{r}_i(t) + f_i(t)) \\
&f_i(t) = S_i(d)\mathbf{y}(t) + (I - R_i(d))\mathbf{u}(t)
\end{aligned}
\tag{4.41}
$$

By construction, it turns out that the hypothetical response of the potential loop (P/C_i) to the virtual reference $\mathbf{r}_i(t)$ would coincide precisely with the measured input–output sequence $(\mathbf{u}, \mathbf{y})$. In other words, the collected data allows one to infer how (P/C_i) would respond to the signal $\mathbf{r}_i(t)$.

However, the computation of $\mathbf{r}_i(t)$ has two associated problems. The first problem arises when the transfer function $T_i(d)$ is non-minimum phase, while the second problem is the invertibility of the nonlinear function $\hat{\varphi}(\cdot)$, which questions the uniqueness of the computed virtual reference. The first problem can be easily circumvented by computing the signal $\beta_i(t) = T_i(d)\mathbf{r}_i(t)$ instead of $\mathbf{r}_i(t)$, thus avoiding the necessity of inverting $T_i(d)$. As for the second problem, when $\mathbf{u}(t)$ belongs to $\hat{\mathbb{U}}^\circ$, i.e. the interior of $\hat{\mathbb{U}}$, the solution is unique and is given by

$$\beta_i(t) = \mathbf{u}(t) - f_i(t) \tag{4.42}$$

On the other hand, when $\mathbf{u}(t)$ belongs to $\partial\hat{\mathbb{U}}$, i.e. the boundary of $\hat{\mathbb{U}}$, many solutions are possible. A reasonable choice amounts to selecting, among all the signals $\beta_i(t)$ satisfying (4.41), the one which is closest to the true signal $\hat{\beta}_i(t) = T_i(d)\,\mathbf{r}(t)$, which would be generated in case the controller C_i was active. In practice, we solve the following optimisation problem:

$$\begin{aligned} &\underset{\beta_i(t)}{\text{argmin}}\|\beta_i(t) - \hat{\beta}_i(t)\|^2 \\ &s.t.\beta_i(t) = \alpha\mathbf{u}(t) - f_i(t), \alpha \geq 1 \end{aligned} \tag{4.43}$$

where $\|\cdot\|$ is the Euclidean norm and the constraint is imposed to preserve the input direction. The optimisation admits the analytic solution

$$\beta_i^{opt}(t) = \max\left\{1, \frac{\sum_j u^{(j)}\left(\hat{\beta}_i^{(j)} + f_i^{(j)}\right)}{\sum_j (u^{(j)})^2}\right\}\mathbf{u}(t) - f_i(t) \tag{4.44}$$

where the superscript (j) denotes the jth component. Notice that the signal $\beta_i(t)$ is used *only* in the computation of the test functionals, as given in (4.45), and hence the non-smoothness of the solution of (4.43) does not affect control performance.

Consider now the vector $\hat{\zeta}_i = (\mathbf{y}, \mathbf{u}, \hat{\mathbf{v}}_i)$ with $\hat{\mathbf{v}}_i = \beta_i + f_i$, consisting of the signals generated by the potential loop (P/C_i) in response to β_i as shown in Figure 4.16. Then, the discrepancy between (P/C_i) and (M_i/C_i) can be measured by computing the outputs $\zeta_i = (\mathbf{y}, \mathbf{u}, \mathbf{v}_i)$ of the reference loop (M_i/C_i) in response to the same input β_i, and then considering the test functionals

$$J_i(t) = \max_{\tau \leq t} \frac{\|\Psi(\hat{\zeta}_i - \zeta_i)\|_\tau}{\|\mathcal{F}(\beta_i)\|_\tau + \mu} \tag{4.45}$$

where Ψ is a positive definite weight matrix, $\mathcal{F}$ is a suitable filter and μ is a positive scalar.

Theorem 4.2 Let the switching signal σ be generated by the hysteresis switching logic (4.16) with test functionals (4.45). Suppose that each reference loop (M_i/C_i) is stable, and the filter $\mathcal{F}$ in (4.45) is stable and stably causally invertible. Then, the switched system (P/C_σ) is stable, provided that the ASC problem is feasible.

For the proof of Theorem 4.2, refer to [8].

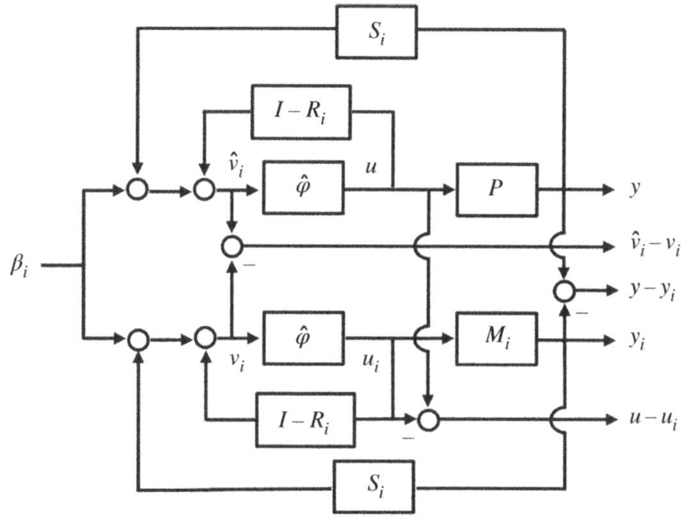

Figure 4.16 Comparison between potential and reference loops both driven by β_i.

4.5.2 The Feasibility Problem

Problem feasibility is a crucial assumption for the stability of any ASC scheme. Hence, in this section, the conditions for the design of a finite family of controllers C to ensure ASC feasibility are presented. Let the true plant belong to a parametric family $S = \{P(\delta) : \mathbf{A} = \mathbf{A}(\delta), \mathbf{B} = \mathbf{B}(\delta), \mathbf{G} = \mathbf{G}(\delta), \delta \in \Delta\}$, and suppose that for any $\delta \in \Delta$ it is possible to design a stabilising controller $C(\delta)$ with transfer function $K(d, \delta) = R(d, \delta)^{-1}[S(d, \delta)T(d, \delta)]$. Then the following result can be stated, which generalises some of the results of [5].

Proposition 4.1 Let Δ be compact. Suppose that for any $\delta \in \Delta$, the feedback loop $(P(\delta)/C(\delta))$ is finite-gain stable relatively to $T(d, \delta)r$. Finally, suppose that the transfer functions $R(d, \delta)$, $S(d, \delta)$ and $T(d, \delta)$ depend continuously on δ. Then, there exists a finite family C of controllers ensuring the feasibility of the ASC problem.

Proposition 4.1 is important because it shows that, for parametric uncertainties on compact domains, the problem of designing a finite family of controllers C ensuring feasibility is always solvable when the design problem for each single plant $P(\delta)$ is solvable (in other words, the ASC scheme does not impose additional restrictions). In this respect, in [17], it is shown that a controller $C(\delta)$ ensuring finite-gain stability of $(P(\delta)/C(\delta))$ always exists when: (i) the plant $P(\delta)$ is neutrally stable, i.e. all the eigenvalues of $\mathbf{A}(\delta)$ are inside the closed unit disk, and those eigenvalues with unitary absolute value are simple; (ii) detectability of $(\mathbf{A}(\delta), \mathbf{G}(\delta))$ and stabilisability of $(\mathbf{A}(\delta), \mathbf{B}(\delta))$ hold. In particular, as shown in [17], a possible choice is an observer-based controller with

$$R(d,\delta) = I - F(\delta)(dI - \mathbf{A}(\delta) + L(\delta)\mathbf{G}(\delta))^{-1}\mathbf{B}(\delta)$$
$$S(d,\delta) = F(\delta)(dI - \mathbf{A}(\delta) + L(\delta)\mathbf{G}(\delta))^{-1}L(\delta) \tag{4.46}$$

where the feedback gain $F(\delta)$ is chosen to ensure passivity, while the observer gain $L(\delta)$ is chosen, for example, by solving the stationary Kalman Filter Riccati equation. When the matrices $\mathbf{A}(\delta)$, $\mathbf{B}(\delta)$, and $\mathbf{G}(\delta)$ depend continuously on δ, such a parametric controller family is continuous in δ and hence satisfies the assumptions of Proposition 4.1.

Example 4.3 Consider a two-carts plant connected mechanically by a spring shown in Figure 4.7 and considered in Example 4.1. Taking as input the force applied to the first cart and as output the velocity of the second cart, the state-space model is

$$\dot{\mathbf{x}}(t) = \begin{bmatrix} 0 & 1 & -1 \\ -\Gamma/m_1 & 0 & 0 \\ \Gamma/m_2 & 0 & 0 \end{bmatrix} \mathbf{x}(t) + \begin{bmatrix} 0 \\ 1/m_1 \\ 0 \end{bmatrix} \varphi(u(t))$$
$$y(t) = \begin{bmatrix} 0 & 0 & 1 \end{bmatrix} \mathbf{x}(t) \tag{4.47}$$

Denoting by $M(s,\Gamma)$ the transfer function of (4.47) from the saturated input $\varphi(u(t))$ to y, consider three models corresponding to the zero-order-hold discretisation of $M(s,\Gamma)$ with sampling time $T_s = 0.1$ seconds and three different stiffness values: $\Gamma_1 = 0.3$ N/m, $\Gamma_2 = 0.7$ N/m and $\Gamma_3 = 1.2$ N/m. To include unmodeled dynamics, the continuous-time plant transfer function used in the simulations is considered as $P(s,\Gamma) = M(s,\Gamma)(1 - s\tau_1)/(1 + s\tau_2)$, with $\tau_1 = 0.2$ and $\tau_2 = 0.4$, where the additional non-minimum phase term can account for the unmodelled actuator/sensor dynamics.

Three simulations for different plant stiffnesses with plant initial condition $y(0) = 0.1$ are performed, case1: $\Gamma = 0.25$ N/m, case2: $\Gamma = 0.5$ N/m and case3: $\Gamma = 1.5$ N/m. The other parameters are hysteresis constant $h = 0.2$, input saturation $\underline{u} = -1.2, \overline{u} = 1.2$. The reference input is a square-wave of amplitude 1.3 m/s and period 105 seonds. The designed controllers are given as follows:

$$R_1(d) = 1 + 0.0366d + 0.1116d^2 + 0.0122d^3 - 0.0019d^4$$
$$S_1(d) = -480.3031 + 1012.1262d - 585.9229d^2 + 55.2913d^3 - 2.7109d^4$$
$$T_1(d) = 1.6 - 0.1757d + 0.0094d^2$$

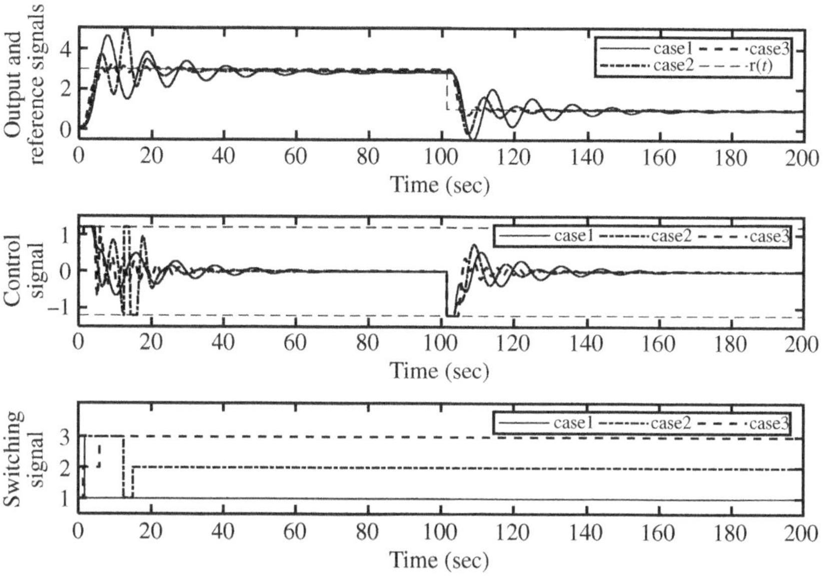

Figure 4.17 LTI case (from top to bottom): reference and plant outputs, control signals and controller selection.

$$R_2(d) = 1 + 0.0854d + 0.1448d^2 + 0.0163d^3 - 0.0025d^4$$

$$S_2(d) = -266.2261 + 563.6374d - 327.9211d^2 + 30.4991d^3 - 1.4713d^4$$

$$T_2(d) = 1.6 - 0.1728d + 0.0091d^2$$

$$R_3(d) = 1 + 0.0099d + 0.0906d^2 + 0.0099d^3 - 0.0015d^4$$

$$S_3(d) = -95.3220 + 200.3013d - 116.8698d^2 + 10.8730d^3 - 0.5246d^4$$

$$T_3(d) = 1.7 - 0.1836d + 0.0097d^2$$

The control loop is initialised with a destabilising controller in every case. As seen in Figure 4.17, the supervisor can select the stabilising controller after a short period of time.

A different simulation is performed by making the stiffness of the plant vary linearly in the simulation horizon from an initial value of 0.25 at time 0 to a final value of 1.4 at time 400. It can be seen from Figure 4.18 that the switching logic can follow the slow parameter variations. Notice that the test functionals are no longer monotone due to the reset logic. The parameters in the reset condition are selected as $\varepsilon_1 = 0.02$ and $M_* = 10$, while the forgetting factor is 0.98.

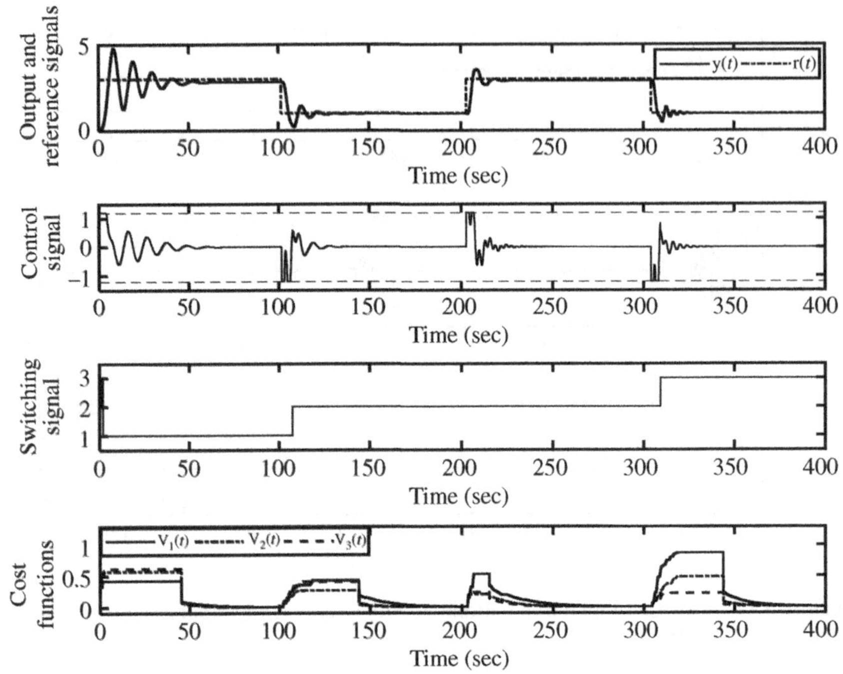

Figure 4.18 Time-varying case (from top to bottom): reference input and plant output, control signal, controller selection and test functionals.

4.5.3 Quadratic Inverse Optimal Control

Derivation of virtual reference signals is central to all unfalsified control strategies. In this section, to implement the MMUCGPC scheme, a quadratic inverse-constrained optimisation problem is proposed and solved to find the inverse GPCs for virtual reference signal calculation.

GPC is stated as an optimal control problem, and the cost function parameters or the weighting matrices are known. The optimal control signal is obtained by solving the optimisation problem. However, in the case of an inverse optimal control problem, the optimal control signal is known, and the cost function parameters must be derived.

Let the *primary* optimisation problem be defined as the case where the cost function parameters are known, and the corresponding optimal control signal is derived through optimising the cost function. Also, let the *secondary* problem be defined as the case where the optimal control signal is known, and it is desired to derive the corresponding unknown cost function parameters that have generated the optimal control signal. A quadratic optimal control scheme is considered to coincide with the GPC optimisation problem.

Define the general quadratic primary problem as follows:

$$\mathcal{J}^{P*} = \min_{U^P}\left(f(U^P) = \sum_{j\in J} b_j^P u_j^P + \frac{1}{2}\sum_{j\in J}\sum_{k\in J} h_{jk}^P u_j^P u_k^P\right)$$

$$\text{s.t.} \sum_{j\in J} a_{ij} u_j^P \le d_i \text{ for all } i \in I \text{ and } j \in J,\ \ I = \{1,2,\dots,m\},\ \ J = \{1,2,\dots,N\}$$

$$U^P = \left[u_1^P\ u_2^P\ \cdots\ u_N^P\right]^T \tag{4.48}$$

where b_j^P and $h_{jk}^P = h_{kj}^P$ are the known weighting parameters, U^P is the control signal and the index P refers to the primary problem and the corresponding optimal values are indicated as U^{P*} and $\mathcal{J}^{P*}$. Next, define the secondary quadratic problem with the same constraints as follows:

$$\mathcal{J}^{S*} = \min_{U^S}\left(f(U^S) = \sum_{j\in J} b_j^S u_j^S + \frac{1}{2}\sum_{j\in J}\sum_{k\in J} h_{jk}^S u_j^S u_k^S\right)$$

$$\text{s.t.} \sum_{j\in J} a_{ij} u_j^S \le d_i \text{ for all } i \in I \text{ and } j \in J,\ \ I = \{1,2,\dots,m\},\ \ J = \{1,2,\dots,N\}$$

$$U^S = \left[u_1^S\ u_2^S\ \cdots\ u_N^S\right]^T \tag{4.49}$$

where b_j^S and $h_{jk}^S = h_{kj}^S$ are the unknown weighting parameters, U^S is the control signal and the index S refers to the secondary problem. By solving this problem, the optimal values should be equal to U^{S*} and $\mathcal{J}^{S*}$ as these are the desired values. Note that in Eqs. (4.48) and (4.49), a_{ij} and d_i are known constraint parameters, and N, m are the control signal prediction horizon and the number of constraints, respectively.

To solve the secondary problem as a constraint optimisation problem, the inequality constraints are converted to equality constraints using the surplus/slack variables q and using these equality constraints with Lagrange multipliers λ, and Lagrangian L can be written as follows:

$$L(U^S, q, \lambda, r) = f(U^S) - \sum_{i\in I}\lambda_i\left(\sum_{j\in J} a_{ij} u_j^S + q_i^2 - d_i\right)$$

$$\sum_{j\in J} a_{ij} u_j^S + q_i^2 = d_i \tag{4.50}$$

and

$$L_j'(u) = b_j^S + \sum_{k\in J} h_{jk}^S u_k^{S*} - \sum_{i\in I}\lambda_i a_{ij} = 0 \tag{4.51}$$

where $L_j'(u)$ is the derivative of L with respect to u_j. Then, as u_j^{S*} and $\mathcal{J}^{S*}$ are known, and using the complementary slackness for the Kuhn–Tucker conditions, the derivative of the Lagrangian is as follows:

$$
\begin{aligned}
L_j'(u) &= b_j^S + \sum_{k\in J} h_{jk}^S u_k^{S*} - \sum_{i\in I} \lambda_i a_{ij} = 0, \text{if } a_{ij} u_j^{S*} = d_i \\
L_j'(u) &= b_j^S + \sum_{k\in J} h_{jk}^S u_k^{S*} = 0, \text{if } a_{ij} u_j^{S*} < d_i
\end{aligned} \tag{4.52}
$$

Also, another constraint is added to the problem as follows:

$$
\mathcal{J}^{S*} = \sum_{j\in J} b_j^S u_j^{S*} + \frac{1}{2} \sum_{j\in J} \sum_{k\in J} h_{jk}^S u_j^{S*} u_k^{S*} \tag{4.53}
$$

The parameters of the cost function of the secondary problem are derived by minimising the distance between the unknown parameters and the parameters of the primary problem with the constraints (4.52) and (4.53) [18]. Therefore, the new problem can be defined as follows:

$$
\min \left(\sum_{j\in J} \left| b_j^S - b_j^P \right| + \sum_{j\in J} \sum_{k\in J} \left| h_{jk}^S - h_{jk}^P \right| \right)
$$

$$
\begin{aligned}
&s.t. \\
&b_j^S + \sum_{k\in J} h_{jk}^S u_k^{S*} - \sum_{i\in I} \lambda_i a_{ij} = 0, \quad \text{if } a_{ij} u_j^{S*} = d_i \\
&L_j'(u) = b_j^S + \sum_{k\in J} h_{jk}^S u_k^{S*} = 0, \quad \text{if } a_{ij} u_j^{S*} < d_i \\
&\quad \mathcal{J}^{S*} = \sum_{j\in J} b_j^S u_j^{S*} + \tfrac{1}{2} \sum_{j\in J} \sum_{k\in J} h_{jk}^S u_j^{S*} u_k^{S*}
\end{aligned} \tag{4.54}
$$

The convergence of this inverse quadratic programming with the optimisation problem (4.54) is proved in [19]. Note that this is not a linear programming (LP) problem. Hence, the following change of variables is introduced to transform (4.54) into a linear optimisation problem:

$$
b_j^S - b_j^P = \alpha_j - \beta_j, h_{jk}^S - h_{jk}^P = \alpha_{jk} - \beta_{jk} \; \alpha_j, \beta_j, \alpha_{jk}, \beta_{jk} \geq 0
$$

$$
b_j^S = b_j^P + \alpha_j - \beta_j, h_{jk}^S = h_{jk}^P + \alpha_{jk} - \beta_{jk} \; \alpha_j, \beta_j, \alpha_{jk}, \beta_{jk} \geq 0
$$

Hence, the transformed problem is as follows:

$$
\min \left(\sum_{j\in J} (\alpha_j + \beta_j) + \sum_{j\in J} \sum_{k\in J} (\alpha_{jk} + \beta_{jk}) \right)
$$

$$
\begin{aligned}
&s.t. \\
&b_j^P + \alpha_j - \beta_j + \sum_{j\in J} \left(h_{jk}^P + \alpha_{jk} - \beta_{jk} \right) u_k^{S*} - \sum_{i\in I} \lambda_i a_{ij} = 0 \text{ if } a_{ij} u_j^{S*} = d_i \\
&b_j^P + \alpha_j - \beta_j + \sum_{j\in J} \left(h_{jk}^P + \alpha_{jk} - \beta_{jk} \right) u_k^{S*} = 0 \text{ if } a_{ij} u_j^{S*} < d_i \\
&\mathcal{J}^{S*} = \sum_{j\in J} \left(b_j^P + \alpha_j - \beta_j \right) u_j^{S*} + \frac{1}{2} \sum_{j\in J} \sum_{k\in J} \left(h_{jk}^{P*} + \alpha_{jk} - \beta_{jk} \right) u_j^{S*} u_k^{S*}
\end{aligned} \tag{4.55}
$$

Note: A special case of the problems (4.54) and (4.55), which will be encountered in the UASC with GPC, is that h_{jk}^S is known and convex and $\mathcal{J}^{S^*}$ is unknown, so the problem must be modified as follows:

$$\min \left(\sum_{j \in J} (\alpha_j + \beta_j) + \sum_{j \in J} \sum_{k \in J} (\alpha_{jk} + \beta_{jk}) \right)$$

$$s.t.$$

$$b_j^P + \alpha_j - \beta_j + \sum_{j \in J} \left(h_{jk}^P + \alpha_{jk} - \beta_{jk} \right) u_k^{S*} - \sum_{i \in I} \lambda_i a_{ij} = 0 \text{ if } a_{ij} u_j^{S*} = d_i$$

$$b_j^P + \alpha_j - \beta_j + \sum_{j \in J} \left(h_{jk}^P + \alpha_{jk} - \beta_{jk} \right) u_k^{S*} = 0 \text{ if } a_{ij} u_j^{S*} < d_i$$

$$\mathcal{J}^{S^*} = \sum_{j \in J} \left(b_j^P + \alpha_j - \beta_j \right) u_j^{S*} + \frac{1}{2} \sum_{j \in J} \sum_{k \in J} \left(h_{jk}^{P*} + \alpha_{jk} - \beta_{jk} \right) u_j^{S*} u_k^{S*}$$

$$\alpha_{jk} - \beta_{jk} = h_{jk}^S - h_{jk}^P, j, k \in J \tag{4.56}$$

The algorithm of quadratic inverse optimal control according to the problem (4.56) is as in Algorithm 4.3.

Algorithm 4.3 Quadratic Inverse Optimal Control Problem

Input Let b_j^P, h_{jk}^P, a_{ij} and d_i be known and $i \in I, j \in J$ such that $I = \{1,2, \ldots, m\}$, $J = \{1,2, \ldots, N\}$

Step 1 Form LP cost function due to the $\alpha_j, \beta_j, \alpha_{jk}, \beta_{jk}$ and $j, k \in J$

1.1. Equality constraint of LP: $\mathcal{J}^{S^*}$ is unknown, h_{jk}^S is known and Kuhn–Tucker complementary slackness condition is known.
1.2. Inequality constraints due to the $\alpha_j, \beta_j, \alpha_{jk}, \beta_{jk}$ and $\lambda_i \geq 0$
1.3. Solve LP problem

Step 2 Generate b_j^S such that $b_j^S = \alpha_j - \beta_j + b_j^P$

4.5.4 Multi-Model Unfalsified Constrained Generalised Predictive Control

This section will address switching between constrained model predictive controllers with an unfalsified strategy. The supervisor uses the MMUASC cost function and the UASSC algorithm. Also, GPC is employed in the controller units. Figure 4.19 depicts the MMUCGPC scheme.

According to Figure 4.19, the detection of a stabilising predictive controller is based on the falsification strategy. By using the inverse optimal control approach, the virtual reference signals are calculated. Then, the predictive control tuned loops and the cost functions are calculated by the virtual reference signals.

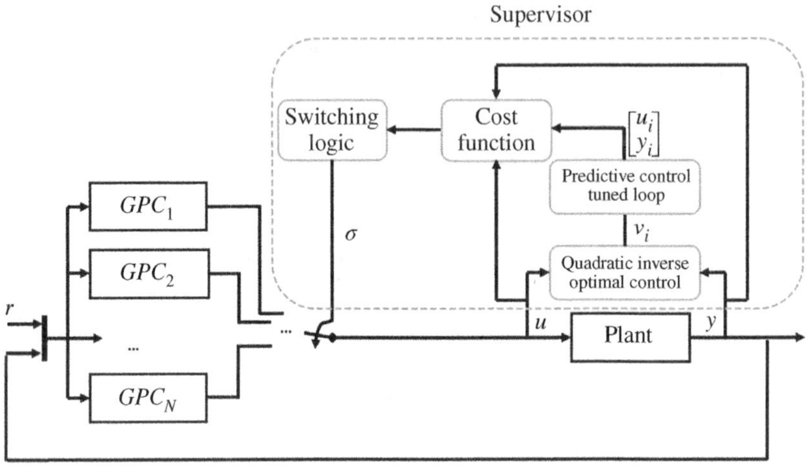

Figure 4.19 The MMUCGPC structure.

The GPC cost function with a constraint on future control increment is expressed as follows [20]:

$$J^* = \min_{U} \left(\sum_{k=N_1}^{N_2} (\hat{y}(t+k \mid t) - r(t+k))^2 + \mu \sum_{k=1}^{N_u} \Delta u(t+k-1)^2 \right)$$

$$s.t. \sum_{i=1}^{M} \sum_{j=1}^{N_u} a_{ij} \Delta u(t+j-1) \leq c_i \tag{4.57}$$

where $\hat{y}(t+k|t)$, $r(t+k)$ and $u(t)$ are the j-step-ahead predictions of the system output, reference sequence and control signal, respectively. The parameters N_1, N_2 and N_u and μ are the minimum and maximum of the prediction horizon, the control horizon and the weighted coefficient, respectively. Also, $\Delta = 1 - d$. It is shown in [21] that the future outputs are related to the control increments in terms of the following matrix algebraic equation:

$$Y = GU + f$$

where

$$Y = \begin{bmatrix} \hat{y}(t+N_1|t) \\ \hat{y}(t+N_1+1|t) \\ \cdots \\ \hat{y}(t+N_2|t) \end{bmatrix}, \quad U = \begin{bmatrix} \Delta u(t) \\ \Delta u(t+k) \\ \cdots \\ \Delta u(t+N_u-1) \end{bmatrix}$$

and f is called the free response of the system, and G is a matrix that can be calculated from the impulse response of the system or using the system model M_i. For a formal definition and derivation of f and G refer to [21]. The above equation can be used to calculate the control action to achieve a specific closed-loop performance. Finally, the GPC optimisation problem can be written as follows:

$$
\begin{aligned}
&J = \min_{U} \left(\frac{1}{2}U^T HU + B^T U + f_0\right)\\
&s.t.\ AU \leq D\\
&H = 2(G^T G + \mu I)\\
&B^T = 2(f - W)^T G\\
&f_0 = (f - W)^T (f - W)\\
&W = [r(t+d+1), r(t+d+2), \dots, r(t+d+N)],\ \mu \text{ is positive constant}
\end{aligned}
\tag{4.58}
$$

In the above optimisation problem, (A, D) are the constraint parameters. Typical constraint examples in practice are the control signals' amplitude limits, the slew rate limits of the actuators and the output signals' amplitude limits.

4.5.5 Virtual Reference Signal in the MMUCGPC Scheme

The inverse optimal control theory is used for the calculation of the virtual reference signal. Consider the primary problem for the active controller and the secondary problem for the candidate controllers in the controller bank. The primary problem can be expressed as follows:

$$
\begin{aligned}
&\mathcal{J}^*_{\sigma(t)} = \min_{U} \left(f(U_{\sigma(t)}) = \frac{1}{2}U^T H_{\sigma(t)} U + B^T_{\sigma(t)} U + f_{0,\sigma(t)}\right)\\
&\quad s.t.\ AU \leq D
\end{aligned}
\tag{4.59}
$$

Since A, D and the control signal U^* (refer to the definition of the virtual reference signal) should be the same for all the controllers, therefore, we can eliminate the P symbol for these parameters. Also, the active controller at time t is represented by the switching signal symbol $\sigma(t)$ instead of P. It can easily be shown that the existence of f_0 in the primary and the secondary problem does not affect the solutions, so for simplicity, it is eliminated in the foregoing equations.

In this optimisation problem, $B^T_{\sigma(t)} = 2(f_{\sigma(t)} - W)^T G_{\sigma(t)}, H_{\sigma(t)} = 2({G_{\sigma(t)}}^T G_{\sigma(t)} + \mu I)$ and its solution is U^*. To obtain a virtual reference signal, the secondary optimisation problem is solved for each candidate controller. Hence, for the ith controller, the problem (4.54) can be written as follows:

$$
\min\left(\|B_i - B_{\sigma(t)}\| + \|H_i - H_{\sigma(t)}\| = \sum_{j=1}^{N_u} |b_{j,i} - b_{j,\sigma(t)}| + \sum_{j\in J}\sum_{k\in J} |h_{jk,i} - h_{jk,\sigma(t)}|\right)
$$

$$
s.t.\begin{cases}
b_{j,i} + \sum_{k=1}^{N_u} h_{jk,i}u^*_k - \sum_{m=1}^{M} \lambda_m a_{mj} = 0, \text{if } a_{mj}u^*_j = d_i\\
b_{j,i} + \sum_{k=1}^{N_u} h_{jk,i}u^*_k = 0, \text{if } a_{mj}u^*_j < d_m\\
\mathcal{J}_i = \sum_{j=1}^{N_u} b_{j,i}u^*_j + \frac{1}{2}\sum_{j=1}^{N_u}\sum_{k=1}^{N_u} h_{jk,i}u^*_j u^*_k
\end{cases}
\tag{4.60}
$$

Now this problem due to the (4.56) is rewritten as:

$$\min\left(\sum_{j=1}^{N_u}(\alpha_{j,i}+\beta_{j,i})+\sum_{j=1}^{N_u}\sum_{k=1}^{N_u}(\alpha_{jk,i}+\beta_{jk,i})\right)$$

$$\text{s.t.}\begin{cases}\alpha_{j,i}-\beta_{j,i}+b_{j,\sigma(t)}+\sum_{k=1}^{N_u}(\alpha_{jk,i}-\beta_{jk,i}+h_{jk,\sigma(t)})u_k^* \\ \quad -\sum_{m=1}^{M}\lambda_m a_{mj}=0,\ \text{if}\ a_{mj}u_j^*=d_i \\ \alpha_{j,i}-\beta_{j,i}+b_{j,\sigma(t)}+\sum_{k=1}^{N_u}(\alpha_{jk,i}-\beta_{jk,i}+h_{jk,\sigma(t)})u_k^*=0,\ \text{if}\ a_{mj}u_j^*<d_m \\ J_i=\sum_{j=1}^{N_u}(\alpha_{j,i}-\beta_{j,i}+b_{j,\sigma(t)})u_j^*+\frac{1}{2}\sum_{j=1}^{N_u}\sum_{k=1}^{N_u}(\alpha_{jk,i}-\beta_{jk,i}+h_{jk,\sigma(t)})u_j^*u_k^* \\ \alpha_{jk}-\beta_{jk}=h_{jk}^S-h_{jk}^P, j,k\in\{1,\dots,N_u\}\end{cases} \tag{4.61}$$

Once this problem is solved, the values of $b_{j,i}$ are calculated as follows:

$$b_{j,i}=\alpha_{j,i}-\beta_{j,i}+b_{j,\sigma(t)} \tag{4.62}$$

Remark 4.2 If $N_u=N, N_1=d+1$ and $N_2=N+d$ then the virtual signal reference is obtained from the following equation:

$$\upsilon_i=f_i-\frac{1}{2}\left(B_i^TG_i^{-1}\right) \tag{4.63}$$

Remark 4.3 Virtual reference signal for the active controller is $\upsilon_i=W$.

Remark 4.4 In the UASC approach, stability analysis is based on the fact that the switching is stopped in a finite time, and the final selected controller guarantees closed-loop stability. To ensure the stability condition from the final controller's cost function, the final controller's virtual reference should converge to the true reference signal. Moreover, as the GPC strategy uses the future reference signal, the virtual reference signal is also related to the future. The direct result of $\upsilon_i=W$ is the convergence of the final virtual reference signal to the real reference signal.

4.5.6 Switching Algorithm in the MMUCGPC

Here the cost function (4.58) is used and u_i and y_i are obtained from the following equations:

$$\begin{aligned}&y_i=\mathrm{M}_i(u_i)(t) && \mathrm{M}_i \text{ denotes the } i^{\text{th}} \textit{ model}\\ &J_i^*=\min_{U_i}\left(\tfrac{1}{2}U_i^TH_iU_i+B_{ii}^TU_i+f_{0ii}\right) && U_i^* \text{ is the optimal solution}\\ &\qquad \text{s.t. } AU_i\le D && u_i=U_i^*(1)\\ &H_i=2\left(G_i{}^TG_i+\mu I\right)\end{aligned}$$

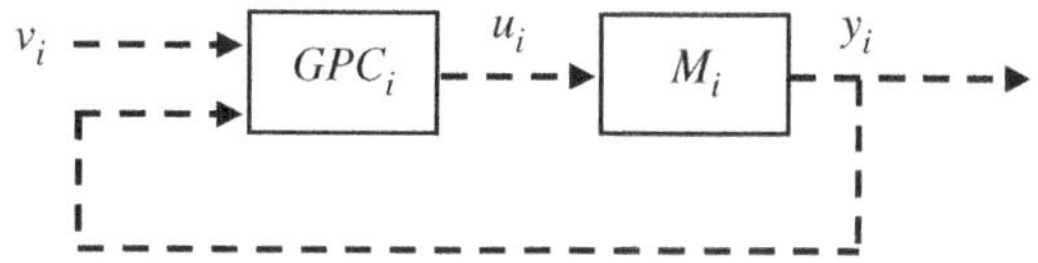

Figure 4.20 The predictive control tuned loop.

$$B_{ii}^T = 2(f_{ii} - \mathbf{v}_i)^T G_i$$
$$f_{0ii} = (f_{ii} - \mathbf{v}_i)^T (f_{ii} - \mathbf{v}_i) \qquad (4.64)$$
$$\boldsymbol{v}_i = [v(t+d+1), v(t+d+2), \dots, v(t+d+N)]^T$$

where $\boldsymbol{v}_i$ is the future virtual reference of the ith model. Figure 4.20 shows the structure of the predictive control tuned loops. By using the virtual reference signal, the signals u_i and y_i are calculated for each controller in this structure.

In the multi-model unfalsified control, the falsification procedure is based on the cost functions' values (4.15) to select the active controller. To calculate the cost function (4.15), the model bank is required along with the input and output of the system. Because of the existence of constraints in the MPC, the controller does not have an explicit representation as (4.5) and (4.11). By solving the inverse optimal control problems, the virtual reference signals are calculated according to (4.54). Utilising the virtual reference signal and using the relations (4.59) and (4.64), the input and output of the predictive control tuned loop are calculated for each controller. The appropriate GPC controller is recognised based on the cost functions and falsification strategy.

The algorithm of MMUCGPC with switching is represented as follows:

Algorithm 4.4 MMUCGPC Algorithm

Input Let $t = 0$, select an arbitrary $GPC_{\sigma(0)}$, and $\varepsilon > 0$ and $h > 0$

Step 1 $t \leftarrow t+1$ collect r, u, y

Step 2 Calculate $\boldsymbol{v}_i$ due to the quadratic inverse optimal problem for each controller

Step 3 Calculate the control candidate loops output: $z_i, i \in \{1, \dots, N\}$

Step 4 Calculate the cost functions $V(GPC_i, z, t), i \in \{1, \dots, N\}$

Step 5 If $V(GPC_{\sigma(t-1)}, z, t) > \min_{C_i \in \mathbb{C}} V(GPC_i, z, t) + h$

$$GPC_{\sigma(t)} \leftarrow \underset{C_i \in \mathbb{C}}{\text{argmin}}\, V(GPC_i, z, t)$$

Else, then

$$GPC_{\sigma(t)} \leftarrow GPC_{\sigma(t-1)}$$

End

Step 6 Go to Step 1

The cost function $V(GPC_i, z, t)$ is given by (4.14) and (4.15). The following theorem is presented and proved in [10] for the closed-loop stability of the MMUCGPC system:

Theorem 4.3 Consider the closed-loop MMUCGPC system shown in Figure 4.19 and Algorithm 4.4. Assume that all the tuned loops shown in Figure 4.4 are stable. Then, the closed-loop switched system is stable with a feasible controller set.

Example 4.4 Consider the 2-cart system in Example 4.1. Let the masses of the carts be $m_1 = m_2 = 1$ kg and uncertain stiffness be $\Gamma \in [0.25, 1.5]$ N/m. Three nominal models corresponding to $\Gamma_1 = 0.3$, $\Gamma_2 = 0.5$, and $\Gamma_3 = 1.00$ are considered:

$$M_1 = \frac{1.25 \times 10^{-06}d + 1.374 \times 10^{-05}d^2 + 1.374 \times 10^{-05}d^3 + 1.25 \times 10^{-06}d^4}{1 - 3.994d + 5.988d^2 - 3.994d^3 + d^4}$$

$$M_2 = \frac{2.083 \times 10^{-06}d + 2.29 \times 10^{-05}d^2 + 2.29 \times 10^{-05}d^3 + 2.083 \times 10^{-06}d^4}{1 - 3.99d + 5.98d^2 - 3.99d^3 + d^4}$$

$$M_3 = \frac{4.164 \times 10^{-06}d + 4.575 \times 10^{-05}d^2 + 4.575 \times 10^{-05}d^3 + 4.164 \times 10^{-06}d^4}{1 - 3.98d + 5.96d^2 - 3.98d^3 + d^4}$$

Also, the plant is fixed with $\Gamma = 0.3$ and the initially active controller is assumed to correspond to M_3 and the control horizon is $N_u = N = 50$. The simulation results are shown in Figures 4.21 and 4.22. The controller's saturation limits are ± 6.

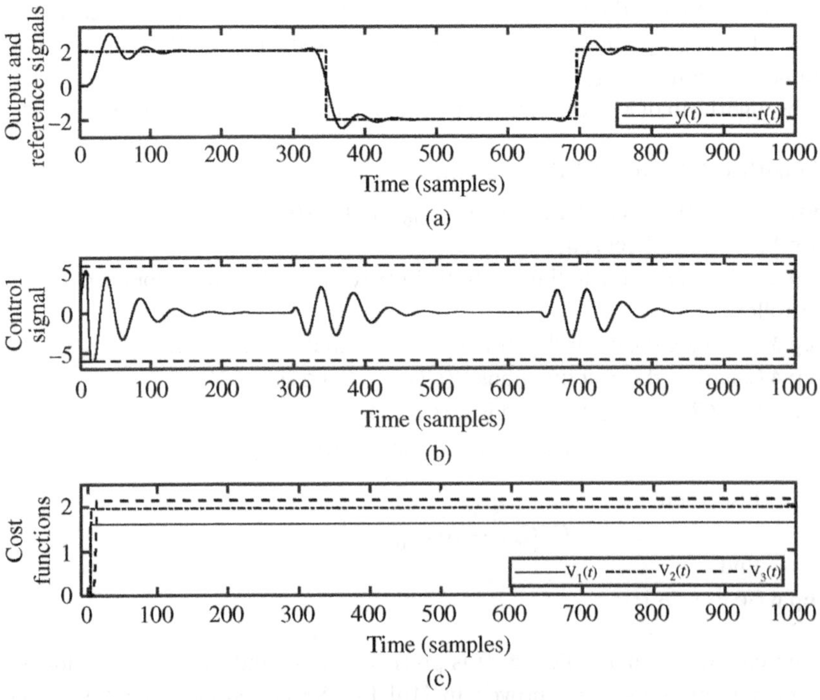

Figure 4.21 (a) Reference and output, (b) control signal, (c) cost value.

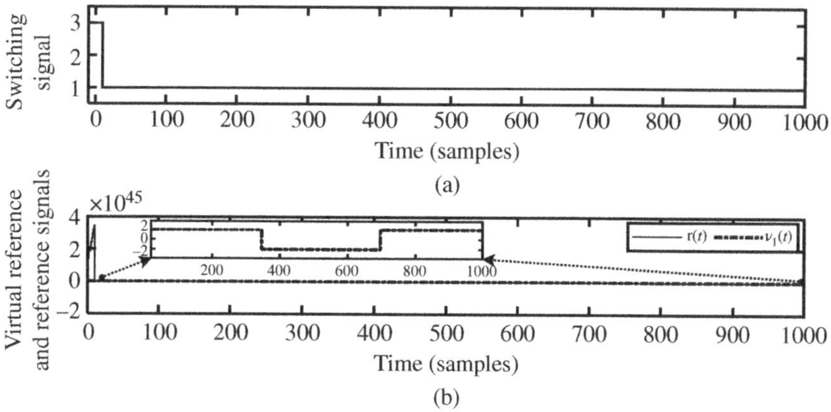

Figure 4.22 (a) Controller selection signal, (b) virtual reference convergence.

In this example, hysteresis is considered as $h = 0.01$. As is evident from the responses shown in Figure 4.21(a) and (b) the supervisor quickly selects the stabilising controller. However, there is a constraint on the control signal. Figure 4.21(c) confirms that the cost function is non-decreasing in time and Figure 4.22(b) shows that the virtual reference signal equals the actual reference signal for the final controller. Figures 4.21 and 4.22 show the appropriate tracking performance of the MMUCGPC and the transient is due to the destabilising controller that is the active controller in the initial samples, and MMUCGPC quickly falsified this controller and replaced it with the stabilising controller.

Remark 4.5 The control horizon affects the closed-loop MMUCGPC performance. It can be shown by simulation studies that a reduction in the control horizon delays the selection of the stabilising controller.

4.6 Conclusion

In this chapter, the MMUASC scheme is addressed. It is discussed that the MMUASC approach takes advantage of both the multi-model and unfalsified control methodologies. Unlike the multi-model methods, the MMUASC does not require a dense model set with the exact match condition to ensure closed-loop stability. The problem feasibility is the only required assumption for closed-loop stability proof. It is shown that the closed-loop performance of the MMUASC approach is superior to the closed-loop performance of the UASC designs. But, in the MMUASC, a model set must be available. The principles of the MMUASC approach are presented in detail, and the closed-loop stability theorem is presented. Different possible cost functions and their specifications are discussed. Finally, GPCs are considered as the underlying designed controllers. In this case, the inverse optimal control approach is utilised for the virtual signal calculations, and the corresponding relationships for the MMUCGPC are derived.

Problems

4.1 Consider the unknown linear system $A(d)y(t) = B(d)u(t)$ and the controller is given by $C_i(d) = S_i(d)/R_i(d)$ corresponding to the linear plant models $M_i(d) = \frac{B_i(d)}{A_i(d)}$ for $i \in \{1, \ldots, N\}$. Also, consider the different closed-loop output data represented as z_i and z, as defined in this chapter. Let the closed-loop characteristic polynomial of the actual plant and the corresponding model with the controller C_i be given by $\alpha_i(d) = A(d)R_i(d) + B(d)S_i(d)$, and $\beta_i(d) = A_i(d)R_i(d) + B_i(d)S_i(d)$, respectively, where $\beta_i(d)$ is designed as a stable polynomial.

a) Show that z_i and z satisfy the following matrix-vector difference equation:

$$\begin{bmatrix} -B_i & A_i \\ R_i & S_i \end{bmatrix} z_i = \begin{bmatrix} 0 & 0 \\ R_i & S_i \end{bmatrix} z \tag{P4.1.1}$$

b) Show that

$$\begin{bmatrix} -B & A \\ R_i & S_i \end{bmatrix} (z - z_i) = \xi_i \tag{P4.1.2}$$

where $\xi_i = \begin{bmatrix} B - B_i & -A + A_i \\ 0 & 0 \end{bmatrix} z_i + \begin{bmatrix} v \\ 0 \end{bmatrix}$ where v is an exogenous disturbance input.

c) Discuss the result of part b) from a stability perspective.

d) Assuming that there exists an i for which $\alpha_i(d)$ is stable, utilising the above results show that there exists $\kappa_0, \kappa_1 \in \mathbb{R}^+$, such that

$$\min_{i \in \{1,\ldots,N\}} J(C_i, z, t) = \frac{\|z - z_i\|_t}{\|z_i\|_t + \varepsilon} \leq \kappa_0 + \kappa_1 \|v\|_\infty, \forall t \in \mathbb{Z}_+$$

Hint: $\|v\|_t^2 \leq \|v\|_\infty^2 (1 - t^2)^{-1}$

4.2 Consider the following system from [22]:

$$P(s) = \frac{k}{s^2 + a_1 s + a_0}$$

where the parameters k, a_1 and a_0 are unknown, and $k \in [-0.5,2]/\{0\}$, $a_1 \in [0.5,2]$, and $a_0 \in [-2,1]$. The reference model is considered as follows:

$$M(s) = \frac{1}{s^2 + 1.4s + 1}$$

Let the reference input be a square wave of unit amplitude and period 10 units of time.

a) Use the multi-model approach to design an appropriate number of dynamic output feedback controllers of the form given by (4.8).
b) Design an MMUASC using Algorithm 4.1 or 4.2.
c) Compare the control systems designed in parts a) and b).

4.3 Consider the following scalar system [23]:

$$y(t+1) = a(t)y(t) + b(t)u(t)$$

where the plant parameter vector $\theta(t) = [a(t), b(t)]^T$ is given for four different operating points as $\theta_1 = [0.8,0.6]^T$, $\theta_2 = [1.8,-0.4]^T$, $\theta_3 = [-0.4,0.8]^T$ and $\theta_4 = [-1.1,-1.2]^T$, over 100 intervals of 100 units. Also, $\theta(t)$ is periodic with period 400. It is desired to track the following reference input: $r(t) = 0.5\left[\cos\left(\frac{2\pi t}{150}\right) + \sin\left(\frac{2\pi t}{50}\right)\right]$. Design an MMUASC using Algorithm 4.1. Study the effect of different control parameters on the closed-loop performance.

4.4 Consider the robot arm model given in [24] with the following transfer function from motor current I to motor angular velocity ω:

$$G_p(s) = \frac{k_m(J_a s^2 + ds + k)}{J_a J_m s^3 + d(J_a + J_m)s^2 + k(J_a + J_m)s}$$

where $J_a \in [0.0001,0.02]$, $J_m = 0.002$, $d = 0.0001$, $k = 100$, $k_m = 0.5$. A simplified model with neglecting the robot arm elasticity is considered as

$$G_m(s) = \frac{k_m}{Js}$$

where $J = J_a + J_m$ is the total moment of inertia and k_m is the current motor gain. Two nominal models are considered with $J = 0.0021$ and $J = 0.02$. The controller equation is $R(d)u(t) = T(d)r(t) + S(d)y(t)$, and the candidate controller set is

$$S_1(d) = -0.0471 + 0.0005d$$

$$R_1(d) = 1 + 1.1138d - 0.0112d^2$$

$$T_1(d) = 0.0476 - 0.0010d$$

$$S_2(d) = -0.0990 + 0.0010d$$

$$R_2(d) = 1 + 0.2300d - 0.0024d^2$$

$$T_2(d) = 0.1000 - 0.0020d$$

For $r(t) = \sin(0.03\ t)$, sampling time $T_s = 0.1$, the hysteresis parameter $h = 0.1, \Psi$ and F as identity, and $\mu = 1$:

a) Simulate the MMUASC algorithm with the cost function (4.45), $\lambda = 1$, $J_a = 0.00015$ and the initial controller C_2.
b) Simulate the MMUASC algorithm with the cost function (4.45), $\lambda = 1$ $J_a = 0.018$ and the initial controller C_1.
c) Simulate the MMUASC algorithm with the cost function (4.45), $\lambda = 1$, J_a linearly varying in time from $J_a = 0.001$ to $J_a = 0.02$, and the initial controller C_2.

4.5 Consider the following uncertain plant [25]:

$$P(s) = \frac{a_3}{s^2 + a_2 s + a_1}$$

where $a_1 \in [-2,2]$, $a_2 \in [-0.6,3.4]$, $a_3 \in [0.5,2]$. The four nominal models are derived as follows:

$$P_1(s) = \frac{0.875}{s^2 + 1.4s - 1}, P_2(s) = \frac{1.625}{s^2 + 1.4s - 1},$$
$$P_3(s) = \frac{1.625}{s^2 + 1.4s + 1}, P_4(s) = \frac{0.875}{s^2 + 1.4s + 1}$$

Let the plant's parameter vary as

$$\begin{cases} 0 \le t \le 20 : a_1 = 2, a_2 = 3.4, a_3 = 2 \\ 20 \le t \le 40 : a_1 = 1.5, a_2 = -0.56, a_3 = 0.75 \\ 40 \le t \le 60 : a_1 = 0, a_2 = 0, a_3 = 1.5 \\ 60 \le t \le 80 : a_1 = 1.5, a_2 = 1.5, a_3 = 0.6 \\ 80 \le t \le 100 : a_1 = 2, a_2 = 1, a_3 = 0.5 \end{cases}$$

Four PID controllers designed for the above four nominal plants are as follows:

$$K_i(s) = k_{Pi}E(s) + k_{Ii}I(s) + k_{Di}D(s), i = 1,2,3,4$$

where k_{Pi}, k_{Ii}, k_{Di}, are the proportional, integral and derivative tuning parameters for the four controllers and are given in the following table:

	k_P	k_I	k_D
$K_1(s)$	13.5	4.7	4.3
$K_2(s)$	1.5	0	0.25
$K_3(s)$	1	0	0.05
$K_4(s)$	2	1	0

a) Check the validity of the four models and the designed controllers using simulation results. Discuss the results.
b) Design an MMUASC using Algorithm 4.1 or 4.2. Compare the results with part a) without controller switching.
c) Study the effect of different control parameters on the closed-loop performance.

References

1 A. S. Morse, "Supervisory control of families of linear set-point controllers-part I. Exact matching," *IEEE Transactions on Automatic Control*, vol. 41, no. 10, pp. 1413–1431, 1996.

2 S. Baldi, G. Battistelli, E. Mosca, and P. Tesi, "Multi-model unfalsified adaptive switching supervisory control," *Automatica*, vol. 46, no. 2, pp. 249–259, 2010.

3 S. Baldi, G. Battistelli, E. Mosca, and P. Tesi, "Multi-model unfalsified adaptive switching control: test functionals for stability and performance," *International Journal of Adaptive Control and Signal Processing*, vol. 25, no. 7, pp. 593–612, 2011.

4 S. Baldi, G. Battistelli, D. Mari, E. Mosca, and P. Tesi, "Multi-model unfalsified switching control of uncertain multivariable systems," *International Journal of Adaptive Control and Signal Processing*, vol. 26, no. 8, pp. 705–722, 2012.

5 G. Battistelli, J. P. Hespanha, E. Mosca, and P. Tesi, "Model-free adaptive switching control of time-varying plants," *IEEE Transactions on Automatic Control*, vol. 58, no. 5, pp. 1208–1220, 2013.

6 M. Nouri Manzar and A. Khaki-Sedigh, "Self-falsification in multimodel unfalsified adaptive switching control," *International Journal of Adaptive Control and Signal Processing*, vol. 31, no. 11, pp. 1723–1739, 2017.

7 G. Battistelli, D. Mari, D. Selvi, A. Tesi, and P. Tesi, "Adaptive disturbance attenuation via logic-based switching," *Systems & Control Letters*, vol. 73, pp. 48–57, 2014.

8 M. N. Manzar, G. Battistelli, and A. Khaki-Sedigh, "Input-constrained multi-model unfalsified switching control," *Automatica*, vol. 83, pp. 391–395, 2017.

9 S. I. Habibi, A. Khaki-Sedigh, and M. N. Manzar, "Performance enhancement of unfalsified adaptive control strategy using fuzzy logic," *International Journal of Systems Science*, vol. 50, no. 15, pp. 2752–2763, 2019.

10 B. Sadeghi Forouz, M. Nouri Manzar, and A. Khaki-Sedigh, "Multiple model unfalsified adaptive generalized predictive control based on the quadratic inverse optimal control concept," *Optimal Control Applications & Methods*, vol. 42, no. 3, pp. 769–785, 2021.

11 K. S. Narendra, J. Balakrishnan, and M. K. Ciliz, "Adaptation and learning using multiple models, switching, and tuning," *IEEE Control Systems Magazine*, vol. 15, no. 3, pp. 37–51, 1995.

12 J. Hespanha, D. Liberzon, A. Stephen Morse, B. D. Anderson, T. S. Brinsmead, and F. De Bruyne, "Multiple model adaptive control. Part 2: Switching," *International Journal of Robust and Nonlinear Control: IFAC-Affiliated Journal*, vol. 11, no. 5, pp. 479–496, 2001.

13 A. S. Morse, D. Q. Mayne, and G. C. Goodwin, "Applications of hysteresis switching in parameter adaptive control," *IEEE Transactions on Automatic Control*, vol. 37, no. 9, pp. 1343–1354, 1992.

14 G. Battistelli, E. Mosca, and P. Tesi, "Adaptive memory in multi-model switching control of uncertain plants," *Automatica*, vol. 50, no. 3, pp. 874–882, 2014.

15 P. Campo and M. Morari, "Robust control of processes subject to saturation nonlinearities," *Computers & Chemical Engineering*, vol. 14, no. 4–5, pp. 343–358, 1990.

16 M. Soroush and S. Valluri, "Optimal directionality compensation in processes with input saturation non-linearities," *International Journal of Control*, vol. 72, no. 17, pp. 1555–1564, 1999.

17 A. Saberi, A. A. Stoorvogel, and P. Sannuti, *Internal and External Stabilisation of Linear Systems with Constraints*. Springer Science & Business Media, 2012.

18 S. Jain and N. Arya, "An inverse transportation problem with the linear fractional objective function," *Advanced Modelling & Optimisation*, vol. 15, no. 3, pp. 677–687, 2013.

19 J. Zhang and L. Zhang, "An augmented Lagrangian method for a class of inverse quadratic programming problems," *Applied Mathematics and Optimisation*, vol. 61, no. 1, pp. 57–83, 2010.

20 E. F. Camacho, "Constrained generalized predictive control," *IEEE Transactions on Automatic Control*, vol. 38, no. 2, pp. 327–332, 1993.

21 E. F. Camacho and C. B. Alba, *Model Predictive Control*. Springer Science & Business Media, 2013.

22 K. S. Narendra and J. Balakrishnan, "Adaptive control using multiple models," *IEEE Transactions on Automatic Control*, vol. 42, no. 2, pp. 171–187, 1997.
23 K. S. Narendra, O. A. Driollet, M. Feiler, and K. George, "Adaptive control using multiple models, switching and tuning," *International Journal of Adaptive Control and Signal Processing*, vol. 17, no. 2, pp. 87–102, 2003.
24 K. J. Åström and B. Wittenmark, *Adaptive Control*. Courier Corporation, 2013.
25 F. Gao, S. E. Li, D. Kum, and H. Zhang, "Synthesis of multiple model switching controllers using H∞ theory for systems with large uncertainties," *Neurocomputing*, vol. 157, pp. 118–124, 2015.

5

Data-Driven Control System Design Based on the Virtual Reference Feedback Tuning Approach

5.1 Introduction

The control laws in the classical adaptive control systems are based on input–output measurements. The two main approaches in classical adaptive control are the *direct* and *indirect* design methodologies. In the indirect adaptive control techniques, a plant model is identified, and using the identified model any model-based control design technique can be employed for closed-loop control. On the other hand, in the direct adaptive control techniques, the controller parameters are directly recursively estimated using the input–output measurement data. Virtual reference feedback tuning (VRFT) is a general design methodology that directly derives the controller parameters in a *non-recursive* or so-called *one-shot direct* approach. Therefore, VRFT can be considered a direct adaptive control design methodology, and the adaptation is performed offline. VRFT minimises a controlled cost of the 2-norm type by using a batch of input–output data collected from the plant, where the optimisation variables are the controller parameters.

VRFT formulates the controller tuning problem as a controller parameter identification problem by introducing a virtual reference signal. It is therefore called a *data-based* controller design methodology and was introduced in Ref. [1]. In VRFT, a controller class is selected, and then a specific controller is chosen in the class based on the collected data. The selection is performed offline. Initially, a batch of input–output data is collected from the plant. Utilising these data and the virtual reference concept, a controller is designed. The designed controller is then placed in the loop without any further adaptation. Note that the designed controller is verified for stability and performance requirements before its placement in the loop.

The main difficulty with direct data-based controller parameters tuning techniques, where no a priori knowledge of the plant is available, is the computation of the gradient of the model-matching error with respect to the controller

An Introduction to Data-Driven Control Systems, First Edition. Ali Khaki-Sedigh.

parameters. A similar problem is addressed in Chapter 6, where the data-based controller tuning is performed utilising a stochastic approach. The presented method in Chapter 6 relies on the estimation of the gradient of the cost function with respect to the controller tuning parameter by simultaneous perturbation stochastic approximation, and three experiments using test signals are performed in each iteration. This leads to an *iterative stochastic algorithm* strategy. However, one appealing feature of the VRFT methodology is its requirement of only a single record of input–output open-loop measurements. It utilises these data in a noniterative, offline and one-shot approach for direct controller design.

To further elaborate on the one-shot direct data-based nature of VRFT, after the batch input–output plant data collection, VRFT returns a controller without requiring iterations and/or further access to the plant for experiments. This is possible because the *design engine* inside VRFT is intrinsically global, and no gradient-descent techniques are involved. The input signals must be exciting enough, and in the case of poor excitation, the designed controller can be inappropriate with non-satisfactory closed-loop responses. The results obtained by VRFT are related to the information content present in the given batch of input–output data. The one-shot feature of VRFT is practically attractive because [2]:

- It is low-demanding, i.e. access to the plant for multiple experiments is not necessary and therefore, the normal operation of the plant is not halted.
- VRFT does not suffer from local minima and initialisation problems.

5.2 The Basic VRFT Methodology

To present the general structure and the main features of the VRFT methodology, consider a linear time-invariant single-input-single-output (SISO) discrete-time plant described by the rational transfer function $P(z)$, where z is the shift operator. It is assumed that $P(z)$ is unknown, and a set of input–output measurement data is available for design purposes. The controllers are assumed linear in the parameters and belong to the controller class described by the following regression model as $C(z;\,\boldsymbol{\theta}) = \boldsymbol{\beta}^T(z)\boldsymbol{\theta}$, where $\boldsymbol{\beta}(z) = [\beta_1(z)\ \beta_2(z)\ \dots\ \beta_n(z)]^T$ and $\boldsymbol{\theta} = \left[\theta_1\ \theta_2\ \dots\ \theta_n\right]^T$ are the n-dimensional known vector of linear discrete-time transfer functions and the n-dimensional vector of unknown parameters, respectively. Also, a reference model $M(z)$ is given that represents the desired closed-loop transfer function.

The control objective is to minimise the following model-reference type cost function:

$$J_{MR}(\boldsymbol{\theta}) = \left\| \left(\frac{P(z)C(z;\boldsymbol{\theta})}{1 + P(z)C(z;\boldsymbol{\theta})} - M(z) \right) W(z) \right\|_2^2 \tag{5.1}$$

where $W(z)$ is a designer-specified weighting function.

The *virtual reference concept* is fundamental to VRFT.[1] Its basic idea is briefly presented in Ref. [3]. Let in the ideal case, the controller $C(z;\theta)$ be designed such that the negative unity feedback closed-loop system transfer function is equal to the desired reference model $M(z)$. Consider any reference input signal $r(t)$, the model reference output will be $M(z)r(t)$, and a similar output is expected from the actual closed-loop plant. In the conventional model reference adaptive control systems, the tracking error is defined as the error between the model reference output and the actual closed-loop system output for the given reference input. The error is derived to zero by designing appropriate adaptation laws that estimate the controller parameters in the direction of the negative gradient of a defined cost function. Hence, gradient calculations are necessary for the model reference adaptive control strategies, and a plant model is required for such calculations.

Let $\bar{r}(t)$ be any given reference input signal. Then, a necessary condition for the closed-loop system to have the same transfer function as the reference model is that the output of the two systems is the same for $\bar{r}(t)$. The output of the reference model will be $M(z)\bar{r}(t)$. On the other hand, the plant transfer function $P(z)$ is unknown, and only the measured output of the plant $y(t)$ is available. For the given $y(t)$, select $\bar{r}(t)$ such that $M(z)\bar{r}(t) = y(t)$. Note that this is a very special selection of the reference input signal, and to determine $\bar{r}(t)$ the reference model $M(z)$ must be *inversely stable*. A selected reference input signal that satisfies this equation is called a *virtual reference input* because it is not, in reality, physically implemented to generate $y(t)$. Notice that $y(t)$, as the output of the reference model, is the desired output of the closed-loop system when the reference signal is $\bar{r}(t)$. Let $e(t) = \bar{r}(t) - y(t)$, be the tracking error for a given virtual reference input signal. Note that $P(z)$ is unknown, but its corresponding input $u(t)$ and output $y(t)$ are available. Also, $u(t) = C(z;\theta)e(t)$, and the controller must be designed to generate $u(t)$. This is equivalent to an identification problem, where the unknown dynamical systems have $e(t)$ and $u(t)$ as their input and output, respectively. That is by defining a cost function of the form $J = \frac{1}{N}\sum_{t=1}^{N}(u(t) - C(z;\theta)e(t))^2$, minimisation of this cost function has a standard least squares solution that gives the unknown parameter θ. However, the minimisation of J will not in general result in a parameter vector close to the solution of (5.1). To overcome this problem a proper filter selection is proposed to bring close the optimisation solutions of J and (5.1). This filter design is discussed in Section 5.2.1. The general structure of a VRFT is shown in Figure 5.1.

The basic VRFT controller design strategy can be implemented by the presented Algorithm 5.1. In this algorithm, data filtering is considered, which will

1 A virtual reference input concept is introduced in an unfalsification context in Chapters 2 and 3.

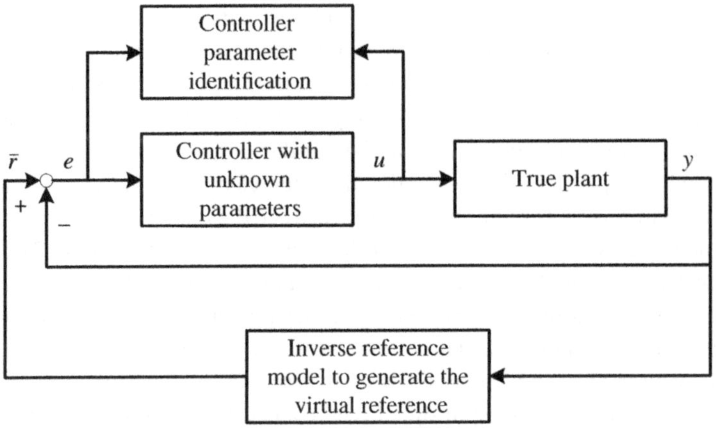

Figure 5.1 The VRFT general structure.

be discussed in the next section. The role of this filter is to attain a near-optimal solution of (5.1).

Algorithm 5.1 The Basic VRFT

Input Select the reference model $M(z) \neq 1$, the filter $L(z)$, data length N and the measured input–output data $\{u(t)\ y(t)\}\ for\, t = 1, \ldots, N$.

Step 1 Calculate the reference input signal $\bar{r}(t)$ such that $M(z)\bar{r}(t) = y(t)$.

Step 2 Calculate the corresponding tracking error $e(t) = \bar{r}(t) - y(t)$.

Step 3 Calculate the filtered tracking error and the filtered input signals as follows:

$$e_L(t) = L(z)e(t), u_L(t) = L(z)u(t)$$

Step 4 Solve the following optimisation problem to derive the controller parameter vector $\boldsymbol{\theta}$:

$$J_{VR}^N(\boldsymbol{\theta}) = \frac{1}{N} \sum_{t=1}^{N} (u_L(t) - C(z; \boldsymbol{\theta})e_L(t))^2 \tag{5.2}$$

Note that by assuming a controller that is linear in its parameters, the controller transfer function can be expressed in the regression model $C(z;\ \boldsymbol{\theta}) = \boldsymbol{\beta}^T(z)\boldsymbol{\theta}$, by defining $\boldsymbol{\varphi}_L(t) = \boldsymbol{\beta}(z)e_L(t)$, the cost function (5.2) can be written as follows:

$$J_{VR}^N(\boldsymbol{\theta}) = \frac{1}{N} \sum_{t=1}^{N} \left(u_L(t) - \boldsymbol{\varphi}_L^T(t)\boldsymbol{\theta}\right)^2 \tag{5.3}$$

This is a standard least squares problem and provided that $\sum_{t=1}^{N} \varphi_L(t)\varphi_L^T(t)$ is a full rank matrix, its solution is given by [4]

$$\hat{\theta}_N = \left[\sum_{t=1}^{N} \varphi_L(t)\varphi_L^T(t)\right]^{-1} \sum_{t=1}^{N} \varphi_L(t)u_L(t) \tag{5.4}$$

Hence, the solution to the optimisation problem (5.2) is given by (5.4). This provides a one-shot design for the controller parameter vector given by $\hat{\theta}_N$.

5.2.1 Filter Design

The VRFT design formulation is stated in terms of the cost function defined in (5.1), $J_{MR}(\theta)$. However, the algorithm was presented using the cost function defined in (5.2), $J_{VR}^N(\theta)$. While $J_{MR}(\theta)$ and $J_{VR}^N(\theta)$ are two different cost functions, it can be shown that by selecting an appropriate filter $L(z)$, their minimum arguments, which represent the unknown parameter vector θ, can converge closely. This implies the key point that the actual VRFT model reference control problem can be solved by utilizing $J_{VR}^N(\theta)$ [3].

Note that for a discrete-time linear transfer function $G(z)$, the 2-norm is defined as [5]

$$\|G(z)\|_2 \triangleq \left(\frac{1}{2\pi}\int_{-\pi}^{\pi} |G(e^{j\omega})|^2 d\omega\right)^{\frac{1}{2}}$$

or in the compact form

$$\|G(z)\|_2 \triangleq \left(\frac{1}{2\pi}\int_{-\pi}^{\pi} |G|^2 d\omega\right)^{\frac{1}{2}}$$

Hence, $J_{MR}(\theta)$ in (5.1) can be rewritten as follows:

$$J_{MR}(\theta) = \frac{1}{2\pi}\int_{-\pi}^{\pi} \left|\frac{PC(\theta)}{1 + PC(\theta)} - M\right|^2 |W|^2 d\omega \tag{5.5}$$

Let $C_0(z)$ be the ideal controller that results in $J_{MR}(\theta) = 0$ and exactly satisfies the model-matching equation. However, this ideal controller would be impractical for the following reasons:

- The ideal controller may be an improper transfer function.
- It may not generally belong to the parametrised controller set.
- It may result in an internally unstable closed-loop system.

Hence, $C_0(z)$ is only used for system analysis and satisfies,

$$\frac{P(z)C_0(z)}{1 + P(z)C_0(z)} = M(z) \tag{5.6}$$

Assume that the input–output measured data are stationary and ergodic stochastic processes. This implies that the statistical properties of the unknown process do not change over time and the collected measured input–output data represent the average statistical properties of the unknown process. By direct substitution of (5.6) in (5.5), it can be written as follows:

$$J_{MR}(\theta) = \frac{1}{2\pi}\int_{-\pi}^{\pi} \frac{|P|^2|W|^2}{|1+PC(\theta)|^2}\frac{|C(\theta)-C_0|^2}{|1+PC_0|^2}d\omega \tag{5.7}$$

It is shown that for ergodic stochastic processes as $N \to \infty$,

$$J_{VR}^N(\theta) \to J_{VR}(\theta) = E\left[(u_L(t) - C(z;\theta)e_L(t))^2\right] \tag{5.8}$$

That is, the time average tends to be the ensemble average [6]. We can therefore use $J_{VR}(\theta)$ instead of $J_{VR}^N(\theta)$ for further analysis. Note that $u_L - Ce_L = [u - Ce]L$, by substitution for $e(t) = \bar{r}(t) - y(t)$, $y(t) = Pu(t)$, $M(z)\bar{r}(t) = y(t)$, noting that $\frac{M}{1-M} = PC_0$, and some mathematical manipulations, we have

$$u_L - Ce_L = Lu - CPL\frac{1-M}{M}u = \frac{P(M-1)(C-C_0)L}{M}u$$

and by applying the Parseval's theorem,[2] it can be shown that the asymptotic cost function (5.8) can be written as

$$J_{VR}(\theta) = \frac{1}{2\pi}\int_{-\pi}^{\pi} |P|^2|C(\theta)-C_0|^2|1-M|^2\frac{|L|^2}{|M|^2}\Phi_u d\omega \tag{5.9}$$

where Φ_u is the spectral density of $u(t)$.

Assuming $C_0(z) \in \{C(z;\theta)\}$ and that $J_{VR}(\theta)$ has a unique minimum, minimisation of $J_{VR}(\theta)$ irrespective of $L(z)$ gives $C_0(z)$, which is in accordance with (5.8). However, this is improbable in most cases, hence considering a filter that satisfies the following equation:

$$|L|^2 = \frac{|M|^2|W|^2}{|1+PC(\theta)|^2}\frac{1}{\Phi_u}, \qquad \forall\omega \in [-\pi\ \pi] \tag{5.10}$$

gives $J_{VR}(\theta) = J_{MR}(\theta)$, and the VRFT design problem can be solved by minimising the cost function $J_{VR}(\theta)$. As the plant transfer function, $P(z)$ appears in (5.10), it cannot be directly utilised in a data-driven control framework. However, noting that for the optimising parameter vector θ of $J_{MR}(\theta)$, we can assume that $|1+PC(\theta)| \approx |1+PC_0(\theta)|$, and also $1-M = (1-PC_0(\theta))^{-1}$, (5.10) can readily be approximated by

$$|L|^2 = |1-M|^2|M|^2|W|^2\frac{1}{\Phi_u}, \forall\omega \in [-\pi\ \pi] \tag{5.11}$$

2 If $\{w(t)\}$ is a quasi-stationary discrete-time signal with spectral density Φ_w, then for a given stable transfer function $G(z)$, for $s(t) = G(z)w(t)$, $\{s(t)\}$ is also quasi-stationary and $\Phi_s = |G(e^{i\omega})|^2\Phi_w$. Also, the Parseval's theorem states that $\sum_{t=0}^{\infty}|w(t)|^2 = \frac{1}{2\pi}\int_{-\pi}^{\pi}|W(e^{i\omega})|^2 d\omega$, where $W(e^{i\omega})$ is the corresponding Fourier transform of $w(t)$.

Finally, for a designer-specified control signal $u(t)$, the spectral density Φ_u is known. Otherwise, the standard techniques can be used for the Φ_u estimation [7].

Remark 5.1 It is shown in [2] that if $L(z)$ is designed using (5.11), the parameter vectors $\hat{\theta}$ and $\overline{\theta}$ minimising the cost functions $J_{VR}(\theta)$, and $J_{MR}(\theta)$ are closely related. Note that $\overline{\theta}$ minimises the original cost function (5.1), while $\hat{\theta}$ minimises (5.9). Also, the controller $C(z;\hat{\theta})$ is a good approximation to the controller $C(z;\overline{\theta})$.

Example 5.1 Consider the following discrete-time linear transfer function $P(z)$ of a flexible transmission system [8]:

$$P(z) = \frac{z^{-3}(0 \cdot 28261 + 0 \cdot 50666z^{-1})}{1 - 1 \cdot 41833z^{-1} + 1 \cdot 58939z^{-2} - 1 \cdot 31608z^{-3} + 0 \cdot 88642z^{-4}}$$

Let the selected closed-loop reference model be

$$M = \frac{z^{-3}(1-\alpha)^2}{(1-\alpha z^{-1})^2}, \alpha = e^{-T_s\overline{\omega}}, \overline{\omega} = 10,$$

using a sampling time $T_s = 0 \cdot 05$ seconds. The weighting factor is considered $W(z) = 1$, and the class of controllers is

$$C(z;\theta) = \frac{\theta_0 + \theta_1 z^{-1} + \theta_2 z^{-2} + \theta_3 z^{-3} + \theta_4 z^{-4} + \theta_5 z^{-5}}{1 - z^{-1}}$$

A set of data has been obtained by feeding $P(z)$ in open-loop with $N = 512$ samples of zero-mean Gaussian white noise ($\Phi_u = 0 \cdot 01$). The derived controller

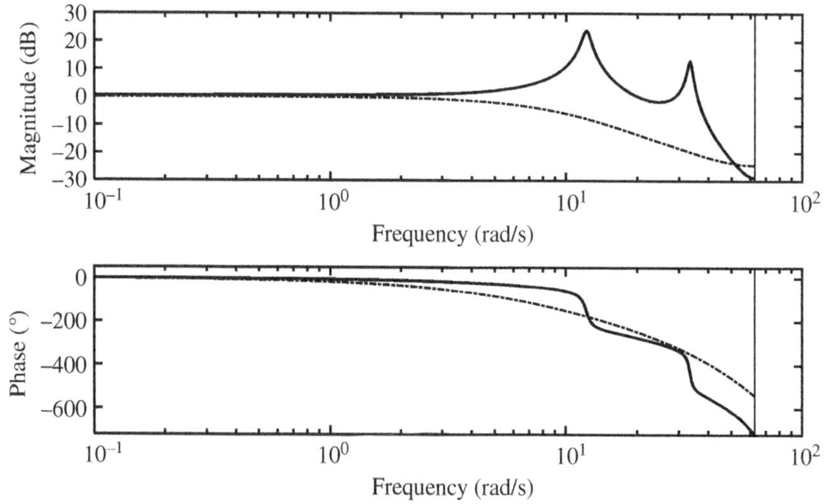

Figure 5.2 Bode magnitude plots: the plant (solid line) and the reference model (dashed line).

parameter vector is as follows:

$$\theta = [0.3256 - 0.5918\ 0.7049 - 0.6461\ 0.4721 - 0.1209]^T$$

Figure 5.2 shows the Bode plots of the open-loop plant and the desired reference model. The closed-loop and reference model step responses are shown in Figure 5.3, and the corresponding Bode plot is depicted in Figure 5.4 to assess the closed-loop plant performance.

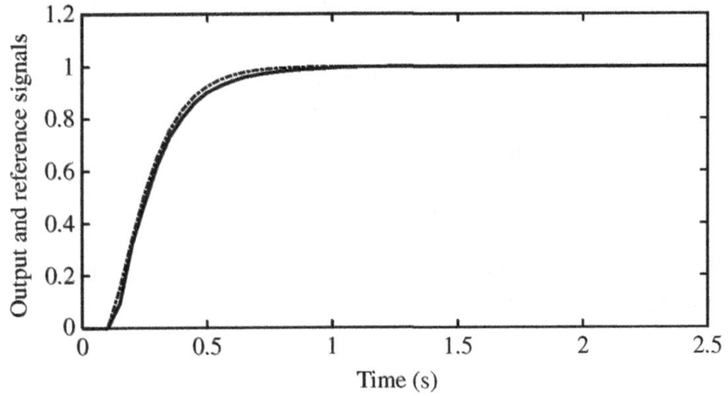

Figure 5.3 Step responses of the control system with θ (solid line) and the reference model (dashed line).

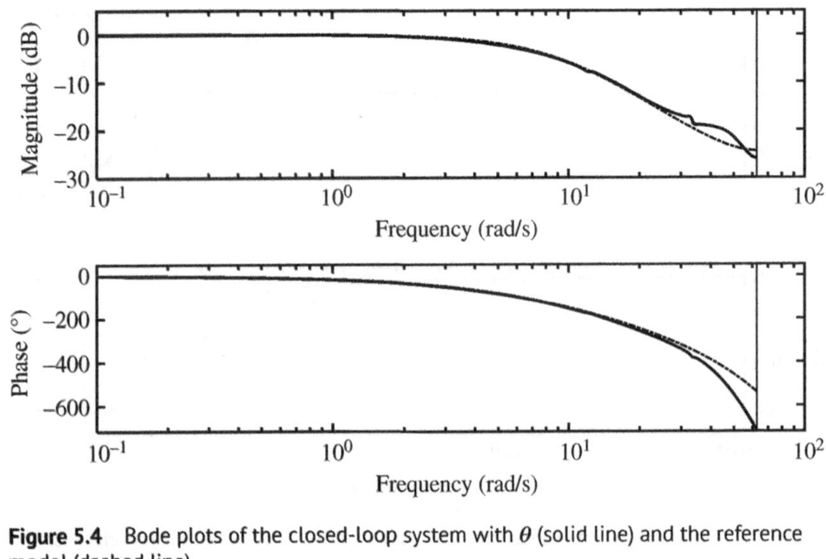

Figure 5.4 Bode plots of the closed-loop system with θ (solid line) and the reference model (dashed line).

5.3 The Measurement Noise Effect

In the presence of sensor or measurement noise, the measured output is given by

$$\tilde{y}(t) = P(z)u(t) + d(t)$$

where $d(t)$ is the measurement noise, it is assumed that $d(t)$ and $u(t)$ are uncorrelated. If the input–output measured data $\{u(t)\ \tilde{y}(t)\}$ *for* $t = 1, \ldots, N$, is employed in Algorithm 5.1, the resulting estimated parameter vector will be *biased*. To show this, the asymptotic cost function utilised in the VRFT algorithm in the face of noisy measurements is as follows:

$$J_{VR}(\theta) = \frac{1}{2\pi}\int_{-\pi}^{\pi} |P|^2|C(\theta) - C_0|^2|1 - M|^2\frac{|L|^2}{|M|^2}\Phi_u + \frac{|C(\theta)|^2}{|P|^2|C_0|^2}|L|^2\Phi_d d\omega \tag{5.12}$$

where Φ_d is the measurement noise spectral density. Clearly, the minimisation of (5.12) will yield different results compared to the minimisation of (5.9) because of the additional term present in (5.12), which introduces estimation bias. The instrumental variable approach briefly presented in the following remark will be used to solve the measurement noise bias problem in the VRFT design.

Remark 5.2 The instrumental variable approach is introduced in the least squares estimation techniques to overcome the problem of biased estimations. Biased estimations occur when the measurement noise is correlated with the regressors. In this remark, the instrumental variables are defined, and the approach is highlighted from Ref. [9]. Let the system be described by the following regression model:

$$y(t) = \psi^T(t)\theta_0 + d(t)$$

Multiplying both sides by $\psi(t)$ and summing over $t = 1, \ldots, N$, gives

$$\frac{1}{N}\sum_{t=1}^{N}\psi(t)y(t) = \left[\frac{1}{N}\sum_{t=1}^{N}\psi(t)\psi^T(t)\right]\theta_0 + \frac{1}{N}\sum_{t=1}^{N}\psi(t)d(t)$$

If $d(t)$ and $\psi(t)$ are uncorrelated, then

$$\frac{1}{N}\sum_{t=1}^{N}\psi(t)d(t) \approx 0$$

And the LS estimate is given by

$$\hat{\theta}_N = \left[\frac{1}{N}\sum_{t=1}^{N}\psi(t)\psi^T(t)\right]^{-1}\frac{1}{N}\sum_{t=1}^{N}\psi(t)y(t)$$

and $\hat{\theta}_N \approx \theta_0$. It can be stated that θ_0 is *correlated out* from the noise using the regressor sequence. In the cases where $d(t)$ and $\psi(t)$ are correlated, a similar idea can be adopted to derive an unbiased estimate of the parameter vector. Let $\zeta(t)$, $t = 1, \ldots, N$, be a vector sequence uncorrelated with $d(t)$ but correlated with $\psi(t)$. Similar to the above, we have

$$\frac{1}{N}\sum_{t=1}^{N}\zeta(t)y(t) = \left[\frac{1}{N}\sum_{t=1}^{N}\zeta(t)\psi^T(t)\right]\theta_0 + \frac{1}{N}\sum_{t=1}^{N}\zeta(t)d(t)$$

That gives

$$\hat{\theta}_N = \left[\frac{1}{N}\sum_{t=1}^{N}\zeta(t)\psi^T(t)\right]^{-1}\frac{1}{N}\sum_{t=1}^{N}\zeta(t)y(t)$$

And it is expected that $\hat{\theta}_N \approx \theta_0$. This estimate is called the *instrumental variable estimate* and $\zeta(t)$ is called the *instrumental variables* or the *instruments*. The main challenge in this technique is the selection of the instruments $\zeta(t)$.

Note that $e(t) = \bar{r} - \tilde{y} = M^{-1}\tilde{y} - \tilde{y} = (M^{-1} - 1)\tilde{y}$, then denote the regressors associated with the noisy output data as follows:

$$\tilde{\varphi}_L = \beta L e = \beta L(M^{-1} - 1)\tilde{y} \tag{5.13}$$

The controller is given by

$$u_L = \tilde{\varphi}_L^T\hat{\theta}$$

Let $\zeta(t)$ be the instrumental variables with the properties characterised in Remark 5.2, then we have

$$\sum_{t=1}^{N}\zeta(t)u_L(t) = \left[\sum_{t=1}^{N}\zeta(t)\tilde{\varphi}_L^T(t)\right]\hat{\theta}_N$$

which gives

$$\hat{\theta}_N = \left[\sum_{t=1}^{N}\zeta(t)\tilde{\varphi}_L^T(t)\right]^{-1}\left[\sum_{t=1}^{N}\zeta(t)u_L(t)\right] \tag{5.14}$$

where $\hat{\theta}_N$ is the *instrumental variable estimate* of the parameter vector.

5.3.1 The Instrumental Variable Selection

Instrumental variables can be selected in different ways, all having the same key characteristics stated in Remark 5.2. Following [3], a method for the instrumental variable selection is presented in this section that does not guarantee the equality of the instrumental variable estimate of the parameter vector with the actual estimate. However, the residual error between the two is expected to be small.

To pursue the derivation, a model of the plant $\widehat{P}(z)$ is identified using the measured input–output data $\{u(t)\ \widetilde{y}(t)\}$ for $t = 1, \ldots, N$. Then a simulated output $\widehat{y}(t) = \widehat{P}(z)u(t)$ is generated, and the instrument variables are selected as follows:

$$\zeta(t) = \beta(z)L(z)(M(z)^{-1} - 1)\widehat{y}(t)$$

Note that any of the standard identification techniques presented in Ref. [4] or [9] can be employed to obtain an estimate of the plant model. The following points are to be noted:

- As there always exists an estimation error corresponding to the identification techniques, the convergence of the instrumental variable estimate of the parameter vector with the actual estimate is not asymptotically guaranteed.
- The modified VRFT method with $\zeta(t)$ is no longer a truly data-driven approach, as it requires an identified model of the plant.
- The identified plant model is only utilised in calculating $\zeta(t)$ and is not involved in the controller design process. Hence, a high-order model can be identified without affecting the controller complexity and ensuring that an underestimated plant model order is not used.

The complete VRFT methodology is presented in the next algorithm.

Algorithm 5.2 VRFT with Measurement Noise

Input Select the reference model $M(z) \neq 1$, the weighting function $W(z)$ in (5.1), the filter $L(z) = (1 - M(z))M(z)W(z)\Phi_u(\omega)$, where $\Phi_u(\omega)$ is the spectral density of $u(t)$, data length N and the measured input–output data $\{u(t)\ \widetilde{y}(t)\}$ *for* $t = 1, \ldots, N$.

Step 1 Compute $u_L(t) = L(z)u(t)$.

Step 2 Compute $\widetilde{\varphi}_L(t)$ from (5.13).

Step 3 Identify a high-order plant model $\widehat{P}(z)$ from the measured input–output data $\{u(t)\ \widetilde{y}(t)\}$.

Step 4 Compute $\zeta(t) = \beta(z)L(z)(M(z)^{-1} - 1)\widehat{P}(z)u(t)$.

Step 5 Compute the estimated controller parameter vector $\widehat{\theta}_N$ from (5.14).

Remark 5.3 Algorithm 5.2 is presented in an open-loop configuration. However, it can also be applied to the measured input–output data collected from a closed-loop plant. In the closed-loop configuration, the procedure in Algorithm 5.2 is implemented using the closed-loop measured input–output data, and the complementary sensitivity function of the closed-loop plant is identified in Step 3 instead of $\widehat{P}(z)$.

Example 5.2 Consider the plant in Example 5.1, where a zero mean white noise corrupts the output signal with the signal-to-noise ratio, SNR = 10. The identified controller parameter vector is determined as follows:

$$\theta = [0.3313, -0.5831, 0.6596, -0.5896, 0.4442, -0.1198]^T$$

Figure 5.5 presents the closed-loop and reference model step responses.

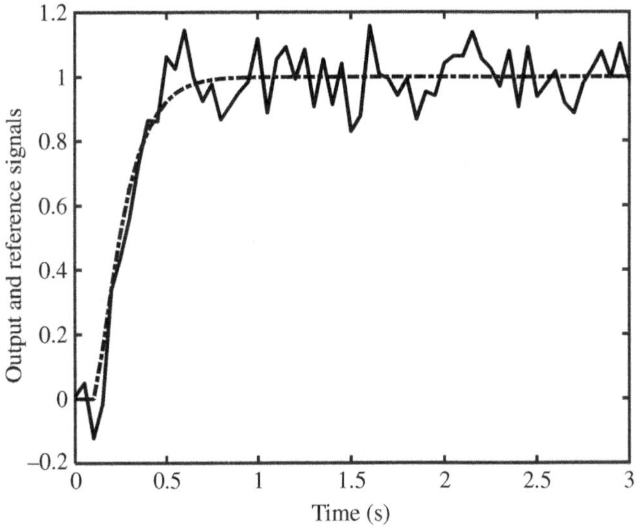

Figure 5.5 Step responses of the closed-loop system with θ (solid line) and the reference model (dashed line).

5.4 The Non-Minimum Phase Plants Challenge in the VRFT Design Approach

Non-minimum phase zeros are a serious issue in the design of data-based controllers. The non-minimum phase zeros in discrete-time systems are zeros outside the unit circle that can be introduced due to the continuous-time right half-plane zeros or can result from the fast sampling of continuous-time minimum phase systems. In particular, it is shortly demonstrated that in the VRFT model reference matching problem, the closed-loop control system can fail in the presence of non-minimum phase plant zeros.

Assume that there exists a controller parameter vector θ^* and a reference model $M(z)$ that satisfies (5.6). Equation (5.6) gives the corresponding controller as follows:

$$C(z;\theta^*) = \frac{M(z)}{(1 - M(z))P(z)} \tag{5.15}$$

Let $P(z) = \frac{n_P(z)}{d_P(z)}$, where $n_P(z)$ and $d_P(z)$ are relatively prime polynomials. As $P(z)$ is assumed to have a nonzero steady-state response,[3] $n_P(1) \neq 0$. Also, let $M(z) = \frac{n_M(z)}{d_M(z)}$, where $n_M(z)$ and $d_M(z)$ are polynomials of appropriate order. Substituting the numerator and denominator polynomials of $P(z)$ and $M(z)$ in (5.15), gives

$$C(z;\theta^*) = \frac{n_M(z)d_P(z)}{(d_M(z) - n_M(z))n_P(z)} \tag{5.16}$$

In the case of a non-minimum phase plant, the following cases are plausible from (5.16):

- The controller transfer function $C(z;\theta^*)$ becomes unstable, and an unstable pole-zero cancellation occurs between the plant and the controller, leading to the closed-loop plant's internal instability.
- The reference model has the same non-minimum phase zero, and a pole-zero cancellation occurs in (5.16), resulting in a stable controller transfer function $C(z;\theta^*)$. Hence, it is concluded that for closed-loop stability, $M(z)$ must include all the non-minimum phase zeros of $P(z)$.

Example 5.3 Consider the following non-minimum phase plant

$$P(z) = \frac{(z - 1.2)(z - 0.8)}{z(z - 0.5)(z - 0.6)}$$

and let the three-term controller be with fixed, stable poles of the following form:

$$C(z,\theta) = \theta^T\beta(z) = [\theta_1\ \theta_2\ \theta_3]\left[\frac{z^2}{z^2 - z}\ \frac{z}{z^2 - z}\ \frac{1}{z^2 - z}\right]^T$$

The desired reference model is selected, ignoring the non-minimum phase open-loop zero as follows:

$$M(z) = \frac{0.7z^2}{z^2(z - 0.3)}$$

3 This assumption is necessary for steady-state output tracking of the reference signal. It is also called the *functional controllability* condition. In the s-domain this implies that for a given transfer function $G(s)$, we have $\lim_{s\to 0} G(s) \neq 0$. In the multivariable plants, this is equivalent to a non-zero determinant of the transfer function matrix.

By applying Algorithm 5.1 for a given random input, the controller parameters are determined as follows:

$$\theta = \begin{bmatrix}0.4651 & 0.1865 & 0.1115\end{bmatrix}^T$$

Finally, the closed-loop transfer function is as follows:

$$T(z) = \frac{0.4651z^4 - 0.7437z^3 + 0.185z^2 - 0.04402z + 0.1071}{z^5 - 1.635z^4 + 0.6563z^3 - 0.115z^2 - 0.04402z + 0.1071}$$

The closed-loop transfer function has a pole outside the unit circle at 1.0847 and four other stable poles. The critical point to note is that the source of the problem is that the model-matching design problem has resulted in an unstable closed-loop system. This is due to ignoring the non-minimum phase zero in the design process.

Hence, the only conceivable approach is to insert the non-minimum phase zeros in the reference model. The non-minimum phase zeros (if any exist) must be included in the reference model before the algorithm steps start. However, this requires precise knowledge of the zero's location(s), which is impractical even for model-based design strategies. There are two different approaches to handling plant non-minimum phase zeros in the context of VRFT-based designs:

- Employ an identification technique to identify the number and locations of the non-minimum phase zeros. Include the identified non-minimum phase zeros in the reference model and continue the VRFT design process [10].
- Add flexibility to the reference model in determining the reference model zeros by free parameters, while the poles are pre-determined before the actual VRFT design [11].

The first proposed approach is impractical as in most real applications, identification of the number and locations of possible non-minimum phase plant zeros is unfeasible. In what follows, based on the results in [11], a reference model with free parameters in the numerator is proposed. The class of parameterised reference models is described by

$$\{M(z;\boldsymbol{\eta}) = \boldsymbol{\eta}^T \boldsymbol{F}(z)\}$$

where $\boldsymbol{\eta} \in \mathbb{R}^q$ is the vector of free parameters and $\boldsymbol{F}(z)$ is a vector of rational functions. That is, the numerator parameters are to be determined, and the denominator is fixed by the designer. Substituting $M(z)$ by $M(z;\boldsymbol{\eta})$ with some simple mathematical manipulations in (5.8), we have the following VRFT cost function:

$$J_{VR}(\boldsymbol{\eta},\boldsymbol{\theta}) = E\left[L(z)\left(u(t) - \left(\frac{1 - M(z;\boldsymbol{\eta})}{M(z;\boldsymbol{\eta})}C(z;\boldsymbol{\theta})\right)y(t)\right)\right]^2 \tag{5.17}$$

Let us assume that the problem is *feasible* and there exists a pair of vector parameters $\begin{bmatrix}\boldsymbol{\theta}^* & \boldsymbol{\eta}^*\end{bmatrix}$ such that

$$C(z;\boldsymbol{\theta}^*) = \frac{M(z;\boldsymbol{\eta}^*)}{(1 - M(z);\boldsymbol{\eta}^*)P(z)} \tag{5.18}$$

In the case of a non-minimum phase transfer function $P(z)$, the global minimum of the proposed flexible cost function (5.17) corresponds to a reference model that includes the non-minimum phase zeros of $P(z)$. Rewriting (5.17) with $C(z;\boldsymbol{\theta})=\boldsymbol{\theta}^T\boldsymbol{\beta}(z)$ and $M(z;\boldsymbol{\eta})=\boldsymbol{\eta}^T\boldsymbol{F}(z)$, gives

$$J_{VR}(\boldsymbol{\eta},\boldsymbol{\theta}) = E[\boldsymbol{\eta}^T \boldsymbol{F}(z)(u_L(t)+\boldsymbol{\theta}^T\boldsymbol{\beta}(z)y_L(t)) - \boldsymbol{\theta}^T\boldsymbol{\beta}(z)y_L(t)]^2 \tag{5.19}$$

As the argument in (5.19) is linear with respect to each of the variables $\boldsymbol{\eta}$ and $\boldsymbol{\theta}$, the optimisation process can be sequentially performed as follows with guaranteed convergence to at least a local minimum [11]:

$$\widehat{\boldsymbol{\eta}}^{(i)} = \text{argmin}_{\eta} J_{VR}(\boldsymbol{\eta},\widehat{\boldsymbol{\theta}}^{(i-1)}) \tag{5.20}$$

$$\widehat{\boldsymbol{\theta}}^{(i)} = \text{argmin}_{\theta} J_{VR}(\widehat{\boldsymbol{\eta}}^{(i)},\boldsymbol{\theta}) \tag{5.21}$$

Where each step conforms to a least square problem with the following solution:

$$\begin{aligned}\widehat{\boldsymbol{\eta}}^{(i)} &= E\{[\boldsymbol{F}(z)(u_L(t)+\boldsymbol{\theta}^T\boldsymbol{\beta}(z)y_L(t))][\boldsymbol{F}(z)(u_L(t)+\boldsymbol{\theta}^T\boldsymbol{\beta}(z)y_L(t))]^T\}^{-1}\\ &\times E\{[\boldsymbol{F}(z)(u_L(t)+\boldsymbol{\theta}^T\boldsymbol{\beta}(z)y_L(t))][(C(z;\widehat{\boldsymbol{\theta}}^{(i-1)})y_L(t))]^T\}\end{aligned} \tag{5.22}$$

$$\begin{aligned}\widehat{\boldsymbol{\theta}}^{(i)} &= E\{[\boldsymbol{\beta}(z)(1-\boldsymbol{\eta}^T\boldsymbol{F}(z))y_L(t)][\boldsymbol{\beta}(z)(1-\boldsymbol{\eta}^T\boldsymbol{F}(z))y_L(t)]^T\}^{-1}\\ &\times E\{[\boldsymbol{\beta}(z)(1-\boldsymbol{\eta}^T\boldsymbol{F}(z))y_L(t)][(M(z;\widehat{\boldsymbol{\theta}}^{(i)})u_L(t))]^T\}\end{aligned} \tag{5.23}$$

The following algorithm for the VRFT design in the context of a two-step procedure is presented as follows:

Algorithm 5.3 VRFT for Non-Minimum Plants

Input Select the reference model $M(z)$, the initial controller, data length N and the measured input–output data $\{u(t)\ y(t)\}$ *for* $t = 1, \ldots, N$.

Step 1 Select the filter $L(z)=M(z)$.

Step 2 Solve the following optimisation problem to derive the controller parameter vector $\boldsymbol{\theta}$ and the free parameter $\boldsymbol{\eta}$:

$$J_{VR}^N(\boldsymbol{\eta},\boldsymbol{\theta}) = \frac{1}{N}\sum_{t=1}^{N}(\boldsymbol{\eta}^T F(z)[u_L(t)+\boldsymbol{\theta}^T\boldsymbol{\beta}(z)y_L(t)] - \boldsymbol{\theta}^T\boldsymbol{\beta}(z)y_L(t))^2$$

Example 5.4 Consider the following discrete-time non-minimum phase plant:

$$P(z) = \frac{(z-1.2)(z-0.4)}{z(z-0.3)(z-0.8)}$$

And the corresponding class of controllers as

$$C(z,\boldsymbol{\theta}) = \boldsymbol{\theta}^T\boldsymbol{\beta}(z) = [\theta_1\ \theta_2\ \theta_3]\begin{bmatrix}\frac{z^2}{z^2-z} & \frac{z}{z^2-z} & \frac{1}{z^2-z}\end{bmatrix}^T$$

The initial controller is selected as

$$C_{init}(z) = \frac{-0.7(z-0.4)(z-0.6)}{z^2 - z}$$

And the desired reference model is

$$M(z) = \frac{0.0706z^2}{(z-0.885)(z^2-0.706z+0.32)}$$

Let the flexible reference model have the same poles as the desired reference model with free numerator parameters to handle the non-minimum phase zeros. That is,

$$M(z,\boldsymbol{\eta}) = \frac{\eta_1 z^2 + \eta_2 z + \eta_3}{(z-0.885)(z^2-0.706z+0.32)}$$

After 30 iterations, Algorithm 5.3 gives

$$M(z,\boldsymbol{\eta}^{(30)}) = \frac{-0.63139(z-1.2)(z-0.4397)}{(z-0.885)(z^2-0.706z+0.32)}$$

$$C(z,\boldsymbol{\theta}^{(30)}) = \frac{-0.61469(z-0.8)(z-0.3231)}{z^2 - z}$$

Figure 5.6 presents the closed-loop and the reference model step responses.

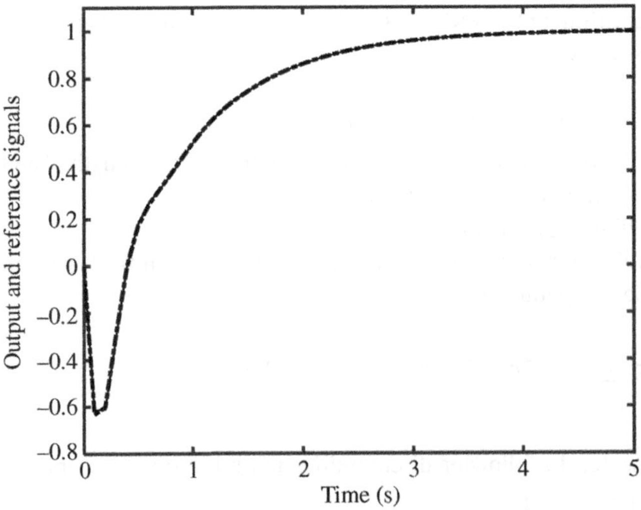

Figure 5.6 The closed-loop step responses of the control system with θ (solid line) and the reference model (dashed line).

5.5 Extensions of the VRTF Methodology to Multivariable Plants

In this section, the VRFT methodology presented in the previous sections for SISO linear time invariant (LTI) plants is extended to multivariable LTI plants. Let $\mathbf{P}(z)$ represent an $m \times m$ square transfer function matrix, and $\mathbf{C}(z;\theta)$ be a corresponding controller matrix transfer function. In general, the multivariable controller class encompasses controllers of the $m \times m$ matrix form

$$\mathbf{C}(z;\theta) = \begin{bmatrix} c_{11}(z,\theta_{11}) & c_{12}(z,\theta_{12}) & \dots & c_{1m}(z,\theta_{1m}) \\ \vdots & \vdots & \ddots & \vdots \\ c_{m1}(z,\theta_{m1}) & c_{m2}(z,\theta_{m2}) & \dots & c_{mm}(z,\theta_{mm}) \end{bmatrix} \tag{5.24}$$

where $\theta = \left[\theta_{11}^T \ \theta_{12}^T \ \dots \ \theta_{m1}^T \ \dots \ \theta_{mm}^T\right]^T$ is the parameter vector. If the controller is linear in its parameters, as it is in most practical designs, it can be written in a regression form. Then, each SISO controller has a linear parameterisation of the following form:

$$c_{ij}(z,\theta_{ij}) = \theta_{ij}^T \bar{\mathbf{c}}_{ij}(z) \tag{5.25}$$

and $\bar{\mathbf{c}}_{ij}(z)$ is a vector of fixed causal rational functions of appropriate dimensions. As a special case, the PID controllers can be written as follows:

$$c_{ij}(z,\theta_{ij}) = \left[K_{Pij} \ K_{Iij} \ K_{Dij}\right] \begin{bmatrix} 1 \\ \dfrac{z}{z-1} \\ \dfrac{z-1}{z} \end{bmatrix}, i,j = 1,\dots,m \tag{5.26}$$

And for a 2×2 PID controller, we will have

$$\mathbf{C}(z;\theta) = \begin{bmatrix} \theta_{P11} & \theta_{P12} \\ \theta_{P21} & \theta_{P22} \end{bmatrix} + \frac{z}{z-1} \begin{bmatrix} \theta_{I11} & \theta_{I12} \\ \theta_{I21} & \theta_{I22} \end{bmatrix} + \frac{z-1}{z} \begin{bmatrix} \theta_{D11} & \theta_{D12} \\ \theta_{D21} & \theta_{D22} \end{bmatrix} \tag{5.27}$$

Also, $\mathbf{M}(z)$ is the $m \times m$ reference matrix transfer function, and the measured input–output data set are m–dimensional vectors $\{\mathbf{u}(t) \ \mathbf{y}(t)\}$ *for* $t = 1,\dots,N$. The m–dimensional *virtual reference input* $\bar{\mathbf{r}}(t)$ is selected such that $\mathbf{M}(z)\bar{\mathbf{r}}(t) = \mathbf{y}(t)$. Hence, the data $\mathbf{y}(t)$ can be considered as the output signal of the feedback control system when the closed-loop transfer function is $\mathbf{M}(z)$ and the reference signal is $\mathbf{r}(t) = \bar{\mathbf{r}}(t)$. Also, $\mathbf{y}(t) = \mathbf{P}(z)\mathbf{u}(t)$. For the controller $\mathbf{C}(z;\theta)$, the control signal is $\mathbf{C}(z;\theta)(\bar{\mathbf{r}}(t) - \mathbf{y}(t))$. In the case of a well-designed model reference controller, $\mathbf{u}(t)$ and $\mathbf{C}(z;\theta)(\bar{\mathbf{r}}(t) - \mathbf{y}(t))$ are approximately equal. Therefore, the model-matching problem can be formulated as a minimisation of the following cost function:

$$J_{VR}^{N}(\theta) = \frac{1}{N}\sum_{t=1}^{N} \|\mathbf{u}(t) - \mathbf{C}(z;\theta)(\mathbf{M}(\mathbf{z})^{-1} - \mathbf{I})\mathbf{y}(t)\|_2^2 \tag{5.28}$$

where for a given vector v, $\|v\|_2 \triangleq \sqrt{v^T v}$, or in the filtered form as

$$J_{VR}^{N}(\theta) = \frac{1}{N}\sum_{t=1}^{N} \|\mathbf{L}(z)\mathbf{u}(t) - \mathbf{C}(z;\theta)(\mathbf{M}(\mathbf{z})^{-1} - \mathbf{I})\mathbf{L}(z)\mathbf{y}(t)\|_2^2 \tag{5.29}$$

where $\mathbf{L}(z)$ is an appropriate $m \times m$ matrix transfer function filter. The SISO–VRFT concept is directly extended to the multivariable plants as presented in [12]. Note that

$$\mathbf{L}(z)\mathbf{y}(t) = \mathbf{L}(z)\mathbf{P}(z)\mathbf{u}(t) \neq \mathbf{P}(z)\mathbf{L}(z)\mathbf{u}(t)$$

And the minimisations of (5.28) and (5.29) will not result in the same results. It is, therefore, necessary in the multivariable plants to restrict the filter structure to filters satisfying the following relation:

$$\mathbf{L}(z)\mathbf{P}(z) = \mathbf{P}(z)\mathbf{L}(z) \tag{5.30}$$

For (5.30) to hold, $\mathbf{L}(z)$ must be diagonal and have similar diagonal elements. Note that $\mathbf{P}(z)$ is assumed unknown, and it is not possible to design a filter that commutes with $\mathbf{P}(z)$. This severely limits the filter selection. A natural choice for the filter selection is $\mathbf{L}(z) = \mathbf{M}(z)$ to avoid matrix inversion of $\mathbf{M}(z)$. However, this restricts the reference model selection to decoupled multivariable models with the same dynamics in all the input–output channels. With this filter choice, (5.29) becomes

$$J_{VR}^{N}(\theta) = \frac{1}{N}\sum_{t=1}^{N} \|\mathbf{M}(z)\mathbf{u}(t) - \mathbf{C}(z;\theta)(\mathbf{I} - \mathbf{M}(\mathbf{z}))\mathbf{y}(t)\|_2^2 \tag{5.31}$$

If the parameters of the controller are described linearly, (5.31) can be rewritten as

$$J_{VR}^{N}(\theta) = \frac{1}{N}\sum_{t=1}^{N} \|\mathbf{M}(z)\mathbf{u}(t) - \boldsymbol{K}(t)\theta\|_2^2 \tag{5.32}$$

where,

$$\boldsymbol{K}(t)\theta = \mathbf{C}(z;\theta)(\mathbf{I} - \mathbf{M}(z))\mathbf{y}(t) \tag{5.33}$$

Algorithm 5.4 The Multivariable VRFT Control

Input Select the appropriate reference model $\mathbf{M}(z)$, data length N and obtain the measured input–output data $\{\mathbf{u}(t)\ \mathbf{y}(t)\}$ *for* $t = 1, \ldots, N$.

Step 1 Calculate $\boldsymbol{K}(t)$ such that (5.33) holds.

Step 2 Determine the following vector and matrix for the solution of the least squares (LS) problem (5.32):

$$\boldsymbol{U} = \begin{bmatrix} \mathbf{Mu}(1) \\ \mathbf{Mu}(2) \\ \vdots \\ \mathbf{Mu}(N) \end{bmatrix}, \boldsymbol{E} = \begin{bmatrix} \boldsymbol{K}(1) \\ \boldsymbol{K}(2) \\ \vdots \\ \boldsymbol{K}(N) \end{bmatrix}$$

Step 3 Compute the estimated controller parameter vector $\hat{\theta}_N$ from

$$\hat{\theta}_N = (\boldsymbol{E}^T\boldsymbol{E})^{-1}\boldsymbol{E}^T\boldsymbol{U} \tag{5.34}$$

The assumption of equal diagonal elements in the decoupled reference matrix is restrictive and a relatively straightforward solution is provided in [12] to handle unequal transfer functions in $\mathbf{M}(z)$. Let $\mathbf{M}(z) = diag\,[M_1(z), \ldots, M_m(z)]$. Then calculate the least common multiple of $M_1(z), \ldots, M_m(z)$ and denote it by $M(z)$. Define $\mathbf{H}(z) = M(z)diag[M_1(z), \ldots, M_m(z)]^{-1}$, and set $\mathbf{L}(z) = \mathbf{M}(z)\mathbf{H}(z)$. In this case, $\mathbf{L}(z)\mathbf{P}(z) = \mathbf{P}(z)\mathbf{L}(z)$, while $\mathbf{M}(z)$ is a decoupled matrix with different transfer functions on its diagonal. Note that if $\mathbf{M}(z)$ has equal elements on its diagonal, then $\mathbf{L}(z) = \mathbf{M}(z)$. For this filter choice, the cost function (5.29) is

$$J_{VR}^N(\theta) = \frac{1}{N}\sum_{t=1}^{N} ||\mathbf{M}(z)\mathbf{H}(z)\mathbf{u}(t) - \boldsymbol{K}(t)\theta||_2^2 \tag{5.35}$$

where

$$\mathbf{K}(t)\theta = \mathbf{C}(z;\theta)(\mathbf{I} - \mathbf{M}(z))\mathbf{H}(z)\mathbf{y}(t) \tag{5.36}$$

This equation can be used in Step 1, and $\mathbf{M}(z)\mathbf{H}(z)\mathbf{u}(t)$ should replace $\mathbf{M}(z)\mathbf{u}(t)$ in Step 2 of Algorithm 5.4. As an example, a realisation is presented for a 2×2 centralised PID controller given by (5.27) [12]. Let us define

$$\begin{bmatrix} p_1(z) & 0 \\ 0 & p_2(z) \end{bmatrix} \begin{bmatrix} y_1(t) \\ y_2(t) \end{bmatrix} = (\mathbf{I} - \mathbf{M}(z))\mathbf{H}(z)\mathbf{y}(t)$$

for some properly defined functions $p_1(z)$ and $p_2(z)$. For the right-hand side of (5.36), using (5.27), we have

$$\begin{aligned} \mathbf{C}(z;\theta) \begin{bmatrix} p_1(z)y_1(t) \\ p_2(z)y_2(t) \end{bmatrix} &= \begin{bmatrix} \theta_{P11} & \theta_{P12} \\ \theta_{P21} & \theta_{P22} \end{bmatrix} \begin{bmatrix} p_1(z)y_1(t) \\ p_2(z)y_2(t) \end{bmatrix} \\ &+ \frac{z}{z-1} \begin{bmatrix} \theta_{I11} & \theta_{I12} \\ \theta_{I21} & \theta_{I22} \end{bmatrix} \begin{bmatrix} p_1(z)y_1(t) \\ p_2(z)y_2(t) \end{bmatrix} \\ &+ \frac{z-1}{z} \begin{bmatrix} \theta_{D11} & \theta_{D12} \\ \theta_{D21} & \theta_{D22} \end{bmatrix} \begin{bmatrix} p_1(z)y_1(t) \\ p_2(z)y_2(t) \end{bmatrix} \end{aligned} \tag{5.37}$$

By defining the parameter vector as

$$\boldsymbol{\theta} = [\theta_{P11}\ \theta_{P12}\ \theta_{P21}\ \theta_{P22}\ \theta_{I11}\ \theta_{I12}\ \theta_{I21}\ \theta_{I22}\ \theta_{D11}\ \theta_{D12}\ \theta_{D21}\ \theta_{D22}]^T$$

the first term in (5.37) can be realised as follows:

$$\begin{bmatrix} \theta_{P11} & \theta_{P12} \\ \theta_{P21} & \theta_{P22} \end{bmatrix} \begin{bmatrix} p_1(z)y_1(t) \\ p_2(z)y_2(t) \end{bmatrix} = \begin{bmatrix} p_1(z)y_1(t) & p_2(z)y_2(t) & 0 & 0 \\ 0 & 0 & p_1(z)y_1(t) & p_2(z)y_2(t) \end{bmatrix} \times \begin{bmatrix} \theta_{P11} \\ \theta_{P12} \\ \theta_{P21} \\ \theta_{P22} \end{bmatrix}$$

Following the same pattern for the other two terms, we have

$$\boldsymbol{K}(t) = \begin{bmatrix} p_1(z)y_1(t) & p_2(z)y_2(t) & 0 & 0 \\ 0 & 0 & p_1(z)y_1(t) & p_2(z)y_2(t) \end{bmatrix} \begin{bmatrix} \mathbf{I}_4 & \frac{z}{z-1}\mathbf{I}_4 & \frac{z-1}{z}\mathbf{I}_4 \end{bmatrix} \tag{5.38}$$

where $\boldsymbol{I}_4$ is the 4×4 identity matrix. Also, in the case of a 2×2 decentralised PID controller, the proportional, integral and derivative coefficients matrices will be diagonal as follows:

$$\mathbf{C}(z; \boldsymbol{\theta}) = \begin{bmatrix} \theta_{P11} & 0 \\ 0 & \theta_{P22} \end{bmatrix} + \frac{z}{z-1} \begin{bmatrix} \theta_{I11} & 0 \\ 0 & \theta_{I22} \end{bmatrix} + \frac{z-1}{z} \begin{bmatrix} \theta_{D11} & 0 \\ 0 & \theta_{D22} \end{bmatrix} \tag{5.39}$$

The corresponding $\boldsymbol{K}(t)$ for the parameter vector $\boldsymbol{\theta} = [\theta_{P11}\ \theta_{P22}\ \theta_{I11}\ \theta_{I22}\ \theta_{D11}\ \theta_{D22}]^T$ is

$$\boldsymbol{K}(t) = \begin{bmatrix} p_1(z)y_1(t) & 0 \\ 0 & p_2(z)y_2(t) \end{bmatrix} \begin{bmatrix} \mathbf{I}_2 & \frac{z}{z-1}\mathbf{I}_2 & \frac{z-1}{z}\mathbf{I}_2 \end{bmatrix} \tag{5.40}$$

Remark 5.4 The proposed extension has two limitations that can cause implementation problems in real multivariable plants. The first limitation is the assumption that all the input–output channels must have the same reference performance, which was handled in this section. In [13], an approach to multivariable VRFT is presented that allows different closed-loop performances for each input–output channel. An unbiased estimate for the ideal controller is determined if it is in the controller class set, and if the ideal controller is not in the controller set, a filter can be used to approximate the minima of the VRFT and the model reference criteria. The second limitation in the above derivations is the noise-free assumption of the measured data. In Ref. [13], for the noise-corrupted signals, an instrumental variable approach is implemented for the multivariable VRFT.

Remark 5.5 The control objective in the VRFT approach is to minimise a model-reference type cost function, as presented in (5.1) for SISO LTI plants. The following equation gives the equivalent cost function for a multivariable plant:

$$J_{MR}(\boldsymbol{\theta}) = \|\mathbf{M}(z) - (\mathbf{I} + \mathbf{P}(z)\mathbf{C}(z;\boldsymbol{\theta}))^{-1}\mathbf{P}(z)\mathbf{C}(z;\boldsymbol{\theta})\|_2^2 \tag{5.41}$$

where the filter $\boldsymbol{W}(z)$ in (5.1) is considered as the identity matrix. However, (5.41) is a non-convex optimisation problem with respect to the control parameter vector $\boldsymbol{\theta}$. It is possible to obtain a convex approximation of (5.41) if, in addition to the previous assumptions, the following two assumptions are imposed [14]:

- The desired sensitivity function $\mathbf{I} - \mathbf{M}(z)$ is close to the actual sensitivity function $(\mathbf{I} + \mathbf{P}(z)\mathbf{C}(z;\widehat{\boldsymbol{\theta}}))^{-1}$ for a $\widehat{\boldsymbol{\theta}}$ that minimises (5.1).
- The controller is linearly parametrised, and a fixed denominator is common to all the transfer functions.

The following approximation of (5.41) is derived from the above assumptions:

$$J_{MR}(\boldsymbol{\theta}) = \|\mathbf{M}(z) - (\mathbf{I} - \mathbf{M}(z)))^{-1}\mathbf{P}(z)\mathbf{C}(z;\boldsymbol{\theta})\|_2^2 \tag{5.42}$$

Note that an ideal controller matrix transfer function that gives $J_{MR}(\boldsymbol{\theta}) = 0$ is generally of a very high order and may be non-causal. Hence, $\mathbf{C}(z;\widehat{\boldsymbol{\theta}})$ is only an approximation of the ideal controller matrix, and the difference between these two controller matrices must be made as small as possible. This strategy is followed in [14], and a multivariable data-driven control design approach based on the VRFT philosophy is proposed by employing the instrumental variables technique in Remark 5.2 and a pre-filter choice.

Example 5.5 Consider the following discrete-time multivariable plant and the desired decoupled reference model as follows:

$$\mathbf{P}(z) = \begin{bmatrix} \dfrac{0.744}{z - 0.9419} & \dfrac{-0.8789}{z - 0.9535} \\ \dfrac{0.5786}{z - 0.9123} & \dfrac{-1.302}{z - 0.9329} \end{bmatrix}$$

and

$$\mathbf{M}(z) = \begin{bmatrix} \dfrac{0.3}{z - 0.7} & 0 \\ 0 & \dfrac{0.5}{z - 0.5} \end{bmatrix}$$

Applying Algorithm 5.4, the following PI controller matrix is obtained:

$$\mathbf{C}(z) = \begin{bmatrix} \dfrac{0.8633z - 0.8176}{z-1} & \dfrac{-0.963z + 0.8894}{z-1} \\ \dfrac{0.3427z - 0.3267}{z-1} & \dfrac{-0.7689z + 0.717}{z-1} \end{bmatrix}$$

The closed-loop step responses are shown in Figure 5.7.

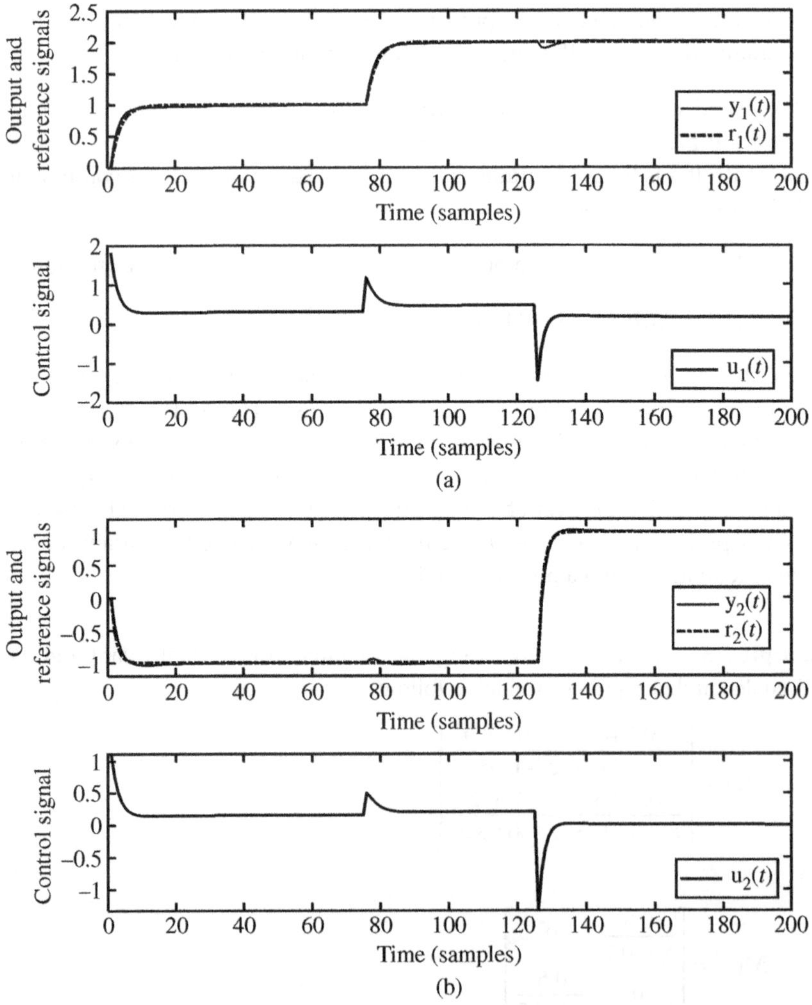

Figure 5.7 Closed-loop output responses and control signals of the decentralised multivariable PI controller for the first and second outputs and control signals.

To show the effectiveness of the centralised PI control in comparison to the decentralised PI control in handling multivariable interactive plants, Algorithm 5.4 is applied to the above multivariable plant, and the following control is designed:

$$\mathbf{C}(z) = \begin{bmatrix} \dfrac{-0.05866z + 0.06328}{z-1} & 0 \\ 0 & \dfrac{-0.1073z + 0.1036}{z-1} \end{bmatrix}$$

The reference model is considered as

$$\mathbf{M}(z) = \begin{bmatrix} \dfrac{0.1}{z-0.9} & 0 \\ 0 & \dfrac{0.1}{z-0.9} \end{bmatrix}$$

The closed-loop step responses for the decentralised multivariable PI controller are shown in Figure 5.8. Figures 5.7 and 5.8 clearly depict that both controllers designed by the VRFT-based data-driven scheme stabilise the closed-loop plant and provide closed-loop steady-state set-point tracking. However, the centralised multivariable PI controller outperforms the decentralised multivariable PI controller in handling interactions and provides superior closed-loop performance.

5.6 Optimal Reference Model Selection in the VRFT Methodology

In the model reference control approach, a reference model is selected to describe the desired closed-loop performance of the plant-controller combination. The model reference in the model-based design techniques is often selected by the designer based on the open-loop plant information and personal experience with the plant under control. Specific guidelines can help the control system designer select an appropriate model reference in the model-based design techniques. In the direct data-driven control techniques, the choice of the reference model is more challenging as the designer must consider the implementation issues in an underlying unknown plant. The closed-loop system should be able to reproduce the desired performance when the designed controller is inserted into the closed loop. In Section 5.2, it is assumed that a reference model is available for the minimum phase plant to be controlled. However, closed-loop stability issues were aroused in the face of SISO plants with non-minimum phase zeros. If the location of non-minimum phase zeros were known, they would be directly included in the reference model. Nevertheless, in the case of unknown plants, a flexible reference model with free numerator coefficients was proposed in Section 5.4. Also, all the numerators, as well as the denominator coefficients, could be free

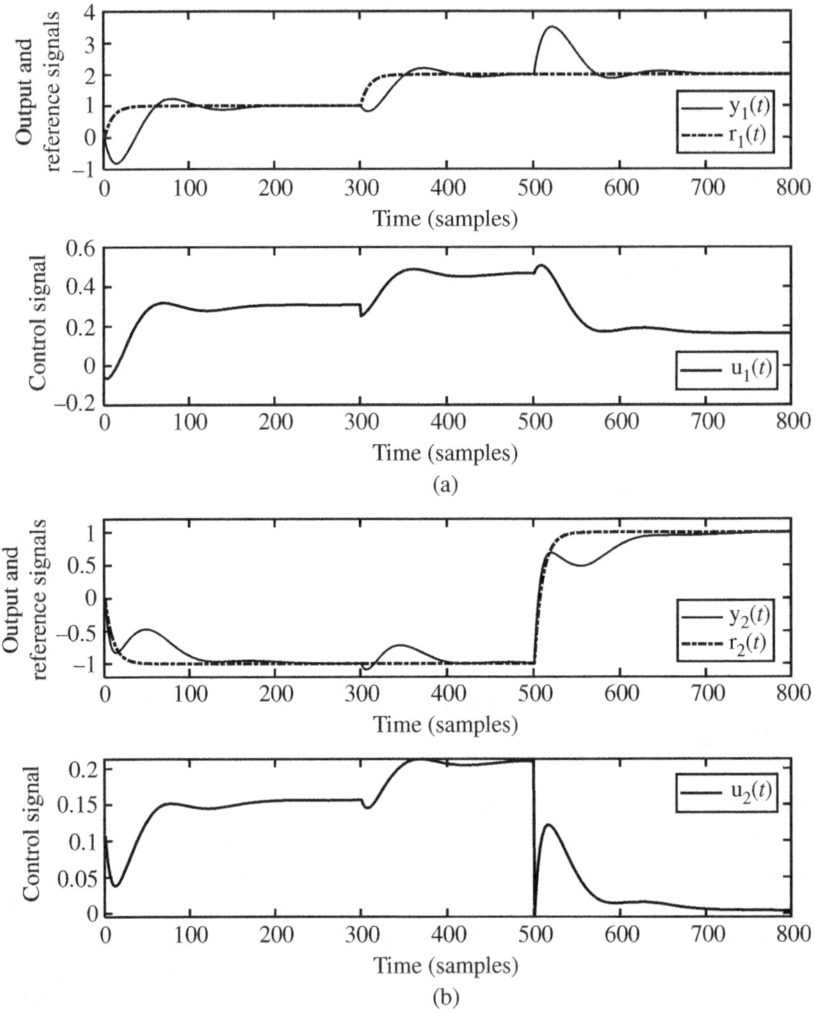

Figure 5.8 Closed-loop output responses and control signals of the decentralised multivariable PI controller for the first and second outputs and control signals.

for an optimised selection. Model reference selection in multivariable plants was addressed in Section 5.5, with a decoupled closed-loop structure.

In this section, based on the results of [15], an alternative method to select the parameters of a flexible reference model for the VRFT design is presented. The resulting reference model is shown to be *optimal* with respect to a cost function reflecting the desired closed-loop performance and control objectives. The reference model selection is stated in terms of an optimisation problem.

Consider the 2 degrees of freedom control structure for a SISO closed-loop plant as follows:

$$u(t) = C(z, \theta)(r(t) - y(t)) + c_0(\theta)r(t) \tag{5.43}$$

where the variables are as previously defined and $c_0(\theta)$ is a feedforward term assigned for set-point tracking. If the controller transfer function has an incorporated integrator, set $c_0(\theta)=0$. Let $M(z;\ \eta)$ be the reference model with the parameter vector η to be optimised. Selecting the reference model would mean the derivation of η, and once η is selected, the corresponding controller $C(z, \theta)$ will be determined by the VRFT methodology of Section 5.2. In this case, θ is a function of η. To further clarify the dependence of various matrices and variables on these parameter vectors and formally state the optimisation problem, we have $C(z;\ \theta^*(\eta))$, $c_0(\theta^*(\eta))$ as the optimal controller set for a given η, which is determined by the VRFT approach:

$$\theta^*(\eta) = \text{argmin}_\theta \sum_{t=1}^{N} (u(t) - \overline{u}(t; \eta, \theta))^2 \tag{5.44}$$

where the virtual input $\overline{u}(t; \eta, \theta)$ is given as follows:

$$\overline{u}(t; \eta, \theta) = C(z; \theta)[\overline{r}(t; \eta) - y(t)] + c_0(\theta)\overline{r}(t; \eta)$$

And $\overline{r}(t; \eta)$ comes from the following equation, and is the corresponding virtual reference for a given parameter set η and the reference model $M(z;\ \eta)$:

$$y(t) = M(z; \eta)\overline{r}(t; \eta)$$

However, to obtain the *optimal reference model*, the optimal reference model parameter vector η^* is selected as

$$\eta^* = \text{argmin}_\eta J(\eta) \tag{5.45}$$

where

$$J(\eta) = \frac{1}{N} \sum_{t=1}^{N} \mathbf{W}(r(t) - y_P(t; \eta))^2 + \mathbf{Q}(\Delta u_P(t; \eta))^2 + \mathbf{S}(u(t) - \overline{u}(t; \eta))^2 \tag{5.46}$$

where $\mathbf{W}$ is the nonnegative weighting parameter for the reference model tracking error with

$$y_P(t; \eta) = M(z; \eta)r(t)$$

and $\mathbf{Q}$ is the nonnegative weighting parameter for the control effort and

$$\Delta u_P(t; \theta) = u_P(\eta; t) - u_P(\eta; t-1)$$

is the corresponding input increment with

$$u_P(\eta; t) = C(z; \theta^*(\eta))(r(t) - y_P(t; \eta)) + c_0(\theta^*(\eta))r(t)$$

Finally, **S** is the nonnegative weighting parameter for the error between the actual and virtual control inputs, that is, the ability to match the reference model $M(z;\, \boldsymbol{\eta})$ for the active controller in the loop with $C(z;\, \boldsymbol{\theta}^*(\boldsymbol{\eta}))$, $c_0(\boldsymbol{\theta}^*(\boldsymbol{\eta}))$. Note that $y_P(t;\, \boldsymbol{\eta})$ is the reference model output in response to the given set point and $u_P(\boldsymbol{\eta};\, t)$ is the corresponding input.

Remark 5.6 The optimisation problem (5.45) is, in general, nonlinear and non-convex. Note that it involves $\boldsymbol{\theta}^*(\boldsymbol{\eta})$, and is, therefore, a bilevel programming problem with another internal optimisation problem. Random optimisation techniques can be employed to solve this problem. In [15], the particle swarm optimisation (PSO) method is used for the outer optimisation problem.

The general algorithm for solving the VRFT design problem with optimal reference model selection will include the following steps:

- Determine the necessary initial conditions as inputs in the previous algorithms and the parameters required for the optimisation technique.
- Initially, apply the VRFT algorithm for the controller design.
- Apply the optimisation technique to find the reference model.
- Go back to the VRFT algorithm with the new reference model.
- Repeat the above process until a sufficiently good fitness for the reference model is achieved.

5.6.1 The Particle Swarm Optimisation Scheme

In this section, based on the results in [15], PSO is employed to solve the optimisation problem (5.46). PSO is a random optimisation technique utilised in *difficult* optimisation problems [16]. PSO is a simple bio-inspired algorithm that searches for an optimal solution in the solution space. It is not dependent on the gradient or any differential form of the cost function and only requires a cost function that correctly formulates the problem at hand. PSO has a computational scheme to iteratively optimise the given cost function. A population of candidate solutions or *particles* called a *swarm* is provided and moved around in the search space according to a simple mathematical formula. The PSO algorithm starts with a number of random points or particles and guides them to look for the minimum point in random directions. At each step, every particle not only searches around the minimum point it has but also searches around the minimum point found by the entire swarm of particles. The cost function minimum point is determined as the minimum point discovered by this swarm of particles based on a proper stopping rule. For a review of the PSO techniques, refer to [17].

In the proposed VRFT design, PSO is used for the optimal selection of the reference model $M(z;\, \boldsymbol{\eta})$. It is, therefore, necessary to present $M(z;\, \boldsymbol{\eta})$ in a parametrised form appropriate for the optimisation process. Hence, the following parametrisation is adopted:

$$M(z;\boldsymbol{\eta}) = K\frac{\prod_{l=1}^{n_{zr}}(z-z_l)\prod_{l=n_{zr}+1}^{n_{zr}+n_{zc}}(z-z_l)\left(z-z_l^*\right)}{\prod_{l=1}^{n_{pr}}(z-p_l)\prod_{l=n_{pr}+1}^{n_{pr}+n_{pc}}(z-p_l)\left(z-p_l^*\right)} \tag{5.47}$$

where * denotes complex conjugate, and n_{zr}, n_{zc}, n_{pr}, n_{pc} denote the number of real zeros, complex conjugate zeros, real poles and complex conjugate poles, respectively. In defining (5.47), $\prod_{l=l_1}^{l_2} = 1$ if $l_1 > l_2$. The parameter vector $\boldsymbol{\eta}$ entails the parameters to be optimised and includes all the real zeros, complex conjugate zeros, real poles and complex conjugate poles of (5.47). The reference model gain K is not an optimisation parameter as it must ensure $M(1;\boldsymbol{\eta}) = 1$ for ideal steady-state tracking.

In the PSO algorithm, the number of particles N_{part} is determined by the designer. In this case, each particle is represented by a parameter vector $\boldsymbol{\eta}^i$. Then, $\theta^*(\boldsymbol{\eta}^i)$ associated with each particle is computed through the VRFT algorithm. At each iteration, using the common rules in PSO, the parameter vector is updated. The process ends within a pre-defined number of iterations k_{max}. Finally, the optimal parameter vector $\boldsymbol{\eta}^*$ is selected as the parameter vector corresponding to the minimum value of $J(\boldsymbol{\eta}^i)$, $i = 1, \ldots, N_{part}$. It is obvious that a reference model must be stable. However, as stated in Section 5.2, in the VRFT approach, it must also be inversely stable or non-minimum phase. That is, z_l for $l = 1, \ldots, n_{zr} + n_{zc}$ and p_l for $l = 1, \ldots, n_{pr} + n_{pc}$ must lie within the unit circle at each iteration in the optimisation process.

This imposes inequality constraints in the optimisation problem. To replace the inequality constraints with a penalising term in the cost function, a *barrier function* $b(\cdot): \mathbb{R} \to \mathbb{R}$ should be defined, which significantly simplifies the optimisation problem. Let $h(\cdot): \mathbb{R} \to \mathbb{R}$ be a function in terms of $\boldsymbol{\eta}$ that characterises the poles and zeros of the reference model as follows:

$$h_l^z(\boldsymbol{\eta}) = |z_l|^2 - 1, \quad l = 1, \ldots, n_{zr} + n_{zc}$$

$$h_l^p(\boldsymbol{\eta}) = |z_p|^2 - 1, \quad l = 1, \ldots, n_{pr} + n_{pc}$$

It is obvious that for $h < 0$, reference model zeros and poles are inside the unit circle and would otherwise lie outside the unit circle. Hence, the following barrier function is defined:

$$b(h) = \begin{cases} 0 & \textit{if } h < 0 \\ \sqrt{k}h & \textit{if } 0 \le h < 1 \\ \sqrt{k}h^2 & \textit{if } h \ge 1 \end{cases} \tag{5.48}$$

where k is the PSO iteration. The cost function given by (5.46) will be replaced with the cost function below by incorporating the barrier function (5.48):

$$J(\boldsymbol{\eta}^i) = \frac{1}{N}\sum_{t=1}^{N} \mathbf{W}(r(t) - y_P(t;\boldsymbol{\eta}^i))^2 + \mathbf{Q}(\Delta u_P(t;\boldsymbol{\eta}^i))^2 + \mathbf{S}(u(t) - \overline{u}(t;\boldsymbol{\eta}^i))^2$$
$$+ \sum_{l=1}^{n_{zr}+n_{zc}} b\left(h_l^z(\boldsymbol{\eta}^i)\right) + \sum_{l=1}^{n_{pr}+n_{pc}} b\left(h_l^p(\boldsymbol{\eta}^i)\right) \tag{5.49}$$

Algorithm 5.5 VRFT with Optimal Reference Model Selection

Input Select the number of particles N_{part}, the number of iterations k_{max}, the weights $\mathbf{W}$, $\mathbf{Q}$ and $\mathbf{S}$, the barrier function $b(\cdot)$, the initial parameter vector $\boldsymbol{\eta}^i$ for $i = 1, \ldots, N_{part}$ satisfying the constraints of poles and zeros inside the unit circle, the filter $L(z)$, data length N and the measured input–output data $\{u(t)\ y(t)\}$ *for* $t = 1, \ldots, N$.

Step 1 Parametrise $M(z; \boldsymbol{\eta}^i)$ as in (5.47) with the appropriate gain.

Step 2 Compute $C(z; \boldsymbol{\theta}^*(\boldsymbol{\eta}^i))$, $c_0(\boldsymbol{\theta}^*(\boldsymbol{\eta}^i))$ through the VRFT algorithm.

Step 3 Select the best particle based on the computed cost function (5.49) and update the particles.

Step 4 Go to step 1 until the stop rule is reached and $\boldsymbol{\theta}^*(\boldsymbol{\eta}^*)$ is derived.

In the next section, the closed-loop stability guarantee of the VRFT design is discussed. A modified algorithm that ensures closed-loop stability is presented in Ref. [15].

Example 5.6 Consider the plant and the controller class given in Example 5.1. Let $\mathbf{W} = 1$, $\mathbf{Q} = 100$ and $\mathbf{S} = 0.01$, and the flexible reference model be given as

$$M(z, \boldsymbol{\eta}) = \frac{z^{-3}(1-\alpha_1)(1-\alpha_2)}{(1-\alpha_1 z^{-1})(1-\alpha_2 z^{-1})}$$

where $\boldsymbol{\eta} = [\alpha_1, \alpha_2]^T$. Running Algorithm 5.5, we have

$$\boldsymbol{\eta}^* = [0.8082, 0.8055]^T$$

And the corresponding controller parameters are

$$\boldsymbol{\theta}^*(\boldsymbol{\eta}^*) = [0.0740\ -0.0772\ 0.0768\ -0.0455\ 0.0190\ 0.0310]^T$$

The closed-loop response, the reference input and the control signal are shown in Figure 5.9.

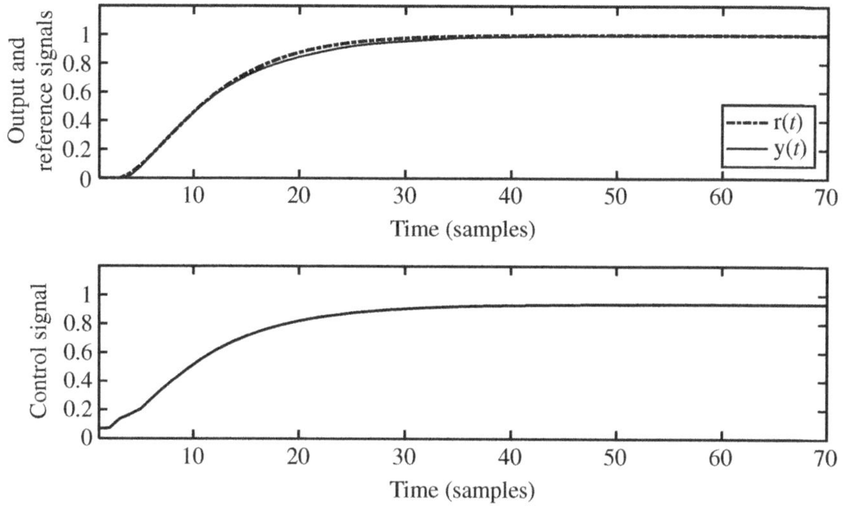

Figure 5.9 The closed-loop performance with the optimal controller parameters.

5.7 Closed-loop Stability of the VRFT-Based Data-Driven Control Systems

5.7.1 An Identification-Based Approach

Stability checks in the model-based approaches are regularly performed by analysing the mathematical closed-loop equations preceding actual implementations. In the presence of uncertainties or time-varying plants, robust stability theorems provide necessary and sufficient conditions for closed-loop stability under certain plant assumptions. Lyapunov functions are also widely used for closed-loop stability assurance, given that detailed plant information is available. However, a closed-loop stability guarantee is one of the main concerns in data-driven control systems design approaches. In the case of unknown plants and no prior information about the plant, since no mathematical model is available, closed-loop stability cannot be typically guaranteed before the implementation of the controller. Hence, some verification processes for closed-loop stability must be performed that test the controller for stability before implementing it. In Ref. [18], an *invalidation* test step based on the available data before implementing the controller is proposed. This invalidation test checks if the number of the free controller parameters (the controller flexibility) and the approximations indicated in (5.2) are consistent with the reference model matching problem or if they result in an unstable closed-loop plant. The idea is similar to the falsification process introduced in Chapter 3. The invalidation process is based on an *identification* approach with two feasible methodologies. The plant model is identified in the first method, and closed-loop stability checks with the designed controller are carried out in a standard setup. Some of the virtual closed-loop transfer functions are directly identified in the second method. To illustrate the second method, let

$$r(t) = C^{-1}(z;\theta)u(t) + y \tag{5.50}$$

be defined as the *virtual reference signal*.[4] Then, the transfer function between $u(t)$ and $r(t)$ is identified. Let this transfer function be $H(\hat{\theta})$. If $H(\hat{\theta})$ has non-minimum phase zeros, it will be inversely unstable, the controller is invalidated and the design process must be repeated with different design parameters.

Remark 5.7 In implementing the identification approach, issues raised in the previous sections and general issues related to the identification technique must be considered for a reliable identified model. For example, the problem of bias and application of instrumental variables, the model structure and order selection, unstable poles and non-minimum zero and the choice of an appropriate identification technique. For further details, refer to [18].

5.7.2 An Unfalsification-Based Approach

This approach is based on the falsification theory introduced in Chapter 3. A stability test is proposed to ensure that the closed-loop plant with the insertion of the VRFT-based designed controller is stable. In Section 5.2, the VRFT data-driven design technique is introduced by defining a reference model $M_r(z)$ and minimising a cost function to ensure closed-loop performance. In Ref. [19] and then [15], another reference model $S_r(s)$ is introduced for stability requirement. The following material is from [15]. Consider the controller transfer function $C(z;\theta)$ and

$$y(t) = P(z)u(t) + d(t) \tag{5.51}$$

where $P(z)$ is assumed unknown, and the variables are as previously defined. Let $r(t)$ be the virtual reference defined in (5.50). From (5.50) and (5.51), we have

$$u(t) = \frac{C(z;\theta)}{1 + P(z)C(z:\theta)}r(t) - \frac{C(z;\theta)}{1 + P(z)C(z;\theta)}d(t)$$

---

4 Note that this definition is similar to the virtual or fictitious input defined in Chapter 3 and is directly derived from the controller equation $u(t) = C(\theta)e(t)$ defined in Section 5.2.

in this case,

$$S(z;\theta) \triangleq \frac{C(z;\theta)}{1+P(z)C(z;\theta)}$$

is the *input sensitivity function*. And, also

$$y(t) = \frac{P(z)C(z;\theta)}{1+P(z)C(z:\theta)}r(t) + \frac{1}{1+P(z)C(z:\theta)}d(t)$$

in this case,

$$M(z;\theta) \triangleq \frac{P(z)C(z;\theta)}{1+P(z)C(z:\theta)}$$

is the *complementary* or *output sensitivity function*. It is desired to minimise the discrepancies between the desired $M_r(z)$ and $S_r(z)$ and the potential $M(z;\theta)$ and $S(z;\theta)$, respectively. To do so, let us define the following variables:

$$u_o(t) \triangleq S_r(z)r(t), \quad y_o(t) \triangleq M_r(z)r(t)$$

and cost functions

$$Z_N(\theta) \triangleq \sum_{t=1}^{N} (u(t) - u_o(t))^2 \tag{5.52}$$

$$V_N(\theta) \triangleq \sum_{t=1}^{N} (y(t) - y_o(t))^2 \tag{5.53}$$

In the case of $d(t) \equiv 0$, we have

$$u(t) - u_o(t) = (S(z;\theta) - S_r(z))r(t)$$

$$y(t) - y_o(t) = (M(z;\theta) - M_r(z))r(t)$$

By direct substitution of (5.50) in equations of $u_o(t)$ and $y_o(t)$, we obtain

$$u(t) - u_o(t) = u(t) - S_r(z)y(t) - C^{-1}(z;\theta)S_r(z)u(t) \tag{5.54}$$

$$y(t) - y_o(t) = [1 - M_r(z)]y(t) - C^{-1}(z;\theta)M_r(z)u(t) \tag{5.55}$$

Equations (5.54) and (5.55) show that to evaluate the cost functions (5.52) and (5.53), the computation of the virtual reference is not required. In addition, if the transfer functions in (5.54) and (5.55) are stable, the recursive computations in these equations will be numerically stable. The transfer functions $M_r(z)$ and $S_r(z)$ are naturally chosen stable. However, if there are non-minimum phase zeros in $C(z;\theta)$, it should be carefully treated to avoid numerical instability in (5.54) and (5.55). If any of the zeros in $C(z;\theta)$ lie outside the unit circle, these zeros must be included in $M_r(z)$ and $S_r(z)$. Hence, describing such zeros by the polynomial $\alpha_u(z)$, that is $C(z;\theta) = \mathrm{C}(z;\overline{\theta})\alpha_u(z)$ where $\mathrm{C}(z;\overline{\theta})$ is a rational function of an appropriate number of poles and zeros with all its zeros inside the unit circle and $\overline{\theta} \subseteq \theta$, we should have

$$M_r(z) = \alpha_u(z)\overline{M}_r(z) \text{ and } S_r(z) = \alpha_u(z)\overline{S}_r(z)$$

where $\overline{M}_r(z)$ and $\overline{S}_r(z)$ are stable transfer functions.

The controller design will involve a multi-objective optimisation problem, minimising a combination of the cost functions $Z_N(\theta)$ and $V_N(\theta)$. The following points are to be noted:

- Minimisation of $V_N(\theta)$ would achieve the desired output performance. But it cannot as a single cost function, ensure internal stability.
- Minimisation of $Z_N(\theta)$ can ensure internal stability (for details, refer to [15]). But it cannot, as a single cost function, ensure desired closed-loop performance.

Finally, it can be shown that the following condition:

$$\left\| S_r(z)\left[C_r^{-1}(z) - C^{-1}(z;\theta)\right] \right\|_\infty < 1 \tag{5.56}$$

where $\|\cdot\|_\infty$ denotes the infinity norm and is defined as the maximum of the absolute values of its components, coming from the small gain theorem is sufficient to ensure the internal stability of the closed-loop system with the controller $C(z;\theta)$ and unknown open-loop transfer function. However, $C_r(z)$ is the ideal controller that, once inserted in the control loop gives the desired reference model for the unknown stable plant. As the plant transfer function is assumed unknown, $C_r(z)$ cannot be directly determined.

It is desired to have an a-posteriori stability test for the closed-loop plant of the unknown transfer function with $C(z;\theta)$ as the controller. To achieve this, it can be shown that for the noise-free case

$$u(t) - u_o(t) = \left[C_r^{-1}(z) - C^{-1}(z;\theta)\right] S_r(z)u(t) \tag{5.57}$$

Equation (5.57) states the transfer function in (5.54) in terms of the input $u(t)$ and the output $u(t) - u_o(t)$. Now employing non-parametric identification techniques such as the *empirical transfer function estimate* (ETFE) method estimates the transfer function by taking the ratios of the Fourier transforms of the corresponding plant output and input data. Hence, for a set of noise-free measured input–output data, (5.57) gives

$$\left\| S_r(z)\left[C_r^{-1}(z) - C^{-1}(z;\theta)\right] \right\|_\infty = \sup_{\omega\in[-\pi,\pi]} \frac{|\hat{u}(\omega) - \hat{u}_o(\omega)|}{|\,\hat{u}(\omega)|} \tag{5.58}$$

where $\hat{u}(\omega)$ and $\hat{u}_0(\omega)$ are the corresponding discrete-time Fourier transforms of $u(t)$ and $u_0(t)$.

Remark 5.8 The ETFE estimates the unknown transfer function $P(z)$ by taking the ratios of the Fourier transforms of its measured output $y(t)$ and input $u(t)$ data, that is

$$\hat{P}(z) = \frac{\text{Fourier Transform of } y(t)}{\text{Fourier Transform of } u(t)}$$

In the case of noisy input–output data, the resulting estimate is also noisy, and the problem cannot be resolved by taking more input–output data, but there are techniques to reduce the variance of the estimate. This ETFE estimate can be implemented using the etfe.m command in the MATLAB system identification toolbox. For further details, refer to [9].

Remark 5.9 An optimal procedure for the determination of the transfer functions $M_r(z)$ and $S_r(z)$ is provided in [15].

5.8 Conclusions

In this chapter, the VRFT approach for the design of data-driven control systems is presented. VRFT is a model-reference–based design methodology that involves direct global minimisation of standard model-reference cost functions with no prior plant information or mathematical models. Hence, the designed controllers are optimal with respect to the proposed cost functions that ensure the desired closed-loop stability and performance. VRFT utilises measured input–output data only and can be used as a data-driven tuning strategy for controllers with predetermined structures, such as the family of proportional integral derivative (PID) controllers. It is a noniterative one-shot design approach with a resulting low-order controller. The basic VRFT ideas are presented for the SISO LTI plants. However, it is shown that these ideas can be extended to multivariable plants. The main challenges of direct data-driven VRFT techniques are the presence of non-minimum phase zeros in the open-loop plant, measurement noise, optimal reference model selection and closed-loop stability guarantee. These issues are tackled in this chapter, algorithms are presented to implement the design strategies and simulation results are provided to show the main features and effectiveness of the design methodologies.

Problems

5.1 Classify the main challenges of the VRFT-based control systems design.

5.2 Consider the feedback control and disturbance rejection feedforward scheme shown in Figure 5.10, where $P(z)$ is the unknown plant, $C(z)$ is the ideal controller, $P_d(z)$ is the unknown disturbance transfer function and $C_f(z)$ is the feedforward controller. Also, let $M(z)$, $N(z)$ and $S(z)$ be the desired closed-loop transfer function, disturbance to the output transfer function and the sensitivity function, respectively.

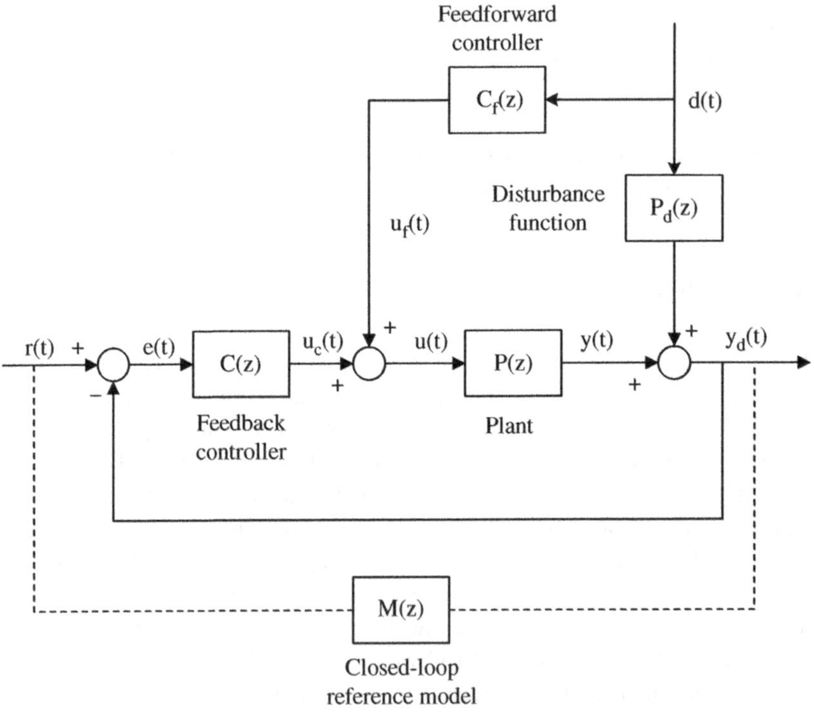

Figure 5.10 A feedback control and disturbance rejection feedforward scheme.

Similar to (5.1) and (5.8) derive a model-reference type cost function for the above control system scheme $J_{MR}(\boldsymbol{\theta}_f, \boldsymbol{\theta}_c)$, and VRFT function $J_{VR}(\boldsymbol{\theta}_f, \boldsymbol{\theta}_c)$, where $\boldsymbol{\theta}_f$ and $\boldsymbol{\theta}_c$ are the feedforward and feedback controller parameter vectors, respectively. (For the complete design procedure and filter design refer to [20].)

5.3 Consider the following plant and reference model:

$$P = \frac{z+1}{z^2 - 0.95z + 0.175}, M_r = \frac{0.2z + 0.2}{z^2 - 0.7z + 0.1}$$

a) Compare the open-loop plant and reference model responses to a given reference input.
b) For the controller given by

$$C(z;\theta) = \frac{\theta_0 + \theta_1 z^{-1} + \theta_2 z^{-2}}{1 - z^{-1}}$$

and a set of appropriately defined input–output data set with $N = 50$, derive the controller parameter vector.
c) Compare the closed-loop plant and reference model responses.
d) Let zero mean white noises with a variance of 0.001 and 0.01 corrupt the output signal. Re-identify the controller parameter vector and show the closed-loop responses in the presence of the measurement noises. Discuss the results.

5.4 Consider the following non-minimum phase plant and reference model:

$$P(z) = \frac{0.5z^2 - 0.8z}{z^2 - 1.5z + 0.56}, M_r(z) = \frac{0.81z^2}{z^2 - 0.2z + 0.01}$$

a) For a square wave reference input, identify the following controller parameters:
b)

$$C(z,\theta) = \theta^T \beta(z) = [\theta_1\ \theta_2\ \theta_3] \left[\frac{z^2}{z^2 - z}\ \frac{z}{z^2 - z}\ \frac{1}{z^2 - z} \right]^T$$

c) Derive the closed-loop transfer function and determine its closed-loop poles. Discuss the results.
d) Consider a reference model with free parameters and by selecting an appropriate initial controller, determine the controller parameter using Algorithm 5.3.
e) Show the closed-loop step responses.

5.5 Consider the following multivariable plant:

$$\mathbf{P}(s) = \begin{bmatrix} \dfrac{0.25}{s^2 + 1.9s + 0.25} & \dfrac{-0.05}{s^2 + 2s + 0.25} \\ \dfrac{-0.5}{s^2 + 3s + 1} & \dfrac{1}{s^2 + 3.2s + 1} \end{bmatrix}$$

a) Discretise the plant with a 0.2-second sampling time and determine the open-loop poles and zeros.

b) For the following decoupled reference model:

$$\mathbf{M}(z) = \begin{bmatrix} \dfrac{0.3}{z(z-0.7)} & 0 \\ 0 & \dfrac{0.5}{z(z-0.5)} \end{bmatrix}$$

Derive the multivariable controller parameters using Algorithm 5.4.

c) Show the corresponding closed-loop responses and discuss the results.

5.6 Consider the plant and the controller class given in Problem 5.3, with possible added terms. Let $\boldsymbol{W} = 1$, $\boldsymbol{Q} = 100$ and $\boldsymbol{S} = 1$ and the flexible reference model be given as

$$M(z, \boldsymbol{\eta}) = \frac{z^{-3}(1-\alpha_1)(1-\alpha_2)}{(1-\alpha_1 z^{-1})(1-\alpha_2 z^{-1})}$$

a) Design an optimal controller using Algorithm 5.5 and derive the optimal controller parameters.

b) Show the closed-loop step response and discuss the results.

References

1 G. O. Guardabassi and S. M. Savaresi, "Virtual reference direct design method: an off-line approach to data-based control system design," *IEEE Transactions on Automatic Control*, vol. 45, no. 5, pp. 954–959, 2000.

2 M. C. Campi and S. M. Savaresi, "Direct nonlinear control design: the virtual reference feedback tuning (VRFT) approach," *IEEE Transactions on Automatic Control*, vol. 51, no. 1, pp. 14–27, 2006.

3 M. C. Campi, A. Lecchini, and S. M. Savaresi, "Virtual reference feedback tuning: a direct method for the design of feedback controllers," *Automatica*, vol. 38, no. 8, pp. 1337–1346, 2002.

4 K. J. Åström and B. Wittenmark, *Adaptive Control*. Courier Corporation, 2013.

5 T. Chen and B. A. Francis, *Optimal Sampled-data Control Systems*. Springer Science & Business Media, 2012.

6 A. Papoulis and S. U. Pillai, *Probability, Random Variables, and Stochastic Processes*. Tata McGraw-Hill Education, 2002.

7 A. Leon-Garcia, *Probability and Rrandom Processes for Electrical Engineering*. Pearson Education India, 1994.

8 I. Landau, D. Rey, A. Karimi, A. Voda, and A. Franco, "A flexible transmission system as a benchmark for robust digital control," *European Journal of Control*, vol. 1, no. 2, pp. 77–96, 1995.

9 L. Ljung, "*System Identification-Theory for the User*," second edition. Upper Saddle River, NJ: Prentice-Hall, 1999.

10 A. Sala and A. Esparza, "Virtual reference feedback tuning in restricted complexity controller design of non-minimum phase systems," *IFAC Proceedings Volumes*, vol. 38, no. 1, pp. 235–240, 2005.

11 L. Campestrini, D. Eckhard, M. Gevers, and A. S. Bazanella, "Virtual reference feedback tuning for non-minimum phase plants," *Automatica*, vol. 47, no. 8, pp. 1778–1784, 2011.

12 M. Nakamoto, "An application of the virtual reference feedback tuning method for multivariable process control," *Transactions of the Society of Instrument and Control Engineers*, vol. 41, no. 4, pp. 330–337, 2005.

13 L. Campestrini, D. Eckhard, L. A. Chia, and E. Boeira, "Unbiased MIMO VRFT with application to process control," *Journal of Process Control*, vol. 39, pp. 35–49, 2016.

14 S. Formentin and S. M. Savaresi, "Noniterative data-driven design of multivariable controllers," in *2011 50th IEEE Conference on Decision and Control and European Control Conference*, 2011: IEEE, pp. 5106–5111.

15 D. Selvi, D. Piga, G. Battistelli, and A. Bemporad, "Optimal direct data-driven control with stability guarantees," *European Journal of Control*, vol. 59, pp. 175–187, 2021.

16 M. Clerc, *Particle Swarm Optimization*. Wiley, 2010.

17 R. Poli, J. Kennedy, and T. Blackwell, "Particle swarm optimization," *Swarm Intelligence*, vol. 1, no. 1, pp. 33–57, 2007.

18 A. Sala and A. Esparza, "Extensions to "virtual reference feedback tuning: a direct method for the design of feedback controllers"," *Automatica*, vol. 41, no. 8, pp. 1473–1476, 2005.

19 G. Battistelli, D. Mari, D. Selvi, and P. Tesi, "Direct control design via controller unfalsification," *International Journal of Robust and Nonlinear Control*, vol. 28, no. 12, pp. 3694–3712, 2018.

20 S. K. Chiluka, A. S. Rao, M. M. Seepana, and G. U. B. Babu, "Design of VRFT based feedback-feedforward controllers for enhancing disturbance rejection on non-minimum phase systems," *Chemical Product and Process Modeling*, vol. 15, no. 2, 2020.

6

The Simultaneous Perturbation Stochastic Approximation-Based Data-Driven Control Design

6.1 Introduction

In this chapter, a multivariate stochastic optimisation technique is utilised in the design of data-driven control systems. In the model-based optimal control approaches, an analytical closed-form solution to the optimisation problem is derived, and the derivation is based on the prior assumption of the availability of almost exact mathematical models of the true plant, either in the state-space form or in the input–output form. However, in the control of real-world complex plants with a corresponding optimisation problem, it is necessary to use a mathematical algorithm that iteratively seeks out the solution [1]. In this context, the *simultaneous perturbation stochastic approximation* (SPSA) method for the control of plants with unknown exact mathematical models has been employed. The control strategy would entail two phases, as shown in Figure 1.5:

- A controller selection whose unknown parameters must be derived for closed-loop stability and performance.
- A cost function optimisation process to derive the controller parameters.

In this chapter, SPSA will be used in the data-driven feedback control framework to obtain the optimised controller parameters. The SPSA algorithm uses the gradient approximation and only requires two measurements of the cost function. Cost function optimisation is performed regardless of the mathematical plant input–output or state-space relationships, order or dynamical characteristics and assumptions. This feature accounts for its control effectiveness and relative ease of implementation.

The control design approaches with a central optimisation problem include a designer-specified scalar-valued cost function that should be minimised with respect to a vector of adjustable controller parameters. Starting from an initial parameter set, the adjustable control parameters are tuned by some derived

An Introduction to Data-Driven Control Systems, First Edition. Ali Khaki-Sedigh.

control law in a step-by-step procedure to reduce the cost function value. In the presence of noisy data, as is the case with many real-world problems, the cost function value reduction will not be uniform in the sampling times that follow. In the model-based control approach, a key assumption is made: The gradient vector associated with the cost function, that is, the gradient of the cost function with respect to the controller parameters, is available at each sampling time using the plant input–output mathematical relationship.

However, recursive optimisation algorithms should be utilised in the data-driven control setting that does not depend on such direct gradient information. Instead, an approximation to the gradient formed from the observed plant noisy data must be used. SPSA has this feature and can be easily and efficiently implemented, as it relies on measurements of the objective function and not on measurements of the gradient of the objective function. Although there are other pioneer stochastic approximation (SA) algorithms in high-dimensional cases, they are very computationally expensive [2].

It is indicated in [1] that, in general, it can be claimed that the gradient-free SA exhibit acceptable convergence properties compared to the gradient-based SA while requiring only cost function measurements without requiring detailed knowledge of the functional relationship between the control parameters and the cost function being minimised. To summarise, the gradient-based algorithms rely on the direct measurements of the gradient of the cost function with respect to the controller parameters, which provide an *estimate* of the gradient in the presence of noisy input–output data. However, direct measurements of the gradient, either in deterministic or stochastic cases, are unavailable. Hence, the approaches based on gradient approximations that only require input–output measurements are of practical interest.

Similar to the discussion in Section 1.2.1, the designer must choose between either implementing the gradient-based or the gradient-free algorithms. As a general rule, it must be stated that the gradient-based algorithms have faster convergence properties than the cost function–based gradient approximations. In a cost-benefit analysis, this is achieved at the expense of the required additional information for the gradient-based algorithms. However, to make an appropriate choice, the following points are to be noted:

- In the case of complex and complex adaptive plants and many complicated plants, it is not possible or practically feasible to derive reliable mathematical descriptions of the system input–output relationships. Hence, the gradient-based algorithms may be either *infeasible* if the system model is unavailable or *unreliable* if the system model is not applicably accurate.
- The computational cost in gradient-based algorithms is generally higher than the gradient-free algorithms. Also, the fulfilment of the required mathematical

assumptions adds to the total cost for effective convergence. Ensuring certain assumptions can lead to further economic and human resources costs.

- Closed-loop control of practical plants must be achieved in finite sampling times. At the same time, the proven convergence rates are only asymptotically valid, and real convergence rates may vary in practical applications. Therefore, there is no guarantee that the gradient-based search algorithm is convergence-wise superior to the gradient approximation-based algorithms.

Finally, it can be concluded that for the control of the complex and complex adaptive plants, the SPSA algorithm type is a practical recommendation, and for simple or certain complicated plants with reliable direct gradient information, it is recommended to utilise this information in a gradient-based optimisation process.

6.2 The Essentials of the SPSA Algorithm

Consider the problem of finding θ^* that minimises the cost function $J(\theta)$, where $J : \mathbb{R}^n \rightarrow \mathbb{R}$ is a function of the parameter vector $\theta \in \mathbb{R}^n$ and the exact mathematical relation between these two is unknown. In general, the measurements of the cost function are assumed to contain noise. The gradient of the cost function with respect to the tuning parameter vector cannot be computed, as their mathematical relationship is assumed unknown. Therefore, an estimation of the gradient should be used. SA is a gradient approximation-based algorithm independent of the direct measurement or calculation of the gradient. SA algorithms start with an initial guess of the optimal parameter vector and then update it with an iterative equation aiming at the convergence of the gradient to zero.

Let $\hat{\theta}_j$ denote the estimated tuning parameter vector at the jth iteration and $\hat{\sigma}_j(\hat{\theta}_j)$ be the corresponding estimated gradient. The general form of the SA is given as follows:

$$\hat{\theta}_{j+1} = \hat{\theta}_j - g_{1_j} \hat{\sigma}_j(\hat{\theta}_j) \tag{6.1}$$

where $\{g_{1_j}\}$ is a gain sequence.

A key step in this class of optimisation algorithms is the gradient estimation procedure. SPSA is a SA algorithm in which the gradient estimation is established through simultaneous stochastic perturbing of all the tuning parameters. To estimate the gradient vector, only two data measurements are required. This shows the superiority of this method over the SA algorithm, i.e. finite-difference SA, which was introduced before the SPSA algorithm. In this algorithm, in contrast to SPSA, two experiments are performed to compute the lth component of the gradient vector estimate by perturbing only the lth component of the parameter vector. Thus, the total number of data measurements to estimate the gradient vector is twice the dimension of the parameter vector. Therefore, when the parameter vector's dimension is high, as in the neural controllers, the reduction of computation burden in SPSA is noticeable.

In the control design applications of the SPSA algorithm, a general nonlinear controlled system with an unknown model is assumed. Then, a closed-loop structure with a controller of a prespecified and known structure and a cost function for control performance evaluation is considered. As the controlled system model is unavailable, the mathematical relation between the performance criterion and the controller parameters is unknown. The goal is to find the optimal parameter vector of the controller such that the cost function is minimised; that is, to set the parameters to values that meet the control objectives of the problem. Therefore, in these problems, SPSA is also called the *tuning algorithm*, and the parameter vector of the controller is called the *tuning parameter vector*. The main specifications of SPSA for data-driven control are shown in Table 6.1.

Each iteration of the SPSA algorithm includes three sampling times or time instants. The parameter vector has a nominal value at the first instant. At the second and third instants, the parameter vector is equal to two perturbed values, obtained by adding and subtracting a random term to and from the nominal value of the corresponding iteration. For these two parameter vectors, two experiments are performed, and the system input–output information is recorded. The reference signal employed in the three experiments is kept constant. Then, using the two sets of information obtained from the last two experiments, the corresponding values of the cost function are obtained, and the corresponding gradient approximation is calculated. Finally, using the updating law, the parameter vector of the next iteration is constructed.

The mathematical relations governing the SPSA algorithm are as follows: Let $\hat{\theta}_j(k-1)$ be the estimated controller parameter vector at instant $k-1$ and jth iteration. The value of this parameter at the second instance of the jth iteration is $\hat{\theta}_j^+(k)$, and the corresponding cost function value is $J_j^+ = J\left(\hat{\theta}_j^+(k)\right)$. The value of the parameter at the third instance of the jth iteration and the corresponding cost function value is given by $\hat{\theta}_j^-(k+1)$ and $J_j^- = J\left(\hat{\theta}_j^-(k+1)\right)$, respectively. Then, the perturbed values of the tuning parameter vector are obtained from the following equations:

$$\hat{\theta}_j^+(k) = \hat{\theta}_j(k-1) + g_{2j}\boldsymbol{\delta}_j \tag{6.2}$$

$$\hat{\theta}_j^-(k+1) = \hat{\theta}_j(k-1) - g_{2j}\boldsymbol{\delta}_j \tag{6.3}$$

where g_{2j} is the value of the perturbation step sequence $\{g_{2_j}\}$ at the jth iteration. In addition, $\boldsymbol{\delta}_j \in \mathbb{R}^n$ is called the *simultaneous perturbation stochastic vector* that includes n mutually independent random variables, which are identically

Table 6.1 SPSA main specifications for data-driven control.

Criterion	SPSA
Controlled system	Nonlinear and unknown
Controller structure	Prespecified and fixed
Using measured input–output data	Online
The gradient employed and its estimation	The gradient of the cost function with respect to the controller tuning parameter is estimated by simultaneous perturbation stochastic approximation
Test signal	Three experiments (using test signals) in each iteration
Computational burden	Low
A systematic method of the proof of Bounded-Input-Bounded-Output stability and error convergence to zero	Parameter convergence to the optimal values is proved
Assumptions of the controlled system	Output controllability in the data-driven framework
Strategy	Identification of the controller parameters with an iterative stochastic algorithm
Application in systems with quickly changing parameters	Not suitable
Algorithm parameters	Parameters of the update step sequence $\{g_{1_j}\}$ and the perturbation step sequence $\{g_{2_j}\}$
Sensitivity to the algorithm parameters	Weak

distributed. This vector should be constructed at every iteration. Then, the estimated gradient at the jth iteration can be obtained using J_j^+ and J_j^- as follows:

$$\hat{\sigma}_j = \begin{bmatrix} \dfrac{J_j^+ - J_j^-}{2g_{2j}\delta_{j_1}} \\ \dfrac{J_j^+ - J_j^-}{2g_{2j}\delta_{j_2}} \\ \vdots \\ \dfrac{J_j^+ - J_j^-}{2g_{2j}\delta_{j_n}} \end{bmatrix} \tag{6.4}$$

where δ_{j_l} is the lth component of the vector δ_j. Finally, the controller parameter vector is updated according to the following equation:

$$\hat{\theta}_{j+1}(k+2) = \hat{\theta}_j(k-1) - g_{1j}\hat{\sigma}_j \tag{6.5}$$

where g_{1j} is the value of the update step sequence $\{g_{1_j}\}$ at the jth iteration.

6.2.1 The Main Theoretical Result of the SPSA Algorithm

The SPSA algorithm is very well theoretically studied in [3, 4]. The standard SPSA algorithm convergence property is utilised in most data-driven control applications, and its convergence conditions are presented in the following theorem. For proof and more details, refer to [3, 4].

Theorem 6.1 In the standard SPSA algorithm, the estimated parameter vector $\hat{\theta}_j$ converges to the optimal parameter values θ^*. That is, we have *almost surely*

$$\lim_{j\to\infty} \hat{\theta}_j = \theta^*$$

subject to the following assumptions:

i. $g_{1_j}, g_{2_j} > 0 \forall j$, $\lim_{j\to\infty} g_{1_j} = 0$, $\lim_{j\to\infty} g_{2_j} = 0$, $\sum_{j=0}^{\infty}\left(\frac{g_{1_j}}{g_{2_j}}\right)^2 < \infty$, and $\sum_{j=0}^{\infty} g_{1_j} = \infty$.
ii. For some α_1, $\alpha_2 > 0$, $\mathrm{E}\left[J(\hat{\theta}_j(k) \pm g_{2j}\delta_j)^2\right] \leq \alpha_1$, which is a condition on the measurements and $\mathrm{E}\left[(\delta_{j_i})^{-2}\right] \leq \alpha_2$ $(i = 1, \dots, n)\forall j$, which is a condition on the allowable perturbations, where $\mathrm{E}[\cdot]$ denotes the expected value.
iii. The cost function $J(\theta)$ must be smooth. That is, it must be three times continuously differentiable and bounded.
iv. $\|\hat{\theta}_j\| < \infty \forall j$.
v. θ^* is an optimal solution of the governing differential equation.
vi. $\hat{\theta}_j$ lies in a closed and bounded (i.e. compact) subset of the domain of attraction for the governing differential equation infinitely often. In the domain of attraction, the solution of the governing differential equation converges to θ^* for any initial condition in the domain of attraction.

Remark 6.1 Condition i can be satisfied by an appropriate selection of the gain and perturbation sequences. It is easily verified that these conditions are all satisfied for the following proposed sequence values. The gain and perturbation sequences are selected as positive terms given as follows:

$$g_{1j} = \frac{a}{(j+1+A)^{\alpha}} \tag{6.6}$$

$$g_{2j} = \frac{c}{(j+1)^{\gamma}} \tag{6.7}$$

where a, A, α, c and γ are nonnegative numbers. The condition $\sum_{j=0}^{\infty}\left(\frac{g_{1_j}}{g_{2_j}}\right)^2 < \infty$ balances the impact of the g_{1_j} and g_{2_j} in the convergence process and can ensure an acceptable transition from $\hat{\theta}_j$ to $\hat{\theta}_{j+1}$, which can be of importance in controller implementations. Also, it prevents the fast decay of g_{2_j} to zero for a better gradient estimate.

Remark 6.2 The boundedness of $\mathrm{E}\left[J(\hat{\theta}_j(k) \pm g_{2j}\delta_j)^2\right]$ and the bounded inverse moment condition of the elements δ_{j_i} in condition ii are essential assumptions in the convergence proof of SPSA. In fact, the simultaneous perturbation stochastic vector can have any symmetric distribution around zero. However, the uniform and normal distributions with zero mean cannot be used as they do not satisfy the finite inverse moment conditions [3, 4]. The Bernoulli ± 1 distribution is usually chosen and has proven to be suitable for control applications.

Remark 6.3 Condition iii is met by the conventional cost functions usually employed in control systems design.

Remark 6.4 The conditions iv, v and vi ensure that the estimated parameter vector $\hat{\theta}_j$ is close enough to the optimal solution θ^*.

Remark 6.5 If the measurements are corrupted by noise, the standard SPSA algorithm has similar theoretical properties and can effectively estimate the unknown parameter vector [3, 4].

The step-by-step procedure of the SPSA algorithm to determine the controller parameter vector is presented as follows. Consider the following general nonlinear multivariable discrete-time system:

$$\mathbf{y}(k+1) = \boldsymbol{f}\left(\mathbf{y}(k), \dots, \mathbf{y}(k-n_y), \mathbf{u}(k), \dots, \mathbf{u}(k-n_u)\right) \tag{6.8}$$

where $\mathbf{y}(k) \in \mathbb{R}^p$ and $\mathbf{u}(k) \in \mathbb{R}^p$ are the system output and input vectors, respectively, n_y and n_u are unknown orders, p is a known integer and $\boldsymbol{f}(\dots) \in \mathbb{R}^p$ is an unknown nonlinear function. Also, the controller parameter vector $\theta(k)$ and the cost function $J(\theta(k))$ are appropriately defined.

Algorithm 6.1 The SPSA Algorithm

Input Select the values of a, A, α, c and γ for the gain sequences $\{g_{1_j}\}$ and $\{g_{2_j}\}$, as given in Eqs. (6.6) and (6.7), and the initial estimated parameter vector $\hat{\theta}_0(0)$.

Step 1 Let $j = 0$, assume that the system output at time instant $k = 0$ is given as $\boldsymbol{y}(0)$, determine $\boldsymbol{u}(0)$ based on $\hat{\theta}_0(0)$, then generate the corresponding output $\boldsymbol{y}(1)$ and evaluate the cost function $J_0(0)$. Now set $k = 1$ and follow the subsequent steps:

Step 2 Given the system output at the time k as $\mathbf{y}(k)$, generate the perturbation vector $\boldsymbol{\delta}_j$ and perturbed value $\hat{\boldsymbol{\theta}}_j^+(k)$ using Eq. (6.2). Determine $\mathbf{u}^+(k)$ based on $\hat{\boldsymbol{\theta}}_j^+(k)$ and generate the output $\mathbf{y}^+(k+1)$, evaluate the cost function $J_j^+(k)$.

Step 3 Given the system output at the time $k+1$ as $\mathbf{y}^+(k+1)$, generate the perturbed value $\hat{\boldsymbol{\theta}}_j^-(k+1)$ using Eq. (6.3). Determine $\mathbf{u}^-(k+1)$ based on $\hat{\boldsymbol{\theta}}_j^-(k+1)$ and generate output $\mathbf{y}^-(k+2)$, evaluate the cost function $J_j^-(k+1)$.

Step 4 Given the system output at the time $k+2$ as $\mathbf{y}^-(k+2)$, approximate the gradient using Eq. (6.4) and obtain $\hat{\boldsymbol{\theta}}_{j+1}(k+2)$ with the updating law (6.5). Then, determine $\mathbf{u}(k+2)$ based on $\hat{\boldsymbol{\theta}}_{j+1}(k+2)$. Generate the corresponding output $\mathbf{y}(k+3)$ and the cost function $J_{j+1}(k+2)$.

Step 5 Set $j=j+1$, $k=k+3$ and return to step 2.

The procedure of generating data in the SPSA algorithm for an unknown system is demonstrated in Figure 6.1. In this figure, p and c represent the plant and the controller, respectively. Note that p generates the outputs and c is denoted by its dependence on its estimated parameter vectors.

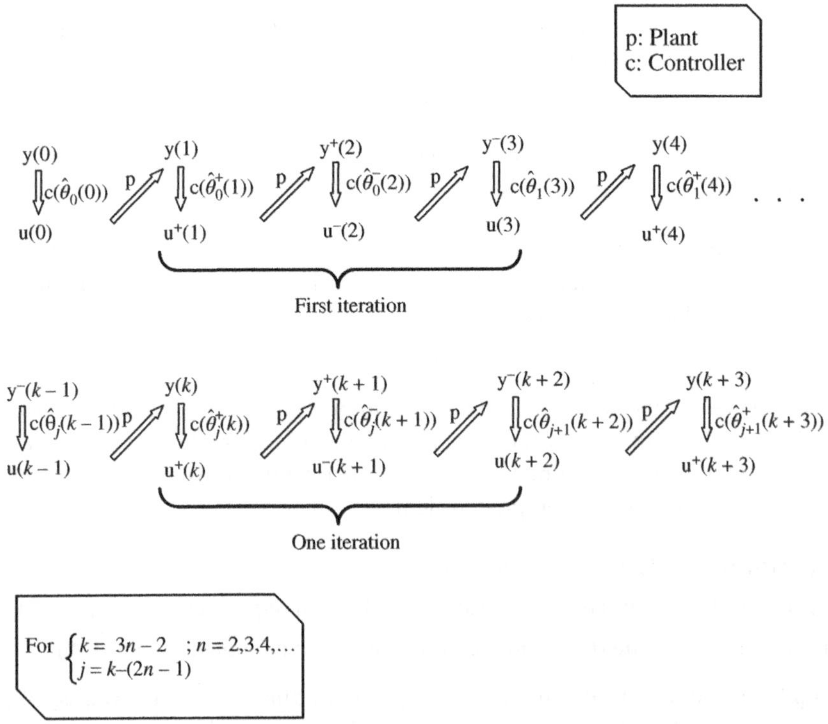

Figure 6.1 The SPSA algorithm design steps.

6.3 Data-Driven Control Design Based on the SPSA Algorithm

In this section, at first, a general scheme of a data-driven controller design based on the SPSA methodology is presented, and after that, two of the data-driven control design techniques based on this method are briefly presented.

The SPSA data-driven control flowchart is demonstrated in Figure 6.2. The controller structure is known and fixed, but its parameters are unknown and must

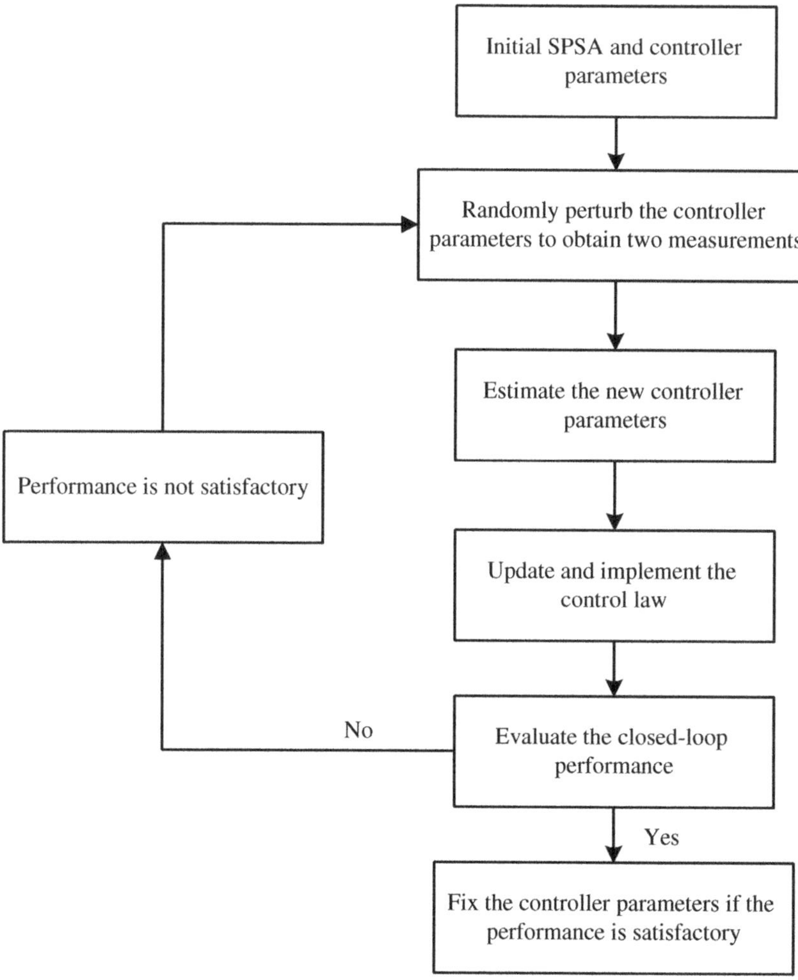

Figure 6.2 The SPSA data-driven control flowchart.

be estimated. An initial controller parameter vector is given that can be randomly selected, and the SPSA algorithm randomly perturbs the given controller parameters. The designer-provided cost function that reflects the desired performance is evaluated using the measured data. The controller parameters are then estimated using the measured data. The new output measurement is recorded by inserting the currently estimated controller parameters in the control law and implementing the control law. This process is repeated until the desired performance is achieved and the optimal controller parameters are determined.

6.3.1 The PID Control

Proportional-Integral-Derivative (PID) is the most widely used control methodology in the industry. The first data-driven control technique of the early last century was PID-based (for further details, refer to Chapter 2). To enhance the PID controller's performance, the model-based versions of PID were later introduced that require an accurate model of the plant to be controlled. Later, with the problems associated with the model-based tuning techniques surfacing, the tuning strategies for PID control parameters based on the input–output data received considerable attention. For a list of the main approaches of the so-called *model-free tuning* strategies, refer to Ahmad et al. [5] and the references therein. Many such strategies are multi-agent-based optimisation, where the computation times per iteration are proportional to the number of agents and therefore require heavy computation time in the design process. In [5], the SPSA-based tuning methodology is introduced to overcome the computational burden with ease of implementation for multivariable plants.

6.3.2 The MPC Approach

Model predictive control (MPC) is the most widely used *advanced* control technique in the industry. MPC has been mainly used in petrochemical industries as early as the 1960s and has since found applications in a wide range of industries with the advent of high-speed computers. The main attractive features of MPC are the ability to control multivariable and over-actuated systems, the possibility of explicitly including constraints in the control system design and the optimisation of a user-defined cost function for optimum output control. The MPC tuning parameters are the weight matrices Q and R in the cost function. Closed-loop stability and performance of plants under MPC are directly related to the tuning of these matrices. The weight matrices are either determined by the professional experience of the designer and trial and error or by analytical and numerical methods. Analytical approaches are very limited in application and require severe assumptions about the plant under control [6]. A numerical method based on random optimisation techniques is used for weight tuning but requires heavy computational effort and is often inappropriate for online applications [7].

In [7], an MPC weight matrix selection algorithm is proposed using SPSA. An evaluation function is also proposed for the selection algorithm. The proposed algorithm utilises the ratios of the control performances and directly uses numerical values of the overshoot and settling time in MPC as the user's requirements in the evaluation function. Therefore, it is possible to directly select the weight matrices that numerically satisfy the user's requirements.

Example 6.1 Consider the plant given by $G(s) = \frac{1}{(s+1)(s+5)}$, and the following PI controller with negative feedback closed-loop structure:

$$u(k) = P(k) + I(k)$$

where, $u(k)$ is the controller output with $P(k)$ the proportional, and $I(k)$ the integral terms, defined as follows:

$$P(k) = k_p e(k)$$
$$I(k+1) = I(k) + k_i T_s e(k)$$

where $e(k) = y_r(k) - y(k)$ is the tracking error and T_s denotes the sampling time. It is assumed that the plant parameters are unknown, and the SPSA algorithm is employed to derive the controller gains k_p and k_i from the input–output data. The control objectives are closed-loop stability and set-point tracking. The controller gain vector is $\zeta = [k_p\ k_i]^T$ and the *tuning parameter vector* is taken as $\theta = \ln(\zeta)$. Note that the proposed tuning parameter vector selection avoids sign changes of gains to prevent instability. Also, the shift from the linear to logarithmic scale increases the search speed of the algorithm for ζ [8]. The cost function is

$$J(\theta(k)) = e^2(k+1)$$

Also, the updating law is modified with a saturation function to avoid large control gain variations as is later given in (6.10), and the parameter δ in (6.10) is chosen as $\delta = 0.1$. For further details, refer to Section 6.4.1. The selected SPSA parameters are $\alpha = 0.6, A = 0.1$ and $a = 80$ in (6.6), and $\gamma = 0.101$ and $c = 0.02$ in (6.7). The plant transfer function is discretised with a sampling time of 0.01 seconds. The initial value of the tuning vector is $\zeta(0) = [1\ 0.2]^T$.

It is clearly shown in Figure 6.3 that the output reached the reference input in about 40 seconds with an appropriate control signal. The controller gains converged to their final values in about 20 seconds, as shown in Figure 6.4, and the cost function decreased to zero, as shown in Figure 6.5.

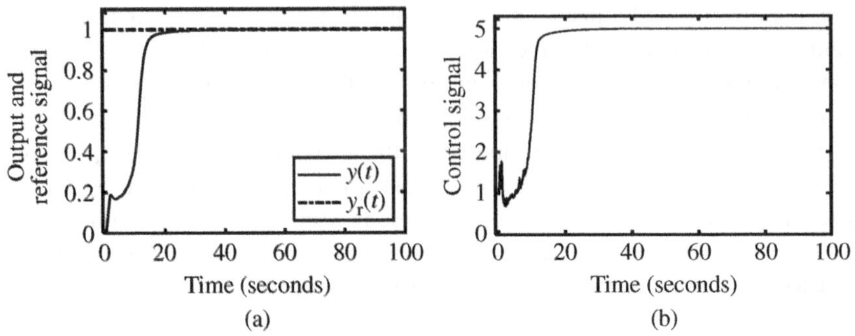

Figure 6.3 (a) The output y, (b) the control effort.

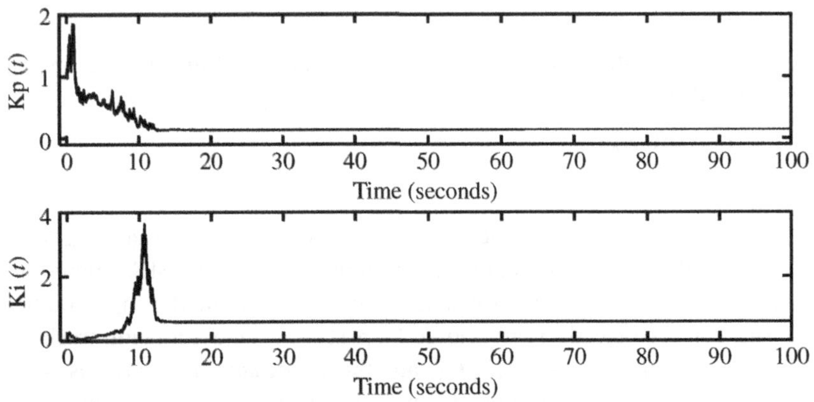

Figure 6.4 PI controller parameters.

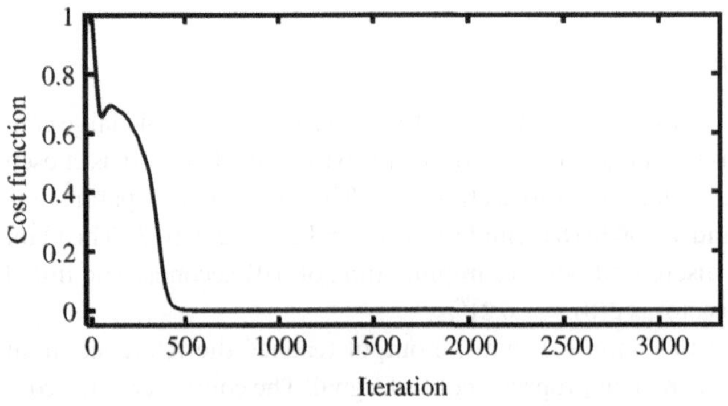

Figure 6.5 The cost function J.

6.4 A Case Study: Data-Driven Control of Under-actuated Systems

There is no general model-based control theory for the systematic analysis and control design of under-actuated systems. This is due to the different structural properties of the under-actuated mechanical systems. Accordingly, such systems are mostly dealt with on a case-by-case basis [9].

In the face of time-varying parameters and coupling, the control of under-actuated mechanical systems becomes even more challenging and to handle such control design problems, there is a tendency towards more complex model-based control algorithms.

However, accurate modelling of under-actuated mechanical systems is, in many cases, non-feasible. For example, unmanned surface vehicles or crane systems such as tower cranes or ship-mounted cranes are subject to sea waves and winds environmental disturbances, and also the parameter uncertainties and simplifications of the physical laws governing them. In the case of crane systems, the frictions may be unknown, and the masses of the trolley and load and the cable length may not be accurately known or can vary in time or under different loading conditions.

To overcome such problems, data-driven control of under-actuated mechanical systems is discussed in [8], using the SPSA-based data-driven control strategy. In this section, the liquid slosh and the ball and beam under-actuated systems are selected to show the procedure and effectiveness of the SPSA-based data-driven controller design.

6.4.1 The Liquid Slosh Example

Any movement of liquid in a partially filled container is called a *slosh*. Liquid slosh leads to additional forces and moments that may degrade system performance. The destructive effects of liquid slosh have been observed in space launch vehicles and long-range projectiles. In these systems, more than 70% of the total weight is dedicated to liquid fuel. Thus, the impact of liquid slosh is severe [10]. Forces and moments caused by the slosh may result in the dangerous overturning of large liquid containers. Examples of such systems are ships or ground vehicles that carry liquid cargo.

The liquid slosh system resembles a moving container and the sloshing liquid in it [10]. The container is mounted on a trolley and is moved by an external force. This process can be appropriately modelled as a set of moving rigid mass and a simple pendulum that moves with reference to the fixed $x-z$ frame. The sloshing part of the liquid is modelled with the pendulum, and the rest of the liquid mass

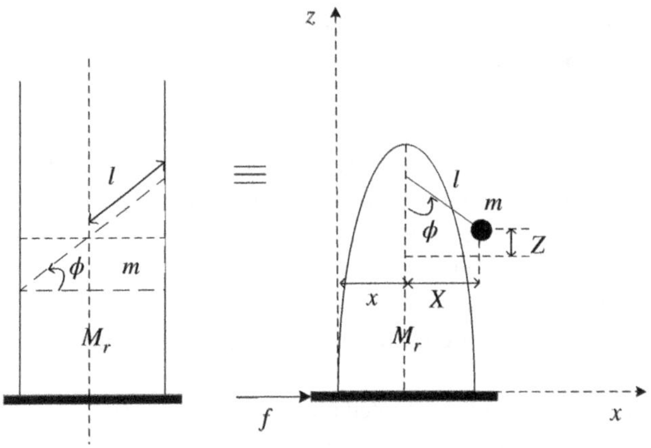

Figure 6.6 The liquid slosh system is modelled by a set of rigid masses and a pendulum.

with the container mass is modelled with a rigid mass. Figure 6.6 represents this schematic modelling.

The model parameters are as follows: M_r is the rigid mass, including masses of the tank, trolley and part of the liquid without slosh (kg), m is the pendulum mass (slosh mass), and M is the total mass that equals the sum of the masses M_r and m. l is the pendulum length (m). f is the force applied to the system for translational motion (N). x is the horizontal displacement of the rigid mass with reference to a fixed $x-z$ frame. X and Z are displacements of m in the horizontal and vertical directions with reference to the fixed $x-z$ frame, respectively. ϕ is the pendulum angle or the slosh angle (rad). The liquid slosh dynamical model is given as follows [10]:

$$M\ddot{x} + ml\cos\phi\ddot{\phi} - ml\dot{\phi}^2\sin\phi = f$$
$$ml\cos\phi\ddot{x} + ml^2\ddot{\phi} + c\dot{\phi} + mgl\sin\phi = 0$$

where g is the gravitational acceleration and c is the viscous damping coefficient between the liquid and the tank. The system has two degrees of freedom, which are the trolley displacement (x) and the slosh angle (ϕ), and one control input, which is the force applied to the trolley (f). The control objective in this system is tracking the desired reference for x and regulating ϕ to zero. Table 6.2 gives the numerical values of the model parameters [8]. The sampling time of the system is chosen as 0.01 seconds. Let $\mathbf{y} = \begin{bmatrix} y_1 \\ y_2 \end{bmatrix} = \begin{bmatrix} x \\ \phi \end{bmatrix}$ in (6.8) and the system state as $\mathbf{x} = \begin{bmatrix} x & \phi & \dot{x} & \dot{\phi} \end{bmatrix}^T$. The initial state is $\mathbf{x}(0) = \begin{bmatrix} 0 & 0 & 0 & 0 \end{bmatrix}^T$ and the reference values are considered as $y_{r_1} = 0.5\,\text{m}$ and $y_{r_2} = 0\,\text{rad}$. The actuator saturation value is 10 N. Figure 6.7 shows the proposed PID control structure that can stabilise the under-actuated systems [8].

Table 6.2 Model parameters of the liquid slosh system [8].

Parameter	Symbol	Value	Unit
The total mass of the rigid part and pendulum	M	6	kg
Pendulum mass	m	1.32	kg
Pendulum length	l	0.052126	m
Gravitational acceleration	g	9.8	m/s^2
Viscous damping coefficient	c	3.049×10^{-4}	kgm^2/s

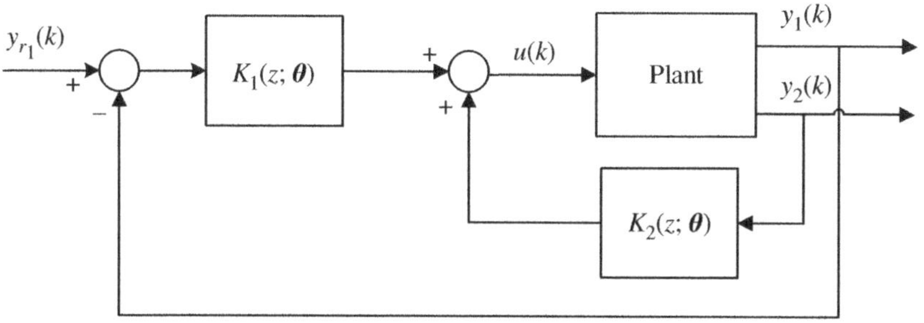

Figure 6.7 The PID control scheme.

The control laws are given by the following equations:

$$v_i(k) = P_i(k) + I_i(k) + D_i(k); \quad \text{for } i = 1, 2$$

where $v_i(k)$ is the output of the ith controller with $P_i(k)$ the proportional, $I_i(k)$ the integral and $D_i(k)$ the derivative term for $i=1, 2$. For $K_1(z;\boldsymbol{\theta})$, these terms are defined as follows:

$$\begin{cases} P_1(k) = k_{p_1} e(k) \\ I_1(k+1) = I_1(k) + k_{i_1} T_s e(k) \\ D_1(k) = k_{d_1} \dfrac{e(k) - e(k-1)}{T_s} \end{cases}$$

where, $e(k) = y_{r_1}(k) - y_1(k)$ is the input of $K_1(z;\boldsymbol{\theta})$ and T_s denotes the sampling time. For $K_2(z;\boldsymbol{\theta})$, the proportional, integral and derivative terms are as follows:

$$\begin{cases} P_2(k) = k_{p_2} y_2(k) \\ I_2(k+1) = I_2(k) + k_{i_2} T_s y_2(k) \\ D_2(k) = k_{d_2} \dfrac{y_2(k) - y_2(k-1)}{T_s} \end{cases}$$

The sum of the controller outputs is the input signal of the actuator, which is specified as $v(k) = v_1(k) + v_2(k)$. By considering the saturation function for the actuator, the control input $u(k)$ is given as follows:

$$u(k) = \text{sat}(v(k))$$

The parameters of the PID controllers in Figure 6.7 are the control gains represented by the vector $\zeta = \left[k_{p_1}\ k_{i_1}\ k_{d_1}\ k_{p_2}\ k_{i_2}\ k_{d_2}\right]^T$. The cost function in the SPSA algorithm is considered as $J(\zeta(k)) = w_1 e^2(k+1) + w_2 {y_2}^2(k+1)$, where $e(k+1)$ is the one-step ahead tracking error of the first output. J is a function of ζ, and the exact mathematical relation between J and ζ is unknown. However, the tuning parameter vector of the algorithm is taken as $\theta = \ln(\zeta)$. As previously stated, this selection avoids sign changes of gains to prevent instability. Also, the shift from the linear to logarithmic scale increases the search speed of the algorithm for ζ [8]. Note that as $J(e^{\theta(k)})$, it is easily observed that J is a function of θ, and it can be written as follows:

$$J(\theta(k)) = w_1 e^2(k+1) + w_2 {y_2}^2(k+1) \tag{6.9}$$

According to Ahmad et al. [11] and to prevent large control gains variations, the updating law given by (6.5) is modified with a saturation function as follows:

$$\hat{\theta}_{j+1}(k+3) = \hat{\theta}_j(k) - \text{sat}_\delta(\mathbf{d}_j); \mathbf{d}_j = g_{1j}\hat{\sigma}_j \tag{6.10}$$

$$\text{sat}_{\delta i}(\mathbf{d}_j) = \begin{cases} \delta & \text{if } \delta < \text{d}_{j_i} \\ \text{d}_{ji} & \text{if} -\delta \le \text{d}_{j_i} \le \delta \\ -\delta & \text{if } \text{d}_{j_i} < -\delta \end{cases}$$

where $\text{sat}_{\delta i}(\mathbf{d}_j)$ and d_{j_i} indicate the ith component of the vector $\text{sat}_\delta(\mathbf{d}_j)$ and $\mathbf{d}_j$, respectively. The weight coefficients in (6.9) are selected as $w_1 = 1$ and $w_2 = 0.5$. Also, the parameter δ in (6.10) is chosen as $\delta = 0.05$. The SPSA parameters are determined as $\alpha = 0.6$, $A = 1$, $a = 10$ in (6.6), $\gamma = 0.101$ and $c = 0.02$ in (6.7).

In the simulation of the SPSA, a simple stop rule is used to prevent the unboundedness of the signals. If the magnitude of the output signal continuously increases during a time interval, the design parameters are reset to their initial values. The initial gain vector is chosen as $\zeta(0) = \left[4\ 0.2\ 1\ 0.5\ 0.01\ 0.1\right]^T$ for which the closed-loop system is stable. The results are depicted in Figures 6.8–6.10.

After about 40 seconds, the trolley position and the slosh angle reach the values 0.5 m and 0 rad, respectively. The control input changes in the range [−2.5, 2] N, which is acceptable. The tuning parameters values have been fixed from about $t = 25$ seconds onwards, which shows the convergence of the algorithm.

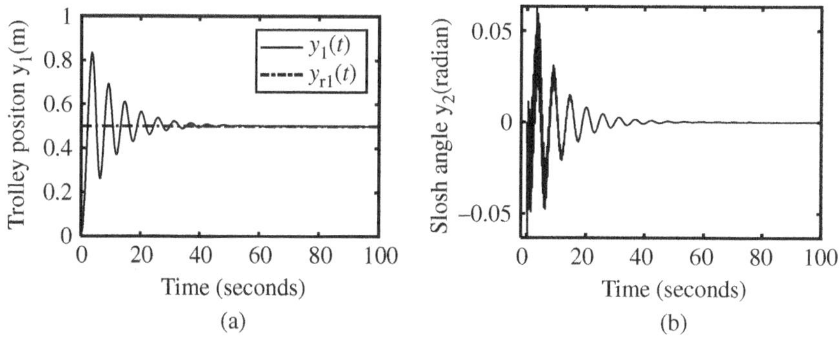

Figure 6.8 (a) y_1 the trolley position, (b) y_2 the slosh angle of the liquid slosh- SPSA algorithm.

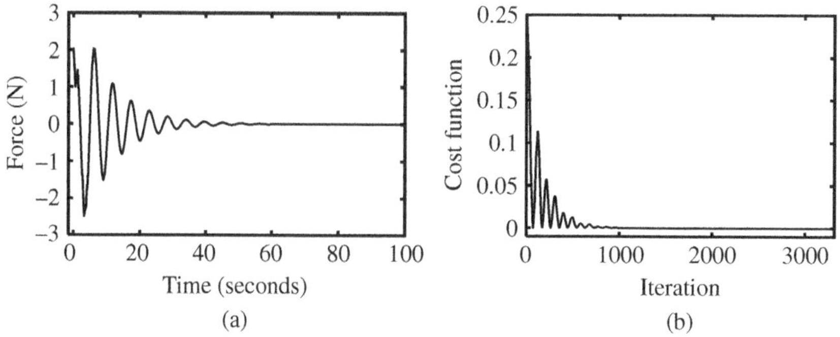

Figure 6.9 (a) Control input u, (b) the cost function J for the liquid slosh-SPSA algorithm.

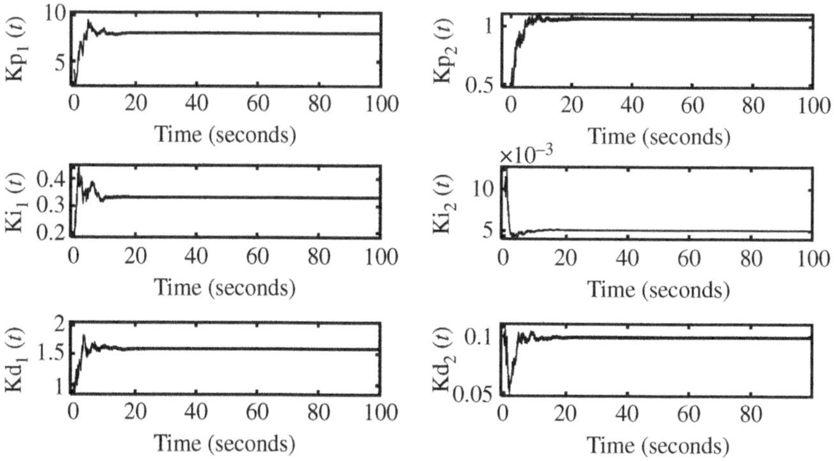

Figure 6.10 The PID controllers' parameters for the liquid slosh.

6.4.2 The Ball and Beam Example

The ball and beam is a well-known benchmark of low-order nonlinear systems, which models small oscillations of platforms, and in particular, the automobile suspension system [12]. The system is composed of a beam set on a base that can rotate in the vertical plane and a ball that moves along the beam. Figure 6.11 shows the ball and beam under-actuated system. System outputs are the ball position and the beam rotation angle, and the only control input is the force applied to the ball. A rotational spring is added, which is represented as two springs on either side of the beam to resemble the automobile suspension system [12].

The model parameters are as follows: m is the ball mass (kg), I_m is the rotational inertia of the ball around its centre of mass (kg m^2) and I_B is the rotational inertia of the beam around the rotation point P. l is the beam length between the points P and a (m). r is the distance of the ball from the point P, and θ is the beam rotational angle with respect to the horizontal axes (rad). k is the stiffness (elastic constant) of the springs (N/m), f is the force applied to the ball (N) and g is the gravitational acceleration constant (kg m/s^2). The dynamic model of the ball and beam is as follows [12]:

$$m\ddot{\theta} - mr\dot{\theta}^2 + mg\sin\theta = f$$
$$(mr^2 + I_T)\ddot{\theta} + 2mr\dot{\theta}\dot{r} + mgrcos\theta + kl^2sin\theta = 0$$

where $I_T = I_m + I_B$ is the total rotational inertia of the system. The control objective in this system is to place the ball in a desired position on the beam while making the beam angle equal to zero. Table 6.3 gives the numerical values of the model parameters [12]. The value of the spring constant is set as $k = 800$ for the ball and beam system. The sampling time of the system is selected as 0.01 seconds.

Let $y_1 = r$ and $y_2 = \theta$ in Figure 6.11 and define the system state vector as $\mathbf{x} = \begin{bmatrix} r & \theta & \dot{r} & \dot{\theta} \end{bmatrix}^T$. The initial state vector is $\mathbf{x}(0) = \begin{bmatrix} 0 & 0 & 0 & 0 \end{bmatrix}^T$ and the reference values are considered as $y_{r_1} = 0.3$ m and $y_{r_2} = 0$ rad. The actuator saturation value is 8 N.

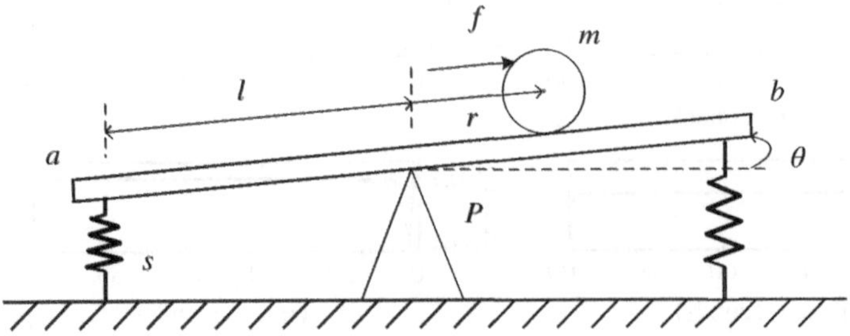

Figure 6.11 The ball and beam system.

Table 6.3 Model parameters of the ball and beam system [12].

Parameter	Symbol	Value	Unit
Ball mass	m	1	kg
Total rotational inertia	I_T	10	kgm^2
Beam length between the points P and a	l	1	m
Gravitational acceleration	g	9.81	m/s^2
Stiffness of the springs	k	800	N/m

The SPSA parameters are determined as $\alpha=0.6$, $A=0.1$, $a=10$ in (6.6), $\gamma=0.101$, and $c=0.02$ in (6.7). The cost function, selected tuning parameter vector and other experimental conditions are the same as in the liquid slosh example. The initial value of the gain vector is chosen as $\zeta(0)=\left[0.5\ 1\ 2\ 5\ 2\ 1\right]^T$ for which the closed-loop system is unstable. The results are depicted in Figures 6.12–6.14.

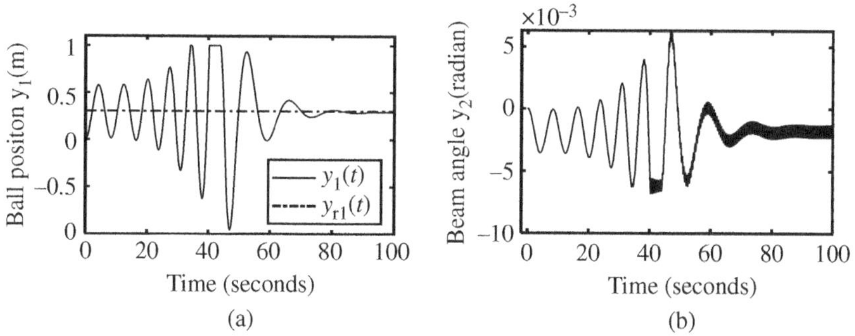

Figure 6.12 (a) y_1 the ball position, (b) y_2 the beam angle of the ball and beam.

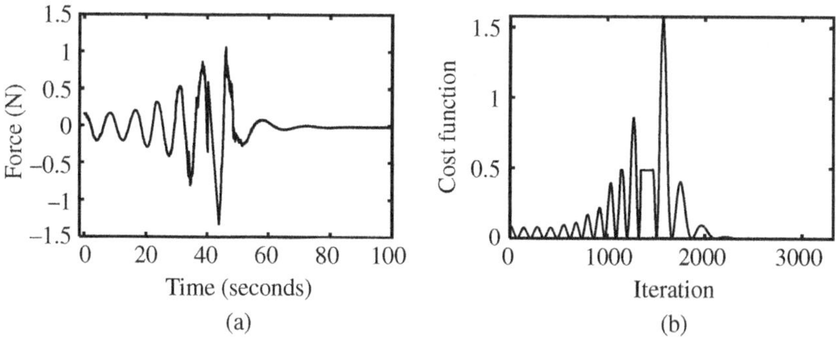

Figure 6.13 (a) Control input u, (b) the cost function J for the ball and beam.

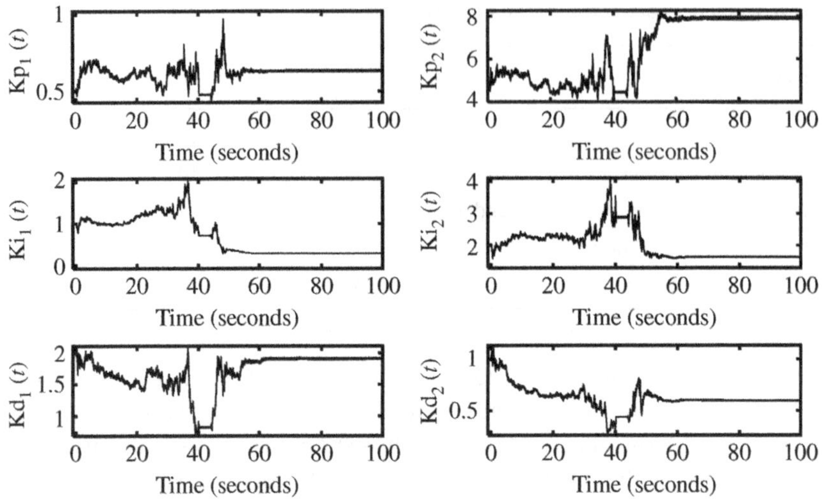

Figure 6.14 PID controllers' parameters for the ball and beam.

A practical limitation of this system is that the ball's distance to the beam rotation centre cannot exceed the beam length. As shown in Figure 6.12, the first output reaches the end of the beam at about $t = 40$ seconds and remains there for about 5 seconds, and then it moves towards its reference value. Similarly, the second output oscillates to zero. The control input remains in the actuator's admissible region. Figure 6.14 indicates that the controllers' gains have converged to their final values in about 70 seconds.

6.5 Conclusions

The optimal control strategies rely on minimising a defined cost function with respect to the controller parameter vector. In the model-based control system design approaches, the required gradient is directly derived using the available mathematical models of the system. However, if the system model is unavailable, the approach would be infeasible, and if the system model is not accurate enough, the resulting design would be unreliable. In this chapter, the SPSA is presented for the cost function optimisation and derivation of the controller parameter vector. It is shown that SPSA can be easily implemented for the controller parameter estimation problem and circumvents the need for cost function gradient calculation. Simulation results are provided to show the effectiveness of the SPSA-based data-driven control approach for PI controller parameter tuning of unknown systems. Also, as a practical case study, closed-loop control of under-actuated systems is effectively achieved for complex unknown systems utilising the SPSA-based data-driven control approach.

Problems

6.1 Consider the smooth Rastrigin test function given as follows:

$$J(\theta) = \sum_{i=1}^{n} \left[\theta_i^2 - 10\cos(2\pi\theta_i) + 10\right]$$

For $n = 100$, $\theta(0) = \begin{bmatrix}0.1 & 0.1 & \cdots & 0.1\end{bmatrix}^T$, $a = 0.002$, $A = 150$, $c = 0.5$, $\alpha = 0.9$ and $\gamma = 1/5$ in (6.6) and (6.7), use the SPSA algorithm to optimise the given cost function with a Bernoulli ±1 distribution for the simultaneous perturbation stochastic vector. Show the convergence of at least two of the parameters and the cost function.

6.2 Consider the Griewank test function given as follows:

$$J(\theta) = \frac{1}{4000}\sum_{i=1}^{n} \left[\theta_i^2 - \prod_{i=1}^{n} \cos\left(\frac{\theta_i}{\sqrt{i}}\right) + 1\right]$$

Assume that the test area is restricted to hypercube $-600 \leq \theta_i \leq 600$, $i = 1, \ldots, n$. For $n = 100$, $\theta(0) = \begin{bmatrix}5 & 5 & \cdots & 5\end{bmatrix}^T$, $a = 700$, $A = 0.05$, $c = 0.5$, $\alpha = 0.6$ and $\gamma = 0.101$ in (6.6) and (6.7), use the SPSA algorithm to optimise the given cost function with a Bernoulli ±1 distribution for the simultaneous perturbation stochastic vector. Show the convergence of at least two of the parameters and the cost function to its global minimum at $J(\theta) = 0$ for $\theta_i = 0$, $i = 1, \ldots, n$.

6.3 Consider the given gain sequences in the SPSA algorithm.

a) Using a sensitivity analysis, prove that g_{1_j} is more sensitive to a change in the parameter a than the parameter A.

b) Consider the liquid slosh plant given in Section 6.4.1. For the given parameters $\alpha = 0.6$, $A = 1$, $a = 10$, $\gamma = 0.101$ and $c = 0.02$, study the effect of the tuning parameters in the SPSA algorithm for the following scenarios and discuss the results. The initial gain vector is given as $\zeta(0) = \begin{bmatrix}k_{p_1} & k_{i_1} & k_{d_1} & k_{p_2} & k_{i_2} & k_{d_2}\end{bmatrix}^T = \begin{bmatrix}4 & 0.2 & 1 & 0.5 & 0.01 & 0.1\end{bmatrix}^T$.

I. Let $a = 1$, keeping the other parameters unchanged, and study the effect of varying a.

II. Let $a = 100$, keeping the other parameters unchanged, and study the effect of varying a.

III. Let $A = 0.01$, keeping the other parameters unchanged, and study the effect of varying A.

IV. Let $A = 1000$, keeping the other parameters unchanged, and study the effect of varying A.

6.4 Consider the following second-order plus dead time plant from Johnson and Moradi [13] and the PI controller is utilised in the unity negative feedback control scheme.

$$G(s) = \frac{(0.75)^2 e^{-0.5s}}{s^2 + 1.2s + (0.75)^2},\ C(s) = k_p + k_i \frac{1}{s}$$

a) For the sampling time $T_s = 0.1$ seconds, select the SPSA parameter set as $\{\alpha = 0.6, A = 0.1, a = 80, \gamma = 0.101 \text{ and } c = 0.02\}$, and the reference value is assumed to be $y_r = 1$. The controllers' parameters vector is considered as $\zeta = \begin{bmatrix} k_p & k_i \end{bmatrix}^T$, and the initial gain vector is $\zeta(0) = \begin{bmatrix} 1 & 0.2 \end{bmatrix}^T$. Tune the PI controller employing the SPSA algorithm.

b) Study the effect of different SPSA parameters on the tuning results.

c) The proposed tuning parameters in [13] are given by $k_p = 1.575$, and $k_i = 0.43$. Compare the proposed tuning results with the tuning parameters given by the SPSA algorithm.

d) Randomly change the plant parameters by 15% and compare the performance of the two derived PI controllers.

6.5 Consider the following multivariable industrial plant [14]:

$$\mathbf{G}(s) = \frac{1}{d(s)} \begin{bmatrix} 1.5s + 1 & 0.15s + 0.2 \\ 0.45s + 0.6 & 0.96s + 0.8 \end{bmatrix}$$

where $d(s) = 2s^4 + 8s^3 + 10.5s^2 + 5.5s + 1$. The closed-loop plant has two PID controllers in a decentralised control structure. It is desired to tune the PID controllers as follows:

$$C_i(s) = k_{Pi} + k_{Ii} \frac{1}{s} + k_{Di},\quad i = 1,2$$

a) For the sampling time $T_s = 0.1$ seconds, select the SPSA parameter as $\{\alpha = 0.6, A = 0.01, a = 10, \gamma = 0.101 \text{ and } c = 0.02\}$, and the reference signal vector is assumed to be $\mathbf{y_r} = \begin{bmatrix} 1 & 0 \end{bmatrix}^T$ and $\mathbf{y_r} = \begin{bmatrix} 0 & 1 \end{bmatrix}^T$. The controllers' parameters vector is considered as $\zeta = \begin{bmatrix} k_{P1} & k_{I1} & k_{D1} & k_{P2} & k_{I2} & k_{D2} \end{bmatrix}^T$, where the subscript i is related to the ith PID controller designed to control the ith output. The initial gain vector is also assumed as $\zeta(0) = \begin{bmatrix} 4 & 0.2 & 1 & 0.5 & 0.01 & 0.1 \end{bmatrix}^T$ Tune the controller parameter vector by employing the SPSA algorithm.

b) Study the effect of different SPSA parameters on the tuning results.

c) The proposed tuning parameters in [14] are given by $\zeta = [6.072\ 1.903\ 0.457\ 6.547\ 1.903\ 0.457]^T$, for the Z-N method and $\zeta = [7.156\ 2.925\ 0.731\ 7.717\ 2.925\ 0.731]^T$ for the characteristic loci approach. Compare the proposed tuning results with the tuning parameters given by the SPSA algorithm.

d) Randomly change the plant parameters by 15% and compare the performance of the two derived PI controllers.

6.6 Consider the following multivariable plant:

$$\mathbf{G}(s) = \frac{1}{(s+1)(s+2)}\begin{bmatrix} 2 & 1 \\ 1 & 1 \end{bmatrix}$$

The plant is unknown and is controlled with two PID controllers in a decentralised control structure given as follows:

$$C_i(s) = k_{Pi} + k_{Ii}\frac{1}{s} + k_{Di}, \quad i = 1,2$$

Determine the controller parameters as $\theta = \begin{bmatrix} k_{P1} & k_{I1} & k_{D1} & k_{P2} & k_{I2} & k_{D2} \end{bmatrix}^T$ using the SPSA algorithm under the following conditions: The sampling time is $T_s = 0.1$ seconds, the SPSA parameter set is $\{\alpha = 0.6, A = 0.2, a = 0.1, \gamma = 0.101$ and $c = 0.02\}$, the reference signal vector is assumed to be $\mathbf{y}_r = \begin{bmatrix} 1 & 0 \end{bmatrix}^T$, and $\mathbf{y}_r = \begin{bmatrix} 0 & 1 \end{bmatrix}^T$. The initial gain vector is $\theta(0) = \begin{bmatrix} 6 & 3 & 1 & 2 & 0.1 & 0.45 \end{bmatrix}^T$ and the actuator saturation value for both the PID controllers is set as 50. The weighting coefficients in the cost function are $w_1 = w_2 = 1$.

References

1 J. C. Spall, "An overview of the simultaneous perturbation method for efficient optimisation," *Johns Hopkins APL Technical Digest*, vol. 19, no. 4, pp. 482–492, 1998.

2 C. Wang, "An overview of SPSA: recent development and applications," *arXiv preprint arXiv:2012.06952*, 2020.

3 J. C. Spall, *Introduction to Stochastic Search and Optimisation: Estimation, Simulation, and Control*. John Wiley & Sons, 2005.

4 J. C. Spall, "Multivariate stochastic approximation using a simultaneous perturbation gradient approximation," *IEEE Transactions on Automatic Control*, vol. 37, no. 3, pp. 332–341, 1992.

5 M. A. Ahmad, S.-i. Azuma, and T. Sugie, "Performance analysis of model-free PID tuning of MIMO systems based on simultaneous perturbation stochastic approximation," *Expert Systems with Applications*, vol. 41, no. 14, pp. 6361–6370, 2014.

6 P. Bagheri and A. K. Sedigh, "Analytical approach to tuning of model predictive control for first-order plus dead time models," *IET Control Theory and Applications*, vol. 7, no. 14, pp. 1806–1817, 2013.

7 S. Cho, M. Otsuki, and T. Kubota, "A new weight selection algorithm using SPSA for model predictive control," *Mechanical Engineering Journal*, vol. 6, no. 5, p. 19-00053, 2019.

8 N. M. Z. A. Mustapha, M. Z. M. Tumari, M. H. Suid, R. M. T. R. Ismail, and M. A. Ahmad, "Data-driven PID tuning for liquid slosh-free motion using memory-based SPSA algorithm," in *Proceedings of the 10th National Technical Seminar on Underwater System Technology 2018*, 2019: Springer, pp. 197–210.

9 A. Choukchou-Braham, B. Cherki, M. Djemaï, and K. Busawon, *Analysis and Control of Underactuated Mechanical Systems*. Springer Science & Business Media, 2013.

10 S. Kurode, S. K. Spurgeon, B. Bandyopadhyay, and P. S. Gandhi, "Sliding mode control for slosh-free motion using a nonlinear sliding surface," *IEEE/ASME Transactions on Mechatronics*, vol. 18, no. 2, pp. 714–724, 2012.

11 M. A. Ahmad, M. Rohani, R. R. Ismail, M. M. Jusof, M. H. Suid, and A. N. K. Nasir, "A model-free PID tuning to slosh control using simultaneous perturbation stochastic approximation," in *2015 IEEE International Conference on Control System, Computing and Engineering (ICCSCE)*, 2015: IEEE, pp. 331–335.

12 I. Fantoni, R. Lozano, and S. Sinha, "Non-linear control for underactuated mechanical systems," *Applied Mechanics Reviews*, vol. 55, no. 4, pp. B67–B68, 2002.

13 M. A. Johnson and M. H. Moradi, *PID control*. Springer, 2005.

14 M. Zhuang and D. Atherton, "PID controller design for a TITO system," *IEE Proceedings-Control Theory and Applications*, vol. 141, no. 2, pp. 111–120, 1994.

7

Data-driven Control System Design Based on the Fundamental Lemma

7.1 Introduction

This chapter presents a class of data-driven controllers based on the Fundamental Lemma. According to this Lemma, persistently exciting data can be used to represent the input–output behaviour of a linear system without the need to identify its matrices [1]. Inspired by the applications of the Fundamental Lemma, a class of data-driven control methods has been developed for unknown systems [2].

The Fundamental Lemma proposed by Willems et al. [3], also referred to as the Willems' Lemma or Willem's Fundamental Lemma, became a foundation for many data-driven analysis and control systems design approaches. The Fundamental Lemma uses the behavioural system theory for discrete-time linear time-invariant (LTI) systems to characterise the system behaviour based on a measured data set from the system. It is stated and reformulated in terms of state-space systems in Refs. [4, 5]. The Fundamental Lemma implies that if the length T input trajectory to an LTI system is *persistently exciting*, that is, the input *excites* all dynamic modes of the system, and the length T measured input–output data spans all possible input–output solutions for a shorter length. This Lemma has led to many positive outcomes on data-driven control systems. A significant immediate implication of the Fundamental Lemma is that a persistently exciting trajectory captures the entire behaviour of the data-generating system. Thus system model identification using subspace methods is possible, and also a data-driven simulation which involves the computation of the system's response to a given reference input can be performed. More importantly, the Fundamental Lemma has played a key role in certain data-driven control systems from a control system design viewpoint. For example, predictive control, optimal control, robust control and linear quadratic regulation are all designed in a data-driven framework using the Fundamental Lemma. There have been attempts to extend the Fundamental Lemma to nonlinear and linear time-varying systems. However, the derived results are so far very restrictive, and there are no general results for

An Introduction to Data-Driven Control Systems, First Edition. Ali Khaki-Sedigh.

nonlinear systems corresponding to those of the Fundamental Lemma for linear systems. For some of the extensions, the interested reader is referred to Refs. [6–8]. A practical application of Willems' Fundamental Lemma to nonlinear systems is through Koopmans' theory [9]. Given that Willems' Fundamental Lemma is defined for a linear time invariant (LTI) system, and the Koopman operator attempts to find a universal LTI model for a nonlinear system, the Koopman operator can be used to apply Willems' Fundamental Lemma for any unknown nonlinear system. The Koopman operator approach is presented in the next chapter.

7.2 The Fundamental Lemma

The Fundamental Lemma was initially formulated and proved in a behavioural system theory framework [3]. Following the Fundamental Lemma, certain conditions are derived under which the state trajectory of a state-space representation spans the whole state-space. This section briefly introduces the Fundamental Lemma in terms of state-space system representations. The results are from Refs. [1, 5]. The Fundamental Lemma states that if the length T input trajectory of a *controllable* LTI system is persistently exciting of a sufficiently high order, then the length T measured input–output data spans all possible input–output solutions of a shorter length.

Consider the following LTI system:

$$\mathbf{x}_{k+1} = \mathbf{A}\mathbf{x}_k + \mathbf{B}\mathbf{u}_k \tag{7.1a}$$

$$\mathbf{y}_k = \mathbf{C}\mathbf{x}_k + \mathbf{D}\mathbf{u}_k \tag{7.1b}$$

where $\mathbf{x}_k \in \mathbb{R}^n$ denotes the state vector, $\mathbf{u}_k \in \mathbb{R}^m$ is the input vector and $\mathbf{y}_k \in \mathbb{R}^p$ is the output vector, $\mathbf{A}$, $\mathbf{B}$, $\mathbf{C}$ and $\mathbf{D}$ are constant matrices of appropriate dimensions. Let

$$\mathbf{u}_{[0,T-1]} \triangleq \left[\mathbf{u}_0{}^T\, \mathbf{u}_1{}^T \ldots \mathbf{u}_{T-1}{}^T\right]^T, \quad \mathbf{y}_{[0,T-1]} \triangleq \left[\mathbf{y}_0{}^T\, \mathbf{y}_1{}^T \ldots \mathbf{y}_{T-1}{}^T\right]^T \tag{7.2}$$

Also, the input–output is restricted to the interval $[0,T-1]$ and (7.2) gives the input–output trajectory of the system (7.1) for this interval. To avoid the notational burden, the sequence of inputs and the corresponding sequence of system outputs for $k = 0,1,\ldots,T-1$ may also be denoted by $\mathbf{u}_{[0,T-1]}$, and $\mathbf{y}_{[0,T-1]}$, respectively. Note that $\mathbf{u}_0, \mathbf{u}_1, \cdots$, and $\mathbf{y}_0, \mathbf{y}_1, \cdots$, are specific instants of the variables, $\mathbf{u}$ and $\mathbf{y}$, respectively.

In the state-space control theory, controllability is defined in terms of a specific state-space realisation of the system. The system given by (7.1a) is *state controllable* if its corresponding controllability matrix $\boldsymbol{\Phi}_c = [\mathbf{B}\,\mathbf{A}\mathbf{B} \cdots \mathbf{A}^{n-1}\mathbf{B}]$ has full rank. This is considered a condition at the *representation level.* Note that the pair $(\mathbf{A}, \mathbf{B})$

is a user realised representation of the system. Controllability in the state-space theory is claimed to be a property of the system, but it is defined as a property of a specific representation of that system. However, it is well-known that one can derive an uncontrollable realisation for a state-controllable system. Hence, the controllability definition in the state-space theory is representation dependent and not a genuine property of the actual system. On the other hand, in the behavioural system theoretical framework, controllability is defined in terms of the system's behaviour or its trajectories. Let $\mathbf{w}_k = \left[\mathbf{u}_{[0,L-1]}^T \ \mathbf{y}_{[0,L-1]}^T\right]^T$ be a trajectory of (7.1). For a given time $T_1 \in \mathbb{N}$, define the past trajectories as $\mathbf{w}_k = \mathbf{w}_k^p$ for $k \leq T_1$, and for a $T_2 \in \mathbb{N}$, define the future trajectories as $\mathbf{w}_k = \mathbf{w}_k^f$ for $k \geq T_1 + T_2$.

Definition 7.1 The system is *controllable* if for any past $\mathbf{w}_k^p$ with T_1 and future trajectories $\mathbf{w}_k^f$ with T_2, there exists an intermediate trajectory $\mathbf{w}_k = \mathbf{w}_k^c$ for $T_1 \leq k \leq T_1 + T_2$ that can patch the past and future trajectories.

As stated by Willems, past trajectories can be considered as trajectories chosen by *nature*. The future trajectories are the desired trajectories imposed by the *designer*, and the controlled trajectory is judiciously selected by the *control engineer* [10]. This general controllability definition is based on the system behaviour and applies to general dynamical systems. However, in the case of LTI systems, it can be related to the conventional state-space controllability definition; however, it should be noted that the two definitions are not equivalent [2]. Figure 7.1 gives a schematic representation of the controllability Definition 7.1.

Definition 7.2 The system input sequences $\mathbf{u}_{[0,T-1]}$ is *persistently exciting of order L* if the following matrix called the *Hankel matrix* $\mathcal{H}_L(\mathbf{u}_{[0,T-1]})$ has full row rank.

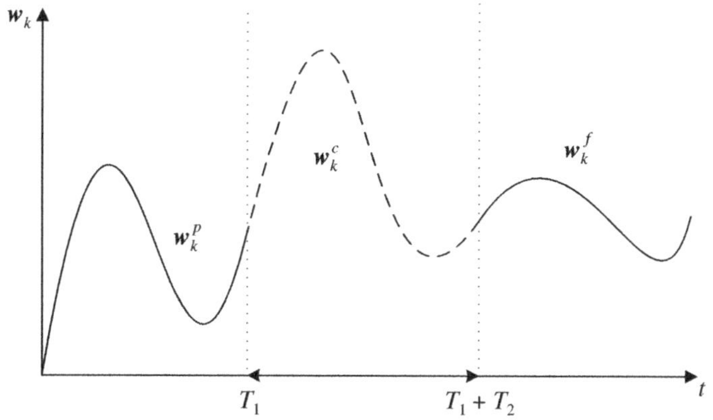

Figure 7.1 A schematic representation of the controllability property.

$$\mathcal{H}_L(\mathbf{u}_{[0,T-1]}) = \begin{bmatrix} \mathbf{u}_0 & \mathbf{u}_1 & \cdots & \mathbf{u}_{T-L} \\ \mathbf{u}_1 & \mathbf{u}_2 & \cdots & \mathbf{u}_{T-L+1} \\ \vdots & \vdots & & \vdots \\ \mathbf{u}_{L-1} & \mathbf{u}_L & \cdots & \mathbf{u}_{T-1} \end{bmatrix} \tag{7.3}$$

The Hankel matrices of input and output sequences are given as follows:

$$\begin{bmatrix} \mathcal{H}_L(\mathbf{u}_{[0,T-1]}) \\ \mathcal{H}_L(\mathbf{y}_{[0,T-1]}) \end{bmatrix} = \begin{bmatrix} \mathbf{u}_0 & \mathbf{u}_1 & \cdots & \mathbf{u}_{T-L} \\ \vdots & \vdots & & \vdots \\ \mathbf{u}_{L-1} & \mathbf{u}_L & \cdots & \mathbf{u}_{T-1} \\ \mathbf{y}_0 & \mathbf{y}_1 & \cdots & \mathbf{y}_{T-L} \\ \vdots & \vdots & & \vdots \\ \mathbf{y}_{L-1} & \mathbf{y}_L & \cdots & \mathbf{y}_{T-1} \end{bmatrix} \tag{7.4}$$

where $L \geq 1$. The columns of the matrix given by (7.4) are the measured inputs–outputs of the system (7.1). The linearity of the system implies that every linear combination of the columns of (7.4) is also an input–output trajectory of (7.1). That is, $\begin{bmatrix} \mathcal{H}_L(\mathbf{u}_{[0,T-1]}) \\ \mathcal{H}_L(\mathbf{y}_{[0,T-1]}) \end{bmatrix} \mathbf{g}$ for any real vector $\mathbf{g} \in \mathbb{R}^{T-L+1}$ is also an input–output trajectory of (7.1).

The essence of the Fundamental Lemma is that for a persistently exciting input $\mathbf{u}_{[0,T-1]}$, any other input–output trajectory of (7.1) of length L such as $\left[\bar{\mathbf{u}}_{[0,L-1]}^T \; \bar{\mathbf{y}}_{[0,L-1]}^T\right]^T$ can be expressed in terms of $\left[\mathbf{u}_{[0,L-1]}^T \; \mathbf{y}_{[0,L-1]}^T\right]^T$ for an appropriate vector $\mathbf{g}$.

Lemma 7.1 ***(The Fundamental Lemma)*** Let

1. $\mathbf{u}_{[0,T-1]}$, $\mathbf{y}_{[0,T-1]}$ be a collected T-samples long trajectory of the LTI system (7.1).
2. The system (7.1) be controllable.
3. The input sequence $\mathbf{u}_{[0,T-1]}$ be persistently exciting of order $L + n$.

Then, any L-samples long trajectory $\bar{\mathbf{u}}_{[0,L-1]}$, $\bar{\mathbf{y}}_{[0,L-1]}$ of (7.1) can be written as a linear combination of the columns of $\begin{bmatrix} \mathcal{H}_L(\mathbf{u}_{[0,T-1]}) \\ \mathcal{H}_L(\mathbf{y}_{[0,T-1]}) \end{bmatrix}$ and any linear combination $\begin{bmatrix} \mathcal{H}_L(\mathbf{u}_{[0,T-1]}) \\ \mathcal{H}_L(\mathbf{y}_{[0,T-1]}) \end{bmatrix} \mathbf{g}$ for any real vector $\mathbf{g} \in \mathbb{R}^{T-L+1}$ is a trajectory of (7.1).

Proof: The Fundamental Lemma is presented and proved in a behavioural approach context in Ref. [1]. Another proof based on the state-space formulation is provided in Ref. [4]. The following proof is based on the proof in Ref. [4]. If the input sequence $\mathbf{u}_{[0,T-1]}$ is persistently exciting of order $L + n$, it is shown in Ref.

[3] that the following Hankel matrix (also see Problem 7.2)

$$\begin{bmatrix} \mathcal{H}_L(\mathbf{u}_{[0,T-1]}) \\ \mathcal{H}_1(\mathbf{x}_{[0,T-L]}) \end{bmatrix}$$

has full rank. Then by applying the Rouché-Capelli theorem[1] (see Ref. [11] for more details), there exists an unknown vector $\mathbf{g} \in \mathbb{R}^{T-L+1}$ such that

$$\begin{bmatrix} \overline{\mathbf{u}}_{[0,L-1]} \\ \overline{\mathbf{x}}_0 \end{bmatrix} = \begin{bmatrix} \mathcal{H}_L(\mathbf{u}_{[0,T-1]}) \\ \mathcal{H}_1(\mathbf{x}_{[0,T-L]}) \end{bmatrix} \mathbf{g} \tag{7.5}$$

where, $\overline{\mathbf{x}}_0$ is the initial state vector of the system (7.1). Direct substitutions in (7.1) give the input–output response of (7.1) over $[0, L-1]$ as

$$\begin{bmatrix} \overline{\mathbf{u}}_{[0,L-1]} \\ \overline{\mathbf{y}}_{[0,L-1]} \end{bmatrix} = \begin{bmatrix} \mathbf{I}_L & \mathbf{0}_{Lm \times n} \\ \mathcal{T}_L & \mathcal{O}_L \end{bmatrix} \begin{bmatrix} \overline{\mathbf{u}}_{[0,L-1]} \\ \overline{\mathbf{x}}_0 \end{bmatrix} \tag{7.6}$$

where

$$\mathcal{T}_L = \begin{bmatrix} \mathbf{D} & \mathbf{0} & \mathbf{0} & \dots & \mathbf{0} \\ \mathbf{CB} & \mathbf{D} & \mathbf{0} & \dots & \mathbf{0} \\ \mathbf{CAB} & \mathbf{CB} & \mathbf{D} & \dots & \mathbf{0} \\ \vdots & \vdots & \vdots & \ddots & \vdots \\ \mathbf{CA}^{L-2}\mathbf{B} & \mathbf{CA}^{L-3}\mathbf{B} & \mathbf{CA}^{L-4}\mathbf{B} & \dots & \mathbf{D} \end{bmatrix}$$

$$\mathcal{O}_L = \begin{bmatrix} \mathbf{C} \\ \mathbf{CA} \\ \vdots \\ \mathbf{CA}^{L-1} \end{bmatrix}$$

are the Toeplitz and observability matrices of order L, respectively. Substitution of (7.5) in (7.6) gives

$$\begin{bmatrix} \overline{\mathbf{u}}_{[0,L-1]} \\ \overline{\mathbf{y}}_{[0,L-1]} \end{bmatrix} = \begin{bmatrix} \mathbf{I}_L & \mathbf{0}_{Lm \times n} \\ \mathcal{T}_L & \mathcal{O}_L \end{bmatrix} \begin{bmatrix} \mathcal{H}_L(\mathbf{u}_{[0,T-1]}) \\ \mathcal{H}_1(\mathbf{x}_{[0,T-L]}) \end{bmatrix} \mathbf{g}$$

which is equal to

$$\begin{bmatrix} \overline{\mathbf{u}}_{[0,L-1]} \\ \overline{\mathbf{y}}_{[0,L-1]} \end{bmatrix} = \begin{bmatrix} \mathcal{H}_L(\mathbf{u}_{[0,T-1]}) \\ \mathcal{H}_L(\mathbf{y}_{[0,T-1]}) \end{bmatrix} \mathbf{g} \tag{7.7}$$

and proves the first result. Also, the substitution of (7.5) and (7.7) in (7.6) gives

$$\begin{bmatrix} \mathcal{H}_L(\mathbf{u}_{[0,T-1]}) \\ \mathcal{H}_L(\mathbf{y}_{[0,T-1]}) \end{bmatrix} \mathbf{g} = \begin{bmatrix} \mathbf{I}_L & \mathbf{0}_{Lm \times n} \\ \mathcal{T}_L & \mathcal{O}_L \end{bmatrix} \begin{bmatrix} \mathcal{H}_L(\mathbf{u}_{[0,T-1]}) \\ \mathcal{H}_1(\mathbf{x}_{[0,T-L]}) \end{bmatrix} \mathbf{g} \tag{7.8}$$

1 The Rouché-Capelli theorem states that a linear system of $L+1$ equations in $T-L+1$ unknowns such as $\boldsymbol{v} = \boldsymbol{S}\mathbf{g}$ has solutions, if and only if the coefficient matrix $\boldsymbol{S}$ of the linear system of equations and its corresponding augmented matrix $(\boldsymbol{S} \mid \boldsymbol{v})$ have the same rank.

By defining

$$\begin{bmatrix} \overline{\mathbf{u}}_{[0,L-1]} \\ \overline{\mathbf{x}}_0 \end{bmatrix} = \begin{bmatrix} \mathcal{H}_L(\mathbf{u}_{[0,T-1]}) \\ \mathcal{H}_1(\mathbf{x}_{[0,T-L]}) \end{bmatrix} \mathbf{g}$$

The first row of (7.8) represents the input sequence of the system (7.1), and the second row gives the output for this input sequence and the given initial condition. This proves the second result. ∎

Remark 7.1 Fundamental Lemma is restated in a state-space framework in Theorem 7.1 of Ref. [5] and an input-state-output trajectory $\mathbf{u}_{[0,T-1]}, \mathbf{x}_{[0,T-1]}, \mathbf{y}_{[0,T-1]}$ of (7.1) is considered. The persistently excitation condition is given as the full rank condition of the following Hankel matrix as it was originally proved in Ref. [3]

$$\begin{bmatrix} \mathcal{H}_1(\mathbf{x}_{[0,T-L]}) \\ \mathcal{H}_L(\mathbf{u}_{[0,T-1]}) \end{bmatrix} = \begin{bmatrix} \mathbf{x}_0 & \mathbf{x}_1 & \dots & \mathbf{x}_{T-L} \\ \mathbf{u}_0 & \mathbf{u}_1 & \dots & \mathbf{u}_{T-L} \\ \vdots & \vdots & & \vdots \\ \mathbf{u}_{L-1} & \mathbf{u}_L & \dots & \mathbf{u}_{T-1} \end{bmatrix} \tag{7.9}$$

The other conditions and conclusions are similar to Lemma 7.1. Note that the full rank condition is on both the input and state matrices that is achievable by injecting a sufficiently exciting input sequence and for a sufficiently large T. This rank condition is essential in data-driven state-space techniques, and for $L = 1$, the full row rank of (7.9) is shown to be instrumental for data-driven state-feedback controller design [4]. To directly verify the rank condition of (7.9), all the states must be accessible. The rank condition cannot be directly verified if the states are not accessible and only input–output data are available. However, it can always be enforced if a sufficiently exciting input signal is applied to the system.

7.3 System Representation and Identification of LTI Systems

7.3.1 Equivalent Data-driven Representations of LTI Systems

State-space and transfer function representations are the two widely used representations for model-based control systems design. Fundamental Lemma provides the framework for an equivalent data-driven representation of (7.1) that can be used in specific data-driven control strategies. In this section, two results are presented from Ref. [4] that give open-loop and closed-loop data-driven system representations. The first result is an open-loop state-space system identification result derived directly from the Fundamental Lemma. The second result is a closed-loop identification result that parametrises the state-feedback vector utilising the input–output data. This provides the basis for further results in deriving data-based control design methodologies.

Theorem 7.1 Let the matrix in (7.9) for $L = 1$ have full rank $m + n$. Then the system (7.1a) has the following equivalent representation:

$$\mathbf{x}_{k+1} = \mathcal{H}_1(\mathbf{x}_{[1,T]}) \begin{bmatrix} \mathcal{H}_1(\mathbf{u}_{[0,T-1]}) \\ \mathcal{H}_1(\mathbf{x}_{[0,T-1]}) \end{bmatrix}^{\dagger} \begin{bmatrix} \mathbf{u}_k \\ \mathbf{x}_k \end{bmatrix} \tag{7.10}$$

where $[\cdot]^{\dagger}$ denotes the pseudo-inverse matrix.

Proof: The proof is from Ref. [4]. Let

$$\boldsymbol{S} := \begin{bmatrix} \mathcal{H}_1(\mathbf{u}_{[0,T-1]}) \\ \mathcal{H}_1(\mathbf{x}_{[0,T-1]}) \end{bmatrix}, \boldsymbol{v} := \begin{bmatrix} \mathbf{u}_k \\ \mathbf{x}_k \end{bmatrix}$$

Then, by the Rouché-Capelli theorem, as in the following linear system of equations

$$\boldsymbol{v} = \boldsymbol{S}\mathbf{g} \tag{7.11}$$

the rank of the coefficient matrix $\boldsymbol{S}$ is equal to the rank of the corresponding augmented matrix $(\boldsymbol{S}\,|\,\boldsymbol{v})$, and all its solutions are given by $\mathbf{g}_p + \bar{\mathbf{g}}$, where $\mathbf{g}_p$ is a particular solution of the system, and $\bar{\mathbf{g}}$ are all the solutions of the associated homogeneous system. Hence, the solutions of (7.11) can be written as follows:

$$\mathbf{g} = \boldsymbol{S}^{\dagger}\boldsymbol{v} + (\mathbf{I} - \boldsymbol{S}^{\dagger}\boldsymbol{S})\boldsymbol{w}, \quad \boldsymbol{w} \in \mathbb{R}^{T} \tag{7.12}$$

Then rewriting (7.1a) as

$$\mathbf{x}_{k+1} = [\mathbf{B}\ \ \mathbf{A}] \begin{bmatrix} \mathbf{u}_k \\ \mathbf{x}_k \end{bmatrix}$$

and using (7.11) gives

$$\mathbf{x}_{k+1} = \left[\mathbf{B}\ \ \mathbf{A}\right]\boldsymbol{S}\mathbf{g}(k) \tag{7.13}$$

for some $\mathbf{g}(k)$. On the other hand, (7.1) easily gives $\left[\mathbf{B}\ \ \mathbf{A}\right]\boldsymbol{S} = \mathcal{H}_1(\mathbf{x}_{[1,T]})$, and from (7.11) and (7.12), we have

$$\mathbf{x}_{k+1} = \mathcal{H}_1(\mathbf{x}_{[1,T]})\mathbf{g}(k) = \mathcal{H}_1(\mathbf{x}_{[1,T]})(\boldsymbol{S}^{\dagger}\boldsymbol{v} + (\boldsymbol{I} - \boldsymbol{S}^{\dagger}\boldsymbol{S})\boldsymbol{w}(k)) = \mathcal{H}_1(\mathbf{x}_{[1,T]})\boldsymbol{S}^{\dagger}\boldsymbol{v} \tag{7.14}$$

which proves the theorem. Note that the last equality in (7.14) results from

$$\mathcal{H}_1(\mathbf{x}_{[1,T]})(\boldsymbol{I} - \boldsymbol{S}^{\dagger}\boldsymbol{S})\boldsymbol{w}(k) = \left[\mathbf{B}\ \ \mathbf{A}\right] \boldsymbol{S}(\boldsymbol{I} - \boldsymbol{S}^{\dagger}\boldsymbol{S})\boldsymbol{w}(k) = \mathbf{0}$$

■

7.3.2 Data-driven State-space Identification

From (7.1a), we have

$$\mathbf{x}_1 = \mathbf{A}\mathbf{x}_0 + \mathbf{B}\mathbf{u}_0 = \begin{bmatrix}\mathbf{B} & \mathbf{A}\end{bmatrix}\begin{bmatrix}\mathbf{u}_0\\ \mathbf{x}_0\end{bmatrix}$$

$$\mathbf{x}_2 = \mathbf{A}\mathbf{x}_1 + \mathbf{B}\mathbf{u}_1 = \begin{bmatrix}\mathbf{B} & \mathbf{A}\end{bmatrix}\begin{bmatrix}\mathbf{u}_1\\ \mathbf{x}_1\end{bmatrix}$$

$$\vdots$$

$$\mathbf{x}_T = \mathbf{A}\mathbf{x}_{T-1} + \mathbf{B}\mathbf{u}_{T-1} = \begin{bmatrix}\mathbf{B} & \mathbf{A}\end{bmatrix}\begin{bmatrix}\mathbf{u}_{T-1}\\ \mathbf{x}_{T-1}\end{bmatrix}$$

and in a compact form, gives

$$\mathcal{H}_1(\mathbf{x}_{[1,T]}) = \begin{bmatrix}\mathbf{B} & \mathbf{A}\end{bmatrix}\begin{bmatrix}\mathcal{H}_1(\mathbf{u}_{[0,T-1]})\\ \mathcal{H}_1(\mathbf{x}_{[0,T-1]})\end{bmatrix}$$

and

$$\begin{bmatrix}\mathbf{B} & \mathbf{A}\end{bmatrix} = \mathcal{H}_1(\mathbf{x}_{[1,T]})\begin{bmatrix}\mathcal{H}_1(\mathbf{u}_{[0,T-1]})\\ \mathcal{H}_1(\mathbf{x}_{[0,T-1]})\end{bmatrix}^{\dagger} \tag{7.15}$$

Note that the right-hand side of (7.15) entails the system inputs and states. Assuming that the states are measurable, (7.15) is a data-based identification equation for the plant state and input matrices.

Remark 7.2 Consider the following least-squares problem:

$$\min_{[\mathbf{B}\ \mathbf{A}]}\left\|\mathcal{H}_1(\mathbf{x}_{[1,T]}) - \begin{bmatrix}\mathbf{B} & \mathbf{A}\end{bmatrix}\begin{bmatrix}\mathcal{H}_1(\mathbf{u}_{[0,T-1]})\\ \mathcal{H}_1(\mathbf{x}_{[0,T-1]})\end{bmatrix}\right\|_F \tag{7.16}$$

where $\|\cdot\|_F$ denotes the Frobenius norm. The solution to the above minimisation problem is

$$\widehat{\begin{bmatrix}\mathbf{B} & \mathbf{A}\end{bmatrix}} = \mathcal{H}_1(\mathbf{x}_{[1,T]})\begin{bmatrix}\mathcal{H}_1(\mathbf{u}_{[0,T-1]})\\ \mathcal{H}_1(\mathbf{x}_{[0,T-1]})\end{bmatrix}^{\dagger} \tag{7.17}$$

where $\begin{bmatrix}\widehat{\mathbf{B}\ \mathbf{A}}\end{bmatrix}$ is the least-squares estimate of $\begin{bmatrix}\mathbf{B} & \mathbf{A}\end{bmatrix}$.

Example 7.1 Consider the discretised version of a batch reactor system using a sampling time of 0.1 second [4]

$$\mathbf{x}_{k+1} = \begin{bmatrix}1.178 & 0.001 & 0.511 & -0.403\\ -0.051 & 0.661 & -0.011 & 0.061\\ 0.076 & 0.335 & 0.560 & 0.382\\ 0 & 0.355 & 0.089 & 0.849\end{bmatrix}\mathbf{x}_k + \begin{bmatrix}0.004 & -0.087\\ 0.467 & 0.001\\ 0.213 & -0.235\\ 0.213 & -0.016\end{bmatrix}\mathbf{u}_k$$

$$\mathbf{y}_k = \mathbf{x}_k$$

The system to be controlled is open-loop unstable. The trajectory data $\mathbf{x}_{[1,T]}$ is generated with zero initial conditions and a random input sequence of length $T = 15$

$$\mathcal{H}_1\left(\mathbf{u}_{[0,T-1]}\right) = \begin{bmatrix} 0.5377 & -2.2588 & \dots & 0.4889 & 0.7269 \\ 1.8339 & 0.8622 & \dots & 1.0347 & -0.3034 \end{bmatrix}$$

and by applying to the system, the following state's trajectory is obtained

$$\mathcal{H}_1(\mathbf{x}_{[0,T-1]}) = \begin{bmatrix} 0 & -0.1574 & -0.4652 & \dots & -10.1253 & -12.2437 \\ 0 & 0.2529 & -0.8701 & \dots & 1.9779 & 2.1207 \\ 0 & -0.3164 & 0.3571 & \dots & 0.4944 & 0.4883 \\ 0 & 0.0852 & -0.4008 & \dots & 1.1974 & 1.2911 \end{bmatrix}$$

Finally, by solving the optimisation problem (7.16), we have

$$\left[\widehat{\mathbf{B}\ \mathbf{A}}\right] = \begin{bmatrix} 0.004 & -0.087 & 1.178 & 0.001 & 0.511 & -0.403 \\ 0.467 & 0.001 & -0.051 & 0.661 & -0.011 & 0.061 \\ 0.213 & -0.235 & 0.076 & 0.335 & 0.560 & 0.382 \\ 0.213 & -0.016 & 0 & 0.355 & 0.089 & 0.849 \end{bmatrix}$$

7.4 Data-driven State-feedback Stabilisation

This section presents a data-dependent closed-loop parametrisation of LTI systems under state feedback. This parametrisation can be used to design various control strategies based only on the measured system data. The following theorem gives the closed-loop parametrisation under state-feedback design and the corresponding data-driven state-feedback matrix.

Theorem 7.2 Let the matrix in (7.9) for $L = 1$ has full rank $m + n$. Then, the closed-loop system with (7.1a) and the state-feedback control $\mathbf{u}_k = \mathbf{K}\mathbf{x}_k$ has the following equivalent representation:

$$\mathbf{x}_{k+1} = \mathcal{H}_1(\mathbf{x}_{[1,T]})\mathbf{G}_\mathbf{K}\mathbf{x}_k \tag{7.18}$$

where $\mathbf{G}_\mathbf{K}$ is a $T \times n$ matrix satisfying

$$\begin{bmatrix} \mathbf{K} \\ \mathbf{I}_n \end{bmatrix} = \begin{bmatrix} \mathcal{H}_1\left(\mathbf{u}_{[0,T-1]}\right) \\ \mathcal{H}_1(\mathbf{x}_{[0,T-1]}) \end{bmatrix} \mathbf{G}_\mathbf{K} \tag{7.19}$$

In particular

$$\mathbf{u}_k = \mathcal{H}_1\left(\mathbf{u}_{[0,T-1]}\right) \mathbf{G}_\mathbf{K}\, \mathbf{x}_k \tag{7.20}$$

Proof: The proof is from Ref. [4]. By applying the Rouché-Capelli theorem, there exists a $\mathbf{G_K}$ that satisfies (7.19). The closed-loop system matrix upon substitution from (7.19) is

$$\mathbf{A}+\mathbf{BK}=\begin{bmatrix}\mathbf{B} & \mathbf{A}\end{bmatrix}\begin{bmatrix}\mathbf{K}\\ \mathbf{I}_n\end{bmatrix}=\begin{bmatrix}\mathbf{B} & \mathbf{A}\end{bmatrix}\begin{bmatrix}\mathcal{H}_1(\mathbf{u}_{[0,T-1]})\\ \mathcal{H}_1(\mathbf{x}_{[0,T-1]})\end{bmatrix}\mathbf{G_K}$$

and from the fact that

$$\begin{bmatrix}\mathbf{B} & \mathbf{A}\end{bmatrix}\begin{bmatrix}\mathcal{H}_1(\mathbf{u}_{[0,T-1]})\\ \mathcal{H}_1(\mathbf{x}_{[0,T-1]})\end{bmatrix}=\mathcal{H}_1(\mathbf{x}_{[1,T]})$$

the closed-loop equation is given by (7.18). And the state-feedback matrix is given by the first row of (7.19), which implies (7.20). ■

Note that (7.19) provides a parametrisation of the closed-loop system with (7.1a) and the state-feedback control $\mathbf{u}_k = \mathbf{K}\mathbf{x}_k$ based on the measured states $\mathbf{x}_{[0,T-1]}$ and the past control inputs $\mathbf{u}_{[0,T-1]}$. The matrix $\mathbf{G_K}$ is a designer-specified matrix that ensures closed-loop stability and must satisfy $\mathbf{I}_n = \mathcal{H}_1(\mathbf{x}_{[0,T-1]})\mathbf{G_K}$. Once $\mathbf{K}$ is determined to be applied, closed-loop stability is checked by examining the eigenvalues of $\mathcal{H}_1(\mathbf{x}_{[1,T]})\,\mathbf{G_K}$ which is equivalent to the closed-loop matrix $\mathbf{A}+\mathbf{BK}$. We know from the Lyapunov stability theorem for discrete-time linear systems that the necessary and sufficient condition for closed-loop stability requires that, for the given $\mathbf{G_K}$, there exists a positive definite symmetric matrix $\mathbf{P} > 0$, such that Ref. [12]:

$$\mathcal{H}_1(\mathbf{x}_{[1,T]})\mathbf{G_K}\mathbf{P}(\mathcal{H}_1(\mathbf{x}_{[1,T]})\mathbf{G_K})^T-\mathbf{P}<\mathbf{0} \tag{7.21}$$

Let $\mathbf{Q} = \mathbf{G_K}\mathbf{P}$, then closed-loop stability requires from (7.21) that, $\mathcal{H}_1(\mathbf{x}_{[1,T]})\mathbf{Q}\mathbf{P}^{-1}\mathbf{Q}^T(\mathcal{H}_1(\mathbf{x}_{[1,T]}))^T - \mathbf{P} < \mathbf{0}$, and from (7.19), we have, $\mathbf{I}_n = \mathcal{H}_1(\mathbf{x}_{[0,T-1]})\mathbf{G_K}$ and multiplying both sides by $\mathbf{P}$ gives $\mathbf{P} = \mathcal{H}_1(\mathbf{x}_{[0,T-1]})\mathbf{Q}$. Similarly, the first row in (7.19) gives $\mathbf{KP} = \mathcal{H}_1(\mathbf{u}_{[0,T-1]})\,\mathbf{Q}$. Recalling the fact that $\mathbf{P} > \mathbf{0}$, and from the basic idea in the linear matrix inequality (LMI) theory [13], we have

$$\begin{bmatrix}\mathcal{H}_1(\mathbf{x}_{[0,T-1]})\mathbf{Q} & \mathcal{H}_1(\mathbf{x}_{[1,T]})\mathbf{Q}\\ (\mathcal{H}_1(\mathbf{x}_{[1,T]})\mathbf{Q})^T & \mathcal{H}_1(\mathbf{x}_{[0,T-1]})\mathbf{Q}\end{bmatrix}>\mathbf{0}. \tag{7.22}$$

Hence, to summarise, any matrix $\mathbf{Q}$ satisfying (7.22) and any state-feedback matrix given by

$$\mathbf{K}=\mathcal{H}_1(\mathbf{u}_{[0,T-1]})\,\mathbf{Q}(\mathcal{H}_1(\mathbf{x}_{[0,T-1]})\mathbf{Q})^{-1} \tag{7.23}$$

stabilises the closed-loop system in a data-based framework assuming that the states are available and measured. Note that in (7.19) and (7.20), only input and state variables data are used, and no knowledge of the state-space matrices is assumed.

Algorithm 7.1 Data-driven State-feedback Control
Input data length T, trajectory data $\mathbf{x}_{[0,T]}$, input data $\mathbf{u}_{[0,T-1]}$.
Step 1 Derive the matrix $\mathbf{Q}$ from (7.22).
Step 2 Obtain the state-feedback gain matrix $\mathbf{K}$ from (7.23).
Step 3 Apply input $\mathbf{u}_k = \mathbf{K}\mathbf{x}_k$ to the system (7.1).

Example 7.2 Consider the discretised version of a batch reactor system given in Example 7.1. The system to be controlled is open-loop unstable. The trajectory data $\mathbf{x}_{[1,T]}$ is generated with random initial conditions and a random input sequence of length $T = 15$. Then the optimisation problem (7.22) is solved, and the following state feedback matrix is obtained:

$$\mathbf{K} = \begin{bmatrix} 0.4893 & -1.2401 & 0.5905 & -1.6244 \\ 2.3250 & 0.3815 & 2.5497 & -1.5430 \end{bmatrix}$$

which stabilises the closed-loop dynamics in agreement with Theorem 7.2. The closed-loop eigenvalues are 0.8334, 0.3712 and $0.0013 \pm 0.2369i$. The open-loop and closed-loop system states are shown in Figure 7.2.

Remark 7.3 The LMI (7.22) is data-dependent, and no parametric model of the system is required in its formulation. The central equation given by (7.19) and the resulting stabilisation algorithm show that persistently exciting data can successfully replace system state-space matrices or transfer function models. Following

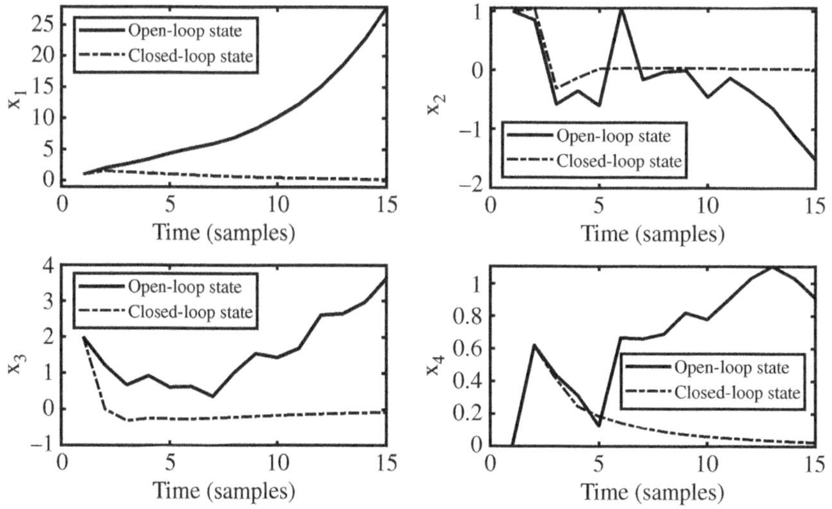

Figure 7.2 The open-loop and closed-loop system states with data-driven state feedback are proposed in Algorithm 7.1.

this LMI-based strategy, a data-driven solution to the linear quadratic regulator (LQR) problem is presented in Ref. [4]. Since LMIs are widely used in many control system design problems [13], the presented approach can promisingly be applied to other model-based approaches for alternative data-driven solutions.

Remark 7.4 In the case of measured state variables corrupted with noise, the data in (7.22) should be replaced with noisy measurements. It is shown in Ref. [4] that there exists a noise level such that a matrix $\mathbf{Q}$ satisfying (7.22) and a stabilisation state-feedback matrix would exist. However, the proven result requires system model assumptions that cannot be verified from measured data. Also, LMIs have shown a certain degree of robustness in the face of various sources of uncertainty (noise is an example), [4] has utilised this property to show that this approach can be extended to the data-driven stabilisation of unstable equilibria of nonlinear systems. Suppose the system can be linearised at an operating point, where the nonlinear system is modelled by a first-order approximation plus higher-order terms. Also, if the input and the state remain sufficiently close to the operating point, the higher-order terms can be viewed as disturbances of small magnitude. Then, the stabilisation result can be assumed to be robust.

7.5 Robust Data-driven State-feedback Stabilisation

In Section 7.4, a data-driven state-feedback stabilisation scheme is presented that does not directly account for noise or disturbances. In this section, based on the results of Ref. [14], the data-driven scheme of Section 7.4 is extended to handle noisy input-state data. Considering the set of systems that are consistent with the measured data and the assumed noise bound, based on results from robust control theory, state-feedback controllers are designed to guarantee closed-loop stability. In this section, the terms disturbance or noise are interchangeably used for any unknown external input affecting an LTI system.

Consider the following controllable LTI system:

$$\mathbf{x}_{k+1} = \mathbf{A}_{tr}\mathbf{x}_k + \mathbf{B}_{tr}\mathbf{u}_k + \mathbf{B}_w\mathbf{w}_k \tag{7.24a}$$

$$\mathbf{y}_k = \mathbf{C}\mathbf{x}_k + \mathbf{D}_w\mathbf{w}_k \tag{7.24b}$$

where $\mathbf{x}_k \in \mathbb{R}^n$ is the state, $\mathbf{w}_k \in \mathbb{R}^{m_w}$ is the disturbance, $\mathbf{u}_k \in \mathbb{R}^m$ is the control input and $\mathbf{y}_k \in \mathbb{R}^p$ is the output to be controlled. The state-feedback controller is given by $\mathbf{u}_k = -\mathbf{K}\mathbf{x}_k$. The matrices $\mathbf{A}_{tr}$ and $\mathbf{B}_{tr}$ are unknown true system matrices of appropriate dimensions. It is assumed that the matrices $\mathbf{B}_w$, $\mathbf{C}$ and $\mathbf{D}_w$ are known, where $\mathbf{B}_w$ is used to model the disturbance influence, $\mathbf{C}$ and $\mathbf{D}_w$ are user-specified for specific performance. We assume that:

- A single open-loop input-state data sequence $\mathbf{u}_{[0,T]}$ and $\mathbf{x}_{[0,T]}$ of (7.24a) is available from tests for some *unknown* noise $\hat{\mathbf{w}}_{[0,T-1]}$.
- $\widehat{\mathbf{W}} = \begin{bmatrix} \hat{\mathbf{w}}_0 & \hat{\mathbf{w}}_1 & \dots & \hat{\mathbf{w}}_{T-1} \end{bmatrix}$ is the unknown disturbance realisation and it is assumed that $\widehat{\mathbf{W}} \in \boldsymbol{W}$, where $\boldsymbol{W}$ is a known set that provides an upper bound for possible disturbances influencing the system's performance. This can be modelled as a quadratic bound on the unknown disturbance sequence. The considered disturbance set provides a framework to model different noise characteristics and is as follows:

$$\boldsymbol{W} = \left\{ \mathbf{W} \in \mathbb{R}^{m_w \times T} \middle| \begin{bmatrix} \mathbf{W} \\ \mathbf{I} \end{bmatrix}^T \begin{bmatrix} \boldsymbol{Q}_w & \boldsymbol{S}_w \\ \boldsymbol{S}_w^T & \boldsymbol{R}_w \end{bmatrix} \begin{bmatrix} \mathbf{W} \\ \mathbf{I} \end{bmatrix} \geq \mathbf{0} \right\}$$

where the matrices $\boldsymbol{Q}_w$, $\boldsymbol{S}_w$ and $\boldsymbol{R}_w > 0$ are assumed known and are of appropriate dimensions. A challenge in this stage would be the proper selection of these matrices. However, many disturbance bounds encountered in practice can be modelled in terms of these matrices.

Consider the measured input-state trajectory $\mathbf{u}_{[0,T-1]}$ and $\mathbf{x}_{[0,T-1]}$ of (7.24a) in the presence of the unknown disturbance realisation $\widehat{\mathbf{W}}$. Define $\Sigma_{\mathcal{H}(\mathbf{u}_{[0,T-1]}),\mathcal{H}_1(\mathbf{x}_{[0,T-1]})}$ as the set of all pairs $(\mathbf{A}\ \mathbf{B})$ that satisfy the system equations for some noise instance $\mathbf{W} \in \boldsymbol{W}$, that is

$$\mathcal{H}_1(\mathbf{x}_{[1,T]}) = \mathbf{A}\mathcal{H}_1(\mathbf{x}_{[0,T-1]}) + \mathbf{B}\mathcal{H}_1\left(\mathbf{u}_{[0,T-1]}\right) + \mathbf{B}_w\mathbf{W} \tag{7.25}$$

Hence, $\Sigma_{\mathcal{H}(\mathbf{u}_{[0,T-1]}),\mathcal{H}_1(\mathbf{x}_{[0,T-1]})}$ parametrises the unknown matrices $\mathbf{A}$ and $\mathbf{B}$ using the data matrices $\mathcal{H}_1(\mathbf{x}_{[0,T-1]})$ and $\mathcal{H}_1(\mathbf{u}_{[0,T-1]})$ via $\mathbf{W}$. It is obvious that $(\mathbf{A}_{tr}\ \mathbf{B}_{tr}) \in \Sigma_{\mathcal{H}(\mathbf{u}_{[0,T-1]}),\mathcal{H}_1(\mathbf{x}_{[0,T-1]})}$. Subsequently, the set of closed-loop matrices $\Sigma^{\mathbf{K}}_{\mathcal{H}(\mathbf{u}_{[0,T-1]}),\mathcal{H}_1(\mathbf{x}_{[0,T-1]})}$ is defined to encompass the closed-loop matrices $\mathbf{A}_{\mathbf{K}} = \mathbf{A} + \mathbf{B}\mathbf{K}$ for $(\mathbf{A}\ \mathbf{B}) \in \Sigma_{\mathcal{H}(\mathbf{u}_{[0,T-1]}),\mathcal{H}_1(\mathbf{x}_{[0,T-1]})}$.

Consider the matrix $\mathbf{G} \in \mathbb{R}^{T \times n}$ and define $\mathcal{A}_{\mathbf{G}}$ as the set of matrices $\mathbf{A}_{\mathbf{G}} \in \mathbb{R}^{n \times n}$ such that

$$\mathbf{A}_{\mathbf{G}} = (\mathcal{H}_1(\mathbf{x}_{[1,T]}) - \mathbf{B}_w\mathbf{W})\mathbf{G} \tag{7.26}$$

for some $\mathbf{W} \in \boldsymbol{W}$ satisfying

$$(\mathcal{H}_1(\mathbf{x}_{[1,T]}) - \mathbf{B}_w\mathbf{W}) \begin{bmatrix} \mathcal{H}_1(\mathbf{x}_{[0,T-1]}) \\ \mathcal{H}_1\left(\mathbf{u}_{[0,T-1]}\right) \end{bmatrix}^{\perp} = \mathbf{0} \tag{7.27}$$

where $[\boldsymbol{.}]^{\perp}$ denotes the matrix containing a basis of the kernel[2] of $[\boldsymbol{.}]$.

Theorem 7.3 If $\mathbf{G} \in \mathbb{R}^{T \times n}$ and $\mathbf{K} \in \mathbb{R}^{m \times n}$ satisfy

$$\begin{bmatrix} \mathcal{H}_1(\mathbf{x}_{[0,T-1]}) \\ \mathcal{H}_1\left(\mathbf{u}_{[0,T-1]}\right) \end{bmatrix} \mathbf{G} = \begin{bmatrix} \mathbf{I} \\ \mathbf{K} \end{bmatrix} \tag{7.28}$$

---

2 The kernel or null space of a given matrix $\mathbf{A}$ consists of the vectors $\mathbf{x}$ such that $\mathbf{A}\mathbf{x} = \mathbf{0}$.

Then $\Sigma^{\mathbf{K}}_{\mathcal{H}(\mathbf{u}_{[0,T-1]}),\mathcal{H}_1(\mathbf{x}_{[0,T-1]})} = \mathcal{A}_{\mathbf{G}}$.

Proof: For proof, refer to Ref. [14].

This theorem gives an exact parameterisation of the uncertain closed-loop system with a state-feedback matrix **K** using a single open-loop trajectory of the uncertain system. The following points must be noted [14]:

- The only required data are $\mathbf{u}_{[0,T-1]}$ and $\mathbf{x}_{[0,T-1]}$, which are obtained by offline tests.
- Disturbance $\mathbf{W} \in \boldsymbol{W}$ must satisfy (7.27). This ensures that the matrices in $\mathcal{A}_{\mathbf{G}}$ contain only those disturbance realisations for which there exists $\begin{pmatrix}\mathbf{A} & \mathbf{B}\end{pmatrix}$ satisfying system dynamics.
- For (7.28), $\mathcal{H}_1(\mathbf{x}_{[0,T-1]})$ must have full rank and the $\mathbf{u}_{[0,T-1]}$ and $\mathbf{x}_{[0,T-1]}$ data need not be necessarily persistently exciting. In this case, a **K** may exist that renders all matrices in $\Sigma^{\mathbf{K}}_{\mathcal{H}(\mathbf{u}_{[0,T-1]}),\mathcal{H}_1(\mathbf{x}_{[0,T-1]})}$ stable; however, with no corresponding **G** satisfying (7.28).
- If the data $\mathbf{x}_{[0,T-1]}$ and $\mathbf{u}_{[0,T-1]}$ are persistently exciting, for any **K**, (7.28) can be solved for **G**. That is, the set of all $\mathbf{A}_{\mathbf{G}} \in \mathcal{A}_{\mathbf{G}}$ with **G** satisfying $\mathcal{H}_1(\mathbf{x}_{[0,T-1]})\mathbf{G} = \mathbf{I}$ equals the set of all possible closed-loop matrices with $\mathbf{u}_k = \mathbf{K}\mathbf{x}_k$.
- The disturbance **W** parametrising $\mathbf{A}_{\mathbf{G}}$ is restricted by $\mathbf{W} \in \boldsymbol{W}$ and must satisfy (7.27). Hence, the construction of $\mathbf{A}_{\mathbf{G}}$ requires the computation of the kernel of $\begin{bmatrix}\mathcal{H}_1(\mathbf{x}_{[0,T-1]})\\ \mathcal{H}_1(\mathbf{u}_{[0,T-1]})\end{bmatrix}$. To overcome the numerical calculation challenge of the computation of the kernel of $\begin{bmatrix}\mathcal{H}_1(\mathbf{x}_{[0,T-1]})\\ \mathcal{H}_1(\mathbf{u}_{[0,T-1]})\end{bmatrix}$, the condition (7.27) is removed by considering the following superset of uncertain closed-loop matrices: $\mathcal{A}^s_{\mathbf{G}} = \{\mathbf{A}_{\mathbf{G}} \mid \mathbf{A}_{\mathbf{G}} = (\mathcal{H}_1(\mathbf{x}_{[1,T]}) - \mathbf{B}_w\mathbf{W})\mathbf{G}, \quad \mathbf{W} \in \boldsymbol{W}\}$. Hence, $\mathbf{A}_{\mathbf{G}} \subseteq \mathcal{A}^s_{\mathbf{G}}$. Note that $\mathcal{A}_{\mathbf{G}}$ considers only those disturbances $\mathbf{W} \in \boldsymbol{W}$, which satisfy (7.27), whereas $\mathcal{A}^s_{\mathbf{G}}$ is in general larger than $\mathcal{A}_{\mathbf{G}}$. Nevertheless, $\mathcal{A}^s_{\mathbf{G}}$ admits a simpler parametrisation and can be translated directly into a standard robust control format.

Linear fractional transformations (LFT) are matrix functions introduced in robust control theory to formulate different control problems in a unified framework. These transformations can then be used as an effective technique to solve them [15]. It is shown that the parameterisation of $\mathcal{A}^s_{\mathbf{G}}$ is equivalent to a particular lower LFT, where a nominal closed-loop system is defined, and a disturbance is considered as the uncertainty term. We have,

$$\mathbf{x}_{k+1} = \mathbf{A}_{\mathbf{G}}\mathbf{x}_k = (\mathcal{H}_1(\mathbf{x}_{[1,T]}) - \mathbf{B}_w\mathbf{W})\mathbf{G}\mathbf{x}_k = \mathcal{H}_1(\mathbf{x}_{[1,T]})\mathbf{G}\mathbf{x}_k + \mathbf{B}_w\mathbf{W}(-\mathbf{G}\mathbf{x}_k)$$

Let $\tilde{\mathbf{z}}_k = -\mathbf{G}\mathbf{x}_k$ and $\tilde{\mathbf{w}}_k = \mathbf{W}\tilde{\mathbf{z}}_k$, then

$$\mathbf{x}_{k+1} = \mathcal{H}_1(\mathbf{x}_{[1,T]})\mathbf{G}\mathbf{x}_k + \mathbf{B}_w\mathbf{W}\tilde{\mathbf{z}}_k = \mathcal{H}_1(\mathbf{x}_{[1,T]})\mathbf{G}\mathbf{x}_k + \mathbf{B}_w\tilde{\mathbf{w}}_k.$$

This can be written as the following lower LFT:

$$\begin{bmatrix} \mathbf{x}(k+1) \\ \tilde{\mathbf{z}}(k) \end{bmatrix} = \begin{bmatrix} \mathcal{H}_1(\mathbf{x}_{[1,T]})\mathbf{G} & \mathbf{B}_w \\ -\mathbf{G} & \mathbf{0} \end{bmatrix} \begin{bmatrix} \mathbf{x}(k) \\ \tilde{\mathbf{w}}(k) \end{bmatrix}. \tag{7.29}$$

Note that Theorem 7.3 implies that if $\mathbf{G}$ satisfies $\mathcal{H}_1(\mathbf{x}_{[0,T-1]})\mathbf{G} = \mathbf{I}$, (7.29) contains all the closed-loop systems with $\mathbf{K} = \mathcal{H}_1\left(\mathbf{u}_{[0,T-1]}\right)\mathbf{G}$. Now by using the results of Ref. [16], we can design a stabilising controller parameter $\mathbf{G}$ for the LFT (7.29) that satisfies all elements of $\Sigma^{\mathbf{K}}_{\mathcal{H}(\mathbf{u}_{[0,T-1]}),\mathcal{H}_1(\mathbf{x}_{[0,T-1]})}$. It is proved in Ref. [14] that if there exists $\aleph > 0$, and a $\mathbf{G}$ such that

$$\mathcal{H}_1(\mathbf{x}_{[0,T-1]})\mathbf{G} = \mathbf{I}$$

and

$$\begin{bmatrix} \mathbf{I} & \mathbf{0} \\ \mathcal{H}_1(\mathbf{x}_{[1,T]})\mathbf{G} & \mathbf{B}_w \\ \mathbf{0} & \mathbf{I} \\ -\mathbf{G} & \mathbf{0} \end{bmatrix}^T \begin{bmatrix} -\aleph & \mathbf{0} & \mathbf{0} & \mathbf{0} \\ \mathbf{0} & \aleph & \mathbf{0} & \mathbf{0} \\ \mathbf{0} & \mathbf{0} & \boldsymbol{Q}_w & \boldsymbol{S}_w \\ \mathbf{0} & \mathbf{0} & \boldsymbol{S}_w^T & \boldsymbol{R}_w \end{bmatrix} \begin{bmatrix} \mathbf{I} & \mathbf{0} \\ \mathcal{H}_1(\mathbf{x}_{[1,T]})\mathbf{G} & \mathbf{B}_w \\ \mathbf{0} & \mathbf{I} \\ -\mathbf{G} & \mathbf{0} \end{bmatrix} < \mathbf{0}. \tag{7.30}$$

Then, $\mathbf{A} + \mathbf{B}\mathbf{K}$ with $\mathbf{K} = \mathcal{H}_1\left(\mathbf{u}_{[0,T-1]}\right)\mathbf{G}$ is stable for all $\left(\mathbf{A}\ \mathbf{B}\right) \in \Sigma_{\mathcal{H}(\mathbf{u}_{[0,T-1]}),\mathcal{H}_1(\mathbf{x}_{[0,T-1]})}$, that is $\mathbf{K}$ robustly stabilises all the elements of $\mathcal{A}^s_{\mathbf{G}}$. ∎

Remark 7.5 Note that (7.30) is not an LMI. It is shown in Ref. [14] that with some mathematical manipulations on (7.30), it leads to the following LMI:

$$\begin{bmatrix} -\boldsymbol{Y} & -\boldsymbol{M}^T\boldsymbol{S}_w^T & \boldsymbol{M}^T\mathcal{H}_1^T(\mathbf{x}_{[1,T]}) & \boldsymbol{M}^T \\ -\boldsymbol{S}_w\boldsymbol{M} & \boldsymbol{Q}_w & \boldsymbol{B}_w^T & \mathbf{0} \\ \mathcal{H}_1(\mathbf{x}_{[1,T]})\boldsymbol{M} & \boldsymbol{B}_w & -\boldsymbol{Y} & \mathbf{0} \\ \boldsymbol{M} & \mathbf{0} & \mathbf{0} & -\boldsymbol{R}_w^{-1} \end{bmatrix} < \mathbf{0} \tag{7.31}$$

in the variables $\boldsymbol{Y} = \aleph^{-1}$, $\boldsymbol{M} = \mathbf{G}\aleph^{-1}$. The stabilising state-feedback gain can then be recovered as $\mathbf{K} = \mathcal{H}_1\left(\mathbf{u}_{[0,T-1]}\right)\boldsymbol{M}\boldsymbol{Y}^{-1}$. The linear equality constraint $\mathcal{H}_1(\mathbf{x}_{[0,T-1]})\mathbf{G} = \mathbf{I}$ must be satisfied with the derived LMI. Multiplying this equation from the right by $\boldsymbol{Y}$ and noting that $\boldsymbol{M} = \boldsymbol{G}\boldsymbol{Y}$ gives the equality constraint $\mathcal{H}_1(\mathbf{x}_{[0,T-1]})\mathbf{M} = \mathbf{Y}$ that must be satisfied when solving the LMI given by (7.31) in Step 1 of the following Algorithm.

Algorithm 7.2 Robust Data-driven State-feedback Control

Input The data length T, the trajectory data $\mathbf{x}_{[0,T]}$, the system matrices $\mathbf{B}_w$, $\mathbf{C}$ and $\mathbf{D}_w$ in (7.24), input data $\mathbf{u}_{[0,T-1]}$, the disturbance bound $\boldsymbol{W}$ and the matrices $\boldsymbol{Q}_w$, $\boldsymbol{S}_w$ and $\boldsymbol{R}_w$.

Step 1 Solve (7.31) to obtain $\mathbf{G} = \boldsymbol{M}\aleph$.

Step 2 Set $\mathbf{K} = \mathcal{H}_1\left(\mathbf{u}_{[0,T-1]}\right)\mathbf{G}$.

Step 3 Apply the input $\mathbf{u}_k = \mathbf{K}\mathbf{x}_k$ to the system (7.24).

Example 7.3 Consider an unstable system described by the following state-space matrices:

$$\mathbf{A}_{tr} = \begin{bmatrix} -0.5 & 1.4 & 0.4 \\ -0.9 & 0.3 & -1.5 \\ 1.1 & 1 & -0.4 \end{bmatrix}, \mathbf{B}_{tr} = \begin{bmatrix} 0.1 & -0.3 \\ -0.1 & -0.7 \\ 0.7 & -1 \end{bmatrix}$$

$$\mathbf{B}_w = \mathbf{I}_3, \mathbf{C} = \mathbf{I}_3, \mathbf{D}_w = \mathbf{0}$$

where it is assumed that $\mathbf{A}_{tr}$ and $\mathbf{B}_{tr}$ are unknown. The data set $\{\mathbf{x}_k, \mathbf{u}_k\}_{k=0}^{T}$ of length, $T = 20$ is generated by sampling the input $\mathbf{u}_k$ uniformly from $[-1,1]$ and the disturbance $\widehat{\mathbf{W}}$ uniformly from the ball $\|\widehat{\mathbf{W}}\|_2 \leq \mathrm{w}$ where $\mathrm{w} = 0.02$. This implies that the disturbance bound $\widehat{\mathbf{W}} \in \boldsymbol{W}$ for $\boldsymbol{Q}_w = -\boldsymbol{I}$, $\boldsymbol{S}_w = \mathbf{0}$, $\boldsymbol{R}_w = \mathrm{w}^2\boldsymbol{I}$. Following the procedure described in Algorithm 7.2, the corresponding controller is obtained as $\mathbf{K} = \begin{bmatrix} -0.8383 & -0.7803 & -0.9290 \\ -0.2213 & 0.7471 & -1.1875 \end{bmatrix}$. Algorithm 7.2 yields a controller with guaranteed performance despite noisy measurements. For larger noise levels $\mathrm{w} \geq 0.04$, the design problem is infeasible because it addresses performance guarantees for all matrices in the respective set, which grows with $\boldsymbol{W}$. Also, it can be shown that the feasibility of (7.31) is enhanced if the data length T increases. However, the computational complexity of the LMI (7.31) scales cubically with the number of data lengths T. The open-loop and closed-loop system states are shown in Figure 7.3.

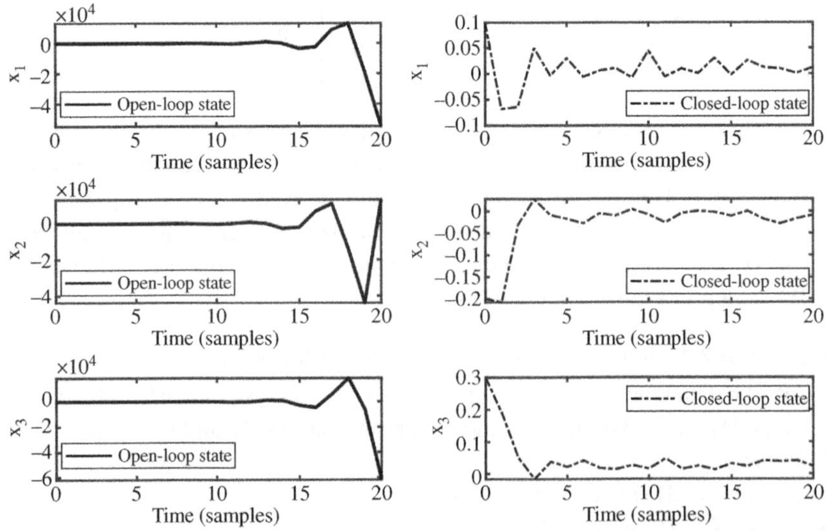

Figure 7.3 The open-loop and closed-loop system states with data-driven robust state feedback proposed in Algorithm 7.2.

7.6 Data-driven Predictive Control

The receding horizon model predictive control (MPC) is a widely used advanced set-point trajectory tracking control in many industrial processes. Its unique characteristics include embedded plant constraints and time delay considerations in the design process. However, the key assumption in the MPC design is the availability of a parametric state-space model of the system with *reasonable* accuracy. This can be a serious limiting factor as the derivation of such models can be a time-consuming and expensive stage in the control system design process for complex practical applications.

In this section, a **D**ata-**e**nabl**e**d **P**redictive **C**ontrol (DeePC) algorithm is presented based on the results in Ref. [17]. The main features of the DeePC algorithm are as follows:

- The DeePC algorithm does not employ a parametric system representation, and it is, therefore, applicable to unknown systems and is superior to the classical MPC and the learning-based control techniques that rely on such representations.
- The initial presentation of the DeePC algorithm is from a behavioural systems theory perspective. Hence, rather than attempting to learn a parametric system model, the aim is to learn the system's *behaviour*.
- The DeePC approach circumvents the need for system identification and state estimation and is therefore much simpler to implement than the learning or model-based predictive control techniques.

The Classical MPC. Consider the discrete-time system given by (7.1). Given a desired reference trajectory $\mathbf{r} = \{\mathbf{r}_0, \mathbf{r}_1, \dots\} \in \mathbb{R}^p$, an input constraint set $\mathbb{U} \subseteq \mathbb{R}^m$ and an output constraint set $\mathbb{Y} \subseteq \mathbb{R}^p$, it is desired to apply the control inputs that optimise a given cost function such that the system output tracks the reference trajectory $\mathbf{r}$ in the presence of constraints. The trajectory tracking receding horizon MPC for a known plant model (7.1) is presented in the following optimisation problem[3]:

$$\begin{aligned}
\min_{\overline{\mathbf{x}},\overline{\mathbf{u}},\overline{\mathbf{y}}} \; & \sum_{k=0}^{N-1} \left(\|\overline{\mathbf{y}}_k(t) - \mathbf{r}_k(t)\|_Q^2 + \|\overline{\mathbf{u}}_k(t)\|_R^2 \right) \\
\text{s.t.} \quad & \overline{\mathbf{x}}_{k+1}(t) = \mathbf{A}\overline{\mathbf{x}}_k(t) + \mathbf{B}\overline{\mathbf{u}}_k(t), \qquad k = 0, \dots, N-1 \\
& \overline{\mathbf{y}}_k(t) = \mathbf{C}\overline{\mathbf{x}}_k(t) + \mathbf{D}\overline{\mathbf{u}}_k(t), \qquad k = 0, \dots, N-1
\end{aligned}$$

3 Note that the plant state space equation given by (7.1) can be equivalently written as $\mathbf{x}(t+1) = \mathbf{A}\mathbf{x}(t) + \mathbf{B}\mathbf{u}(t), \mathbf{y}(t) = \mathbf{C}\mathbf{x}(t)$. In this section, in the defined predictive optimisation problems, this notation is utilised to show that the optimisation is performed at time t and the predictions are obtained for $k = 0, \dots, N-1$ at the specified sampling time.

$$\begin{aligned} &\bar{\mathbf{x}}_0(t) = \hat{\mathbf{x}}_k \\ &\bar{\mathbf{u}}_k(t) \in \mathbb{U}, \qquad k = 0, \ldots, N-1 \\ &\bar{\mathbf{y}}_k(t) \in \mathbb{Y}, \qquad k = 0, \ldots, N-1 \end{aligned} \tag{7.32}$$

where $N \in \mathbb{Z}^+$ is the prediction horizon, $\bar{\mathbf{u}}_k(t), \bar{\mathbf{x}}_k(t), \bar{\mathbf{y}}_k(t), k = 0, \ldots, N-1$ are the predicted control input, predicted state and predicted output, respectively. Also, $\mathbf{r}_k(t) \in \mathbb{R}^p$ is the desired reference trajectory at time $t + k$, where $t \in \mathbb{Z}^+$ is the time at which the optimisation problem should be solved. The norm $\| \cdot \|_*^2$ denotes the standard quadratic form and $*$ is the corresponding weighting or cost matrix of the appropriate dimension. Finally, if the states are unavailable and the system is observable, $\hat{\mathbf{x}}_k$ would denote the estimated state vector, otherwise $\hat{\mathbf{x}}_k = \mathbf{x}_k$. The basic structure of the classical MPC is shown in Figure 7.4. A model is employed to predict the future outputs, based on the past and current values and the predicted future control inputs. The predicted inputs are calculated by solving the optimisation problem (7.32), with the proposed cost function and the plant constraints. Note that only the first input in the resulting predicted control sequence is applied to the plant. It is evident that the plant model plays a decisive role in the presented predictive control structure and should entail the necessary plant characteristics for an accurate output prediction, and in the case of required state estimation, the estimated states must be sufficiently precise.

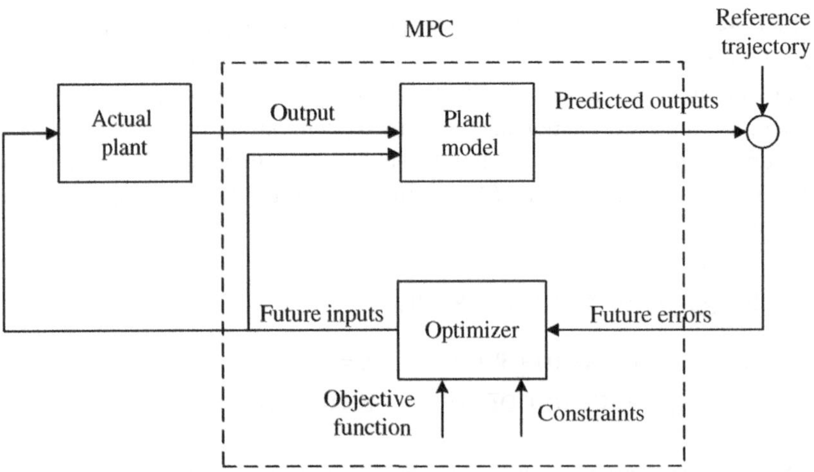

Figure 7.4 The basic structure of a classical MPC.

7.6.1 The Data-enabled Predictive Control (DeePC)

To implement the data-driven predictive control methodology based on the Fundamental Lemma, three steps are presented as follows:

- Input–output data are collected from the unknown plant.
- State estimation and input–output predictions are derived utilising non-parametric representations based on the Fundamental Lemma.
- An optimisation problem is formulated, and the DeePC algorithm is utilised to produce the control signals.

7.6.1.1 Input–Output Data Collection

Consider an unknown controllable LTI system denoted by $\mathfrak{B}$[4] with a minimal input–output-state representation in (7.1). Let the positive integer $T_{ini} \geq lag(\mathfrak{B})$, where $lag(\mathfrak{B})$ is defined as the smallest positive integer L such that the matrix is defined as $\mathcal{O}_L \triangleq [\mathbf{C}^T\ (\mathbf{CA})^T \cdots (\mathbf{CA}^{L-1})^T]^T$ has rank n. Note that in the state-space theory, the observability matrix is denoted by $\mathcal{O}_n$ and the system is observable if $\mathcal{O}_n$ has rank n. Also, the smallest positive integer i for which $\mathcal{O}_i$ has full rank is defined as the observability index. It is well known that the observability index is invariant under state-space similarity transformations and can therefore be considered a behavioural property of the system.

Let $N \in \mathbb{Z}^+$ be such that $T \geq (m+1)(T_{ini} + N + n) - 1$, for a given designer-specified window length T. Let $\mathbf{u}_{[0,T-1]}$ be a sequence of T appropriately selected or designed inputs. Then, $\mathbf{u}_{[0,T-1]}$ is applied to $\mathfrak{B}$, and the corresponding output sequence $\mathbf{y}_{[0,T-1]}$ is collected from the real plant. To use the collected *rich enough* input–output data in the system state estimation and trajectory prediction, it is assumed that $\mathbf{u}_{[0,T-1]}$ is persistently exciting of order $T_{ini} + N + n$. Alternatively, if an exact minimal state-space representation of the plant given by (7.1) is available, the input–output data $\left[\mathbf{u}^T{}_{[0,T-1]}\ \mathbf{y}^T{}_{[0,T-1]}\right]^T$ can be derived from solving (7.1).

7.6.1.2 State Estimation and Trajectory Prediction

In contrast to the MPC process depicted in Figure 7.4, where the predicted outputs are derived from the plant model, the Fundamental Lemma is used to construct any $T_{ini} + N$ length trajectory of the LTI system $\mathfrak{B}$ using the data collected. Define the following Hankel matrices:

---

4 The notations and definitions are from the behavioural approach to the control theory.

$$\mathcal{H}_{T_{ini}+N}\left(\mathbf{u}_{[0,T-1]}\right)=\begin{bmatrix}\mathbf{u}_0 & \mathbf{u}_1 & \cdots & \mathbf{u}_{T-T_{ini}-N}\\ \mathbf{u}_1 & \mathbf{u}_2 & \cdots & \mathbf{u}_{T-T_{ini}-N+1}\\ \vdots & \vdots & \cdots & \vdots\\ \mathbf{u}_{T_{ini}-1} & \mathbf{u}_{T_{ini}} & \cdots & \mathbf{u}_{T-N-1}\\ \mathbf{u}_{T_{ini}} & \mathbf{u}_{T_{ini}+1} & \cdots & \mathbf{u}_{T-N}\\ \vdots & \vdots & \cdots & \vdots\\ \mathbf{u}_{T_{ini}+N-1} & \mathbf{u}_{T_{ini}+N} & \cdots & \mathbf{u}_{T-1}\end{bmatrix}$$

and

$$\mathcal{H}_{T_{ini}+N}\left(\mathbf{y}_{[0,T-1]}\right)=\begin{bmatrix}\mathbf{y}_0 & \mathbf{y}_1 & \cdots & \mathbf{y}_{T-T_{ini}-N}\\ \mathbf{y}_1 & \mathbf{y}_2 & \cdots & \mathbf{y}_{T-T_{ini}-N+1}\\ \vdots & \vdots & \cdots & \vdots\\ \mathbf{y}_{T_{ini}-1} & \mathbf{y}_{T_{ini}} & \cdots & \mathbf{y}_{T-N-1}\\ \mathbf{y}_{T_{ini}} & \mathbf{y}_{T_{ini}+1} & \cdots & \mathbf{y}_{T-N}\\ \vdots & \vdots & \cdots & \vdots\\ \mathbf{y}_{T_{ini}+N-1} & \mathbf{y}_{T_{ini}+N} & \cdots & \mathbf{y}_{T-1}\end{bmatrix}$$

Note that the first T_{ini} rows of the first column of the above matrices correspond to the data up to the time T_{ini}. The first T_{ini} rows are called the *past* data, and the last N rows correspond to the data that are called the *future* data. The past data can be used to estimate the initial state of the plant and the future data can be utilised for the prediction of plants' behaviour. These are denoted as U_p and U_f for past and future input blocks, and Y_p and Y_f for past and future output blocks of the corresponding Hankel matrices. That is, we have for the available measured input–output data the following partitioning of the Hankel matrices:

$$\begin{bmatrix}U_p\\ U_f\end{bmatrix} \triangleq \mathcal{H}_{T_{ini}+N}\left(\mathbf{u}_{[0,T-1]}\right)$$

and

$$\begin{bmatrix}Y_p\\ Y_f\end{bmatrix} \triangleq \mathcal{H}_{T_{ini}+N}\left(\mathbf{y}_{[0,T-1]}\right)$$

According to the Fundamental Lemma, any trajectory such as

$$\begin{bmatrix}\overline{\mathbf{u}}^T_{[-T_{ini},-1]} & \overline{\mathbf{u}}^T_{[0,N-1]} & \overline{\mathbf{y}}^T_{[-T_{ini},-1]} & \overline{\mathbf{y}}^T_{[0,N-1]}\end{bmatrix}^T$$

is an input–output trajectory of the system $\mathfrak{B}$ if and only if there exists $\mathbf{g} \in \mathbb{R}^{T-T_{ini}-N+1}$, such that

$$\begin{bmatrix}U_p\\ Y_p\\ U_f\\ Y_f\end{bmatrix}\mathbf{g}=\begin{bmatrix}\overline{\mathbf{u}}_{[-T_{ini},-1]}\\ \overline{\mathbf{y}}_{[-T_{ini},-1]}\\ \overline{\mathbf{u}}_{[0,N-1]}\\ \overline{\mathbf{y}}_{[0,N-1]}\end{bmatrix} \tag{7.33}$$

Note that the block rows can be interchanged within (7.33) without affecting the final results. Based on this result to group the past and future input and output data, Y_p and U_f rows are interchanged to derive (7.33). Recall that the input–output data $\begin{bmatrix} \bar{\mathbf{u}}^T_{[-T_{ini},-1]} & \bar{\mathbf{y}}^T_{[-T_{ini},-1]} \end{bmatrix}^T$ can be used to derive the initial state $\bar{\mathbf{x}}_{[-T_{ini},-1]}$, and the input–output data $\begin{bmatrix} \bar{\mathbf{u}}^T_{[0,N-1]} & \bar{\mathbf{y}}^T_{[0,N-1]} \end{bmatrix}^T$ is evolved from this initial state. The state $\bar{\mathbf{x}}_{[-T_{ini},-1]}$ is only *fixed* implicitly by $[\bar{\mathbf{u}}^T_{[-T_{ini},-1]} \; \bar{\mathbf{y}}^T_{[-T_{ini},-1]}]^T$. Hence, one can predict future trajectories based on a given initial trajectory $[\bar{\mathbf{u}}^T_{[-T_{ini},-1]} \; \bar{\mathbf{y}}^T_{[-T_{ini},-1]}]^T$, and the past data in $\mathcal{H}_{T_{ini}+N}(\mathbf{u}_{[0,T-1]})$, and $\mathcal{H}_{T_{ini}+N}(\mathbf{y}_{[0,T-1]})$. Indeed, given an initial trajectory $[\bar{\mathbf{u}}^T_{[-T_{ini},-1]} \; \bar{\mathbf{y}}^T_{[-T_{ini},-1]}]^T$ of length $T_{ini} \geq lag(\mathfrak{B})$ and a sequence of future inputs $\mathbf{u}_{[0,T-1]}$, the first three block equations of (7.33) can be solved for $\mathbf{g}$. The sequence of future outputs is then given by $\bar{\mathbf{y}}_{[0,N-1]} = Y_f\mathbf{g}$.

7.6.1.3 The DeePC Algorithm

Let $N \in \mathbb{Z}^+$ be the prediction horizon, $\mathbf{r} = (\mathbf{r}_0, \mathbf{r}_1, \dots) \in \mathbb{R}^p$ be the desired reference trajectory, and the past input–output data be given by $\begin{bmatrix} \bar{\mathbf{u}}^T_{[-T_{ini},-1]} & \bar{\mathbf{y}}^T_{[-T_{ini},-1]} \end{bmatrix}^T$. For practical considerations, the input and output constraint sets are given by $\mathbb{U} \subseteq \mathbb{R}^m$ and $\mathbb{Y} \subseteq \mathbb{R}^p$, respectively. The positive semidefinite and positive definite diagonal output and input cost matrices provided by the designer are $\mathbf{Q} \in \mathbb{R}^p$ and $\mathbf{R} \in \mathbb{R}^p$, respectively. The data-driven predictive control problem can be formulated as the following optimisation problem:

$$\min_{\mathbf{g},\bar{\mathbf{u}},\bar{\mathbf{y}}} \sum_{k=0}^{N-1} \left(\|\bar{\mathbf{y}}_k(t) - \mathbf{r}_k(t)\|^2_{\mathbf{Q}} + \|\bar{\mathbf{u}}_k(t)\|^2_{\mathbf{R}} \right) \tag{7.34a}$$

$$\text{s.t.} \quad \begin{bmatrix} U_p \\ Y_p \\ U_f \\ Y_f \end{bmatrix} \mathbf{g}(t) = \begin{bmatrix} \bar{\mathbf{u}}_{[-T_{ini},-1]}(t) \\ \bar{\mathbf{y}}_{[-T_{ini},-1]}(t) \\ \bar{\mathbf{u}}_{[0,N-1]}(t) \\ \bar{\mathbf{y}}_{[0,N-1]}(t) \end{bmatrix} \tag{7.34b}$$

$$\begin{bmatrix} \bar{\mathbf{u}}_{[-T_{ini},-1]}(t) \\ \bar{\mathbf{y}}_{[-T_{ini},-1]}(t) \end{bmatrix} = \begin{bmatrix} \mathbf{u}_{[t-T_{ini},t-1]} \\ \mathbf{y}_{[t-T_{ini},t-1]} \end{bmatrix} \tag{7.34c}$$

$$\begin{aligned} &\bar{\mathbf{u}}_k(t) \in \mathbb{U}, \qquad k = 0, \dots, N-1 \\ &\bar{\mathbf{y}}_k(t) \in \mathbb{Y}, \qquad k = 0, \dots, N-1 \end{aligned} \tag{7.34d}$$

Note here that $\bar{\mathbf{u}}_{[0,N-1]}(t)$ and $\bar{\mathbf{y}}_{[0,N-1]}(t)$ are not independent decision variables of the optimisation problem. Instead, they are entirely described by the fixed data matrices U_f and Y_f and the decision variable $\mathbf{g}(t)$. Also, $\bar{\mathbf{u}}_{[-T_{ini},-1]}(t)$ and $\bar{\mathbf{y}}_{[-T_{ini},-1]}(t)$ are the actual input–output data of the system that are obtained at each sampling time by deriving the controller for the cost function minimisation, giving it to the system, and measuring the corresponding output, in a range with length T_{ini}. Further, (7.34c) ensures that the internal state of the true trajectory aligns with the internal state of the predicted trajectory at time t. A comparison of the model-based predictive control problem given by (7.32) and the corresponding data-driven predictive control problem given by (7.34) reveals that:

- The cost functions, physical input–output constraints and designer-provided cost matrices are the same.
- The employed models are different. A state-space parametric model is used in (7.32), whereas a non-parametric data-based model is used in (7.34). This difference is tremendous in practical predictive control of complex systems, unknown systems or systems whose corresponding models poorly describe them.
- In (7.34), no state estimation is required, whereas state estimation or complete state measurement is necessary for (7.32).

The DeePC algorithm is presented next.

Algorithm 7.3 Data-enabled Predictive Control (DeePC)

Input $\left[\mathbf{u}^T{}_{[0,T-1]}\ \mathbf{y}^T{}_{[0,T-1]}\right]^T$, reference trajectory $\mathbf{r} \in \mathbb{R}^p$, past input–output data $\left[\overline{\mathbf{u}}^T{}_{[-T_{ini},-1]}\ \overline{\mathbf{y}}^T{}_{[-T_{ini},-1]}\right]^T$, constraint sets $\mathbb{U}$ and $\mathbb{Y}$ and weighting or cost matrices $\mathbf{Q}$ and $\mathbf{R}$.
Step 1 Solve (7.34) for $\mathbf{g}^*$ and the optimal input sequence $\overline{\mathbf{u}}_s^* = U_f\mathbf{g}^*$ for $s = 0, \dots, N-1$.
Step 2 Apply input $\overline{\mathbf{u}}_s^*$ for some $s \leq N-1$.
Step 3 Set t to $t+s$ and update $\overline{\mathbf{u}}_{[-T_{ini},-1]}$ and $\overline{\mathbf{y}}_{[-T_{ini},-1]}$ to the T_{ini} most recent input–output measurements.
Step 4 Return to Step 1.

The MPC problem is *feasible* if it has a solution. Also, the *feasible set* is defined as the set of initial states for which the MPC problem (7.32) with horizon N is feasible. The MPC problem is also *recursively feasible* if it can guarantee that the states remain in a region where the online optimisation problem has a feasible solution. That is, for all feasible initial states, feasibility is guaranteed at every state along the closed-loop trajectory. Feasibility and recursive feasibility is similarly defined for the data-driven predictive control methodologies. It is proved in the next theorem from Ref. [17] that the feasible sets of (7.32) and (7.34) are equivalent.

Theorem 7.4 Consider a controllable LTI system $\mathfrak{B}$ with the minimal input–output-state representation as in (7.1). Consider the MPC and DeePC optimisation problems (7.32) and (7.34). Let $T_{ini} \geq lag(\mathfrak{B})$ and $\left[\overline{\mathbf{u}}^T{}_{[-T_{ini},-1]}\ \overline{\mathbf{y}}^T{}_{[-T_{ini},-1]}\right]^T$ be the most recent input–output measurements from the system (7.1). Assume that the data $\left[\mathbf{u}^T{}_{[0,T-1]}\ \mathbf{y}^T{}_{[0,T-1]}\right]^T$ is such that $\mathbf{u}_{[0,T-1]}$ is persistently exciting of order T_{ini} +

$N + n$, where $T \geq (m+1)(T_{ini} + N + n) - 1$. Then there exists a state estimate $\hat{\mathbf{x}}_k$ in (7.32) such that the feasible sets of (7.32) and (7.34) are equal.

***Proof*. [17]**

The feasible set of (7.34): Since $\mathbf{u}_{[0,T-1]}$ is persistently exciting of order $T_{ini} + N + n$, (7.33) holds by the Fundamental Lemma. Hence, the feasible set of (7.34) equals the set of $(\overline{\mathbf{u}}_{[0,N-1]}, \overline{\mathbf{y}}_{[0,N-1]})$ for which (7.33) is satisfied. The system $\mathfrak{B}$ has an equivalent state-space representation given by (7.1), for which Eq. (7.6) is satisfied. Hence, by the Fundamental Lemma, the feasible set of (7.34) can be written as the set of $(\overline{\mathbf{u}}_{[0,N-1]}, \overline{\mathbf{y}}_{[0,N-1]})$ that satisfies

$$\overline{\mathbf{y}}_{[0,N-1]} = \mathcal{O}_N(\mathbf{A}, \mathbf{C})\overline{\mathbf{x}}_{[-T_{ini},-1]} + \mathcal{T}_N(\mathbf{A}, \mathbf{B}, \mathbf{C}, \mathbf{D})\overline{\mathbf{u}}_{[0,N-1]}$$

where $\overline{\mathbf{x}}_{[-T_{ini},-1]}$ is uniquely determined from $[\overline{\mathbf{u}}^T_{[-T_{ini},-1]} \ \overline{\mathbf{y}}^T_{[-T_{ini},-1]}]^T$, and the matrices $\mathcal{O}_N$ (**A**, **C**) and $\mathcal{T}_N$ (**A**, **B**, **C**, **D**) are as previously defined. Also, by direct substitutions in the state-space constraints, the feasible set of (7.32) is given as follows:

$$\overline{\mathbf{y}}_{[0,N-1]} = \mathcal{O}_N(\mathbf{A}, \mathbf{C})\hat{\mathbf{x}}_k + \mathcal{T}_N(\mathbf{A}, \mathbf{B}, \mathbf{C}, \mathbf{D})\overline{\mathbf{u}}_{[0,N-1]},$$

where $\hat{\mathbf{x}}_k$ is the estimation of the state $\overline{\mathbf{x}}_k$ at instant k. Setting the state estimate $\hat{\mathbf{x}}_k = \overline{\mathbf{x}}_{[-T_{ini},-1]}$ yields equal feasible sets. Note that if all the state variables are measured or when input–output measurements are deterministic, then $\hat{\mathbf{x}}_k = \overline{\mathbf{x}}_{[-T_{ini},-1]}$.

Remark 7.6 The closed-loop trajectories of the MPC (7.32) and the data-driven predictive control strategy (7.34) are proved to coincide if the input and output constraint sets, $\mathbb{U}$ and $\mathbb{Y}$, are convex [17]. In the Euclidean input and output space, these constraint sets are convex, provided the line segment connecting each pair of points in these sets is also within the sets.

Remark 7.7 It was previously noted that T is the number of collected data, and in practice, it is determined by the designer. However, for the input signal to be persistently exciting, the minimum required number of collected data is $(m+1)(T_{ini} + N + n) - 1$. This gives at least $m(T_{ini} + N) + (m+1)n$ variables in (7.34) to be determined. To ensure the persistency of excitation conditions, the plant order n and $lag(\mathfrak{B})$ must be known a priori. It is clear that $lag(\mathfrak{B}) \leq n$, and knowledge of an upper bound for n is sufficient.

7.6.2 LTI Systems with Measurement Noise

Consider the linear discrete-time system given by (7.1) affected by the measurement noise. In this case, we have

$$\mathbf{y}_k = \mathbf{C}\mathbf{x}_k + \mathbf{D}\mathbf{u}_k + \omega_k$$

where $\boldsymbol{\omega}_k$ is some ***unknown measurement noise***. The constraint equation in (7.34) may become inconsistent in the presence of noisy real-time measurements. Hence, safeguards must be introduced to ensure the feasibility of the constraints in (7.34) during the plant operation and online data collection. This is achieved through the application of explicit regularisation by adding some terms or costs to the optimisation problem. The regularised optimal control problem (7.34) is proposed as follows [8, 17]:

$$
\begin{aligned}
&\min_{\mathbf{g},\overline{\mathbf{u}},\overline{\mathbf{y}},\boldsymbol{\sigma}_{\mathbf{y}}} \sum_{k=0}^{N-1} \|\overline{\mathbf{y}}_k(t)-\mathbf{r}_k(t)\|_{\mathbf{Q}}^2+\|\overline{\mathbf{u}}_k(t)\|_{\mathbf{R}}^2+\lambda_{\mathbf{g}}\|\mathbf{g}(t)\|_2+\lambda_{\sigma}\|\boldsymbol{\sigma}_{\mathbf{y}}(t)\|_2\\
&\text{s.t.}\quad \begin{bmatrix}\hat{U}_p\\ \hat{Y}_p\\ \hat{U}_f\\ \hat{Y}_f\end{bmatrix}\mathbf{g}(t)=\begin{bmatrix}\overline{\mathbf{u}}_{[-T_{ini},-1]}(t)\\ \overline{\mathbf{y}}_{[-T_{ini},-1]}(t)\\ \overline{\mathbf{u}}_{[0,N-1]}(t)\\ \overline{\mathbf{y}}_{[0,N-1]}(t)\end{bmatrix}+\begin{bmatrix}0\\ \boldsymbol{\sigma}_{\mathbf{y}}(t)\\ 0\\ 0\end{bmatrix}\\
&\begin{bmatrix}\overline{\mathbf{u}}_{[-T_{ini},-1]}(t)\\ \overline{\mathbf{y}}_{[-T_{ini},-1]}(t)\end{bmatrix}=\begin{bmatrix}\mathbf{u}_{[t-T_{ini},t-1]}\\ \mathbf{y}_{[t-T_{ini},t-1]}\end{bmatrix}\\
&\overline{\mathbf{u}}_k(t)\in\mathbb{U},\qquad k=0,\dots,N-1\\
&\overline{\mathbf{y}}_k(t)\in\mathbb{Y},\qquad k=0,\dots,N-1
\end{aligned}
\tag{7.35}
$$

where $\boldsymbol{\sigma}_{\mathbf{y}} \in \mathbb{R}^{T_{ini}p}$ is an auxiliary slack variable, $\lambda_{\mathbf{g}}, \lambda_{\sigma} \in \mathbb{R}^{+}$ are the regularisation parameters, and $\hat{U}_*(\cdot), \hat{Y}_*(\cdot)$ column matrix in (7.35) denote a low-rank approximation of $U_*(\cdot), Y_*(\cdot)$, column matrix in (7.34b), respectively. The following points must be noted for (7.35):

- The slack variable $\boldsymbol{\sigma}_{\mathbf{y}}$ in the constraint is to ensure the feasibility of the constraint at all times. The slack variable is penalised with a weighted two-norm penalty function by selecting a sufficiently large λ_{σ} such that the implicit initial condition estimation is accurate.
- The cost includes a two-norm penalty on $\mathbf{g}$ as a regularisation term to avoid overfitting in case of noisy data samples. When stochastic disturbances affect the output measurements, a two-norm regularisation of $\mathbf{g}$ coincides with distributional two-norm robustness in the trajectory space [18].

Example 7.4 In this example, the DeePC algorithm in (7.34) is applied to a four-tank system described by an open-loop stable linearised state-space model given in Ref. [19]:

$$
\mathbf{x}_{k+1}=\begin{bmatrix}0.921 & 0 & 0.041 & 0\\ 0 & 0.918 & 0 & 0.033\\ 0 & 0 & 0.924 & 0\\ 0 & 0 & 0 & 0.937\end{bmatrix}\mathbf{x}_k+\begin{bmatrix}0.017 & 0.001\\ 0.001 & 0.023\\ 0 & 0.061\\ 0.072 & 0\end{bmatrix}\mathbf{u}_k
$$

$$\mathbf{y}_k = \begin{bmatrix} 1 & 0 & 0 & 0 \\ 0 & 1 & 0 & 0 \end{bmatrix} \mathbf{x}_k + \omega_k$$

It is assumed that the system matrices are unknown and only the measured input–output data is available. The control goal is tracking the following fixed set-point trajectory:

$$\mathbf{r}_k = \begin{bmatrix} 0.64 & 0.64 & \cdots \\ 0.75 & 0.75 & \cdots \end{bmatrix}$$

In an open-loop experiment, an input–output trajectory of length $T = 500$ is collected, where the input is chosen randomly from the unit interval; that is, the elements of the control input vector are in the interval [−1,1]. The output is subjected to uniformly distributed additive measurement noise with a bound of 0.002. The online measurements used to update the initial conditions in (7.35) are subject to the same noise type. The prediction horizon is $N = 30$, and the design parameters are selected as follows:

$$\mathbf{Q} = 10\boldsymbol{I}_p, \boldsymbol{R} = \mathbf{I}_m, \lambda_{\mathbf{g}} = 50, \lambda_\sigma = 1000, T_{ini} = N.$$

The closed-loop outputs from (7.35) in a N-step DeePC scheme are displayed in Figure 7.5(a). It is shown that a steady-state error is present in the closed-loop responses. For step set-point tracking it may be assumed that the steady-state value of the control signal $\mathbf{u}_{ss}$ is known and equals $[1\ 1]^T$ as in Ref. [19], and the second term in the cost function (7.35) is modified as $\|\bar{\mathbf{u}}_k(t) - \mathbf{u}_{ss}\|^2_{\mathbf{R}}$. Figure 7.5(b) shows that satisfactory set-point trajectory tracking is achieved. However, in many practical applications $\mathbf{u}_{ss}$ is not known a priori. In such cases, $\mathbf{u}_{ss}$ can be considered as an optimisation parameter and is incorporated in (7.35) by modifying the second term as $\|\bar{\mathbf{u}}_k(t) - \mathbf{u}_{ss}\|^2_{\mathbf{R}}$ with an unknown $\mathbf{u}_{ss}$. It is also shown in Figure 7.5(c) that satisfactory set-point trajectory tracking is achieved in this case.

7.6.3 Data-driven Predictive Control for Nonlinear Systems

One of the challenging issues in learning-based and data-driven control strategies is the development of methods to control unknown nonlinear systems with closed-loop guarantees. In the following discussion, this issue is addressed with the solution proposed in Ref. [7]. For a detailed treatment of data-driven predictive control for nonlinear systems, refer to Ref. [20]. Consider the following nonlinear system

$$\begin{aligned} \mathbf{x}_{k+1} &= f(\mathbf{x}_k) + g(\mathbf{x}_k)\mathbf{u}_k \\ \mathbf{y}_k &= h_0(\mathbf{x}_k) + h_1(\mathbf{x}_k)\mathbf{u}_k \end{aligned} \tag{7.36}$$

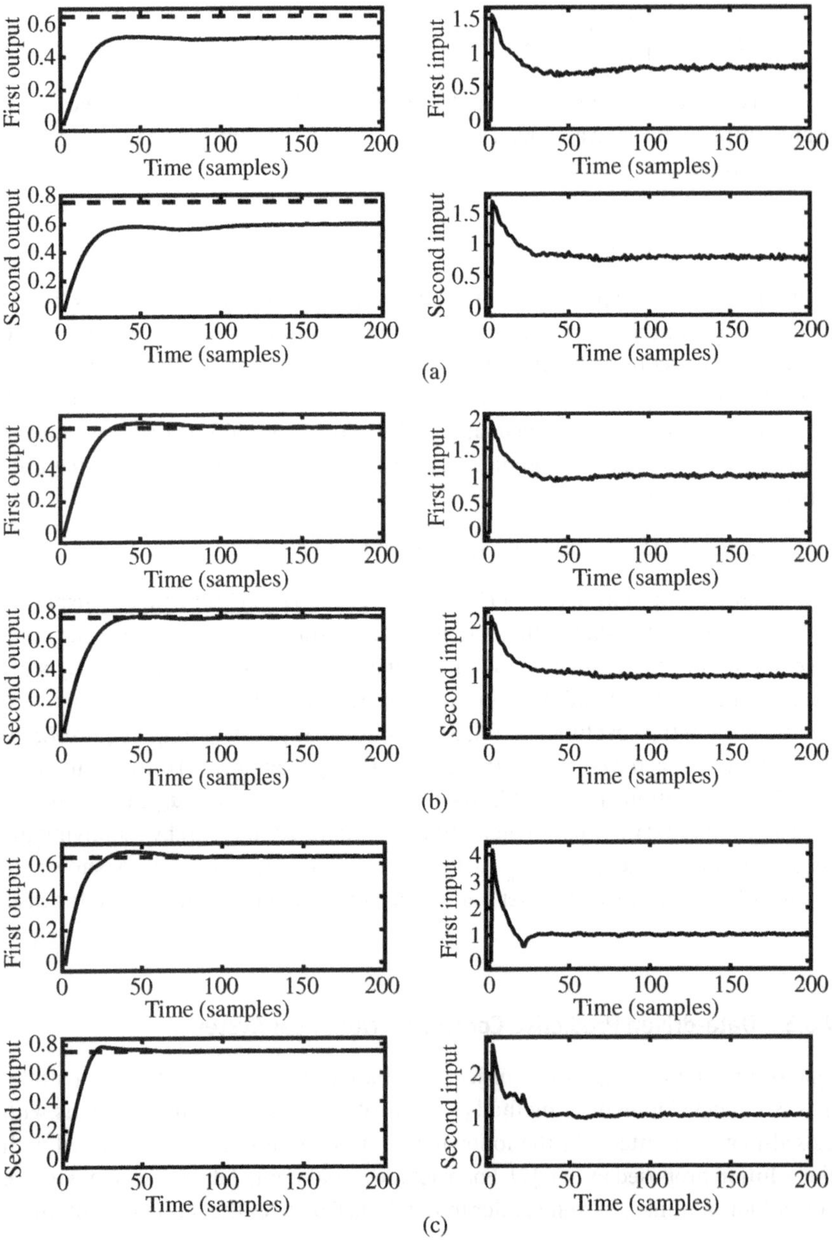

Figure 7.5 *N*-step data-enabled predictive control responses.

where $\mathbf{x}_k \in \mathbb{R}^n$, $\mathbf{u}_k \in \mathbb{R}^m$, $\mathbf{y}_k \in \mathbb{R}^p$ and f, g, h_0, h_1 are unknown sufficiently smooth vector fields of appropriate dimensions. The control objective is to track a desired set-point trajectory $\mathbf{r}$, while satisfying the pointwise-in-time input constraints $\mathbf{u}_k \in \mathbb{U}$ *for* $t \geq 0$ with some convex, compact polytope $\mathbb{U}$.

A measured input–output trajectory is assumed to be available and to account for the local approximation of the nonlinear system (7.36), and the measured data are constantly updated. Let the past T input–output measurements $\left[\mathbf{u}^T{}_{[t-T,t-1\,]}\ \tilde{\mathbf{y}}^T{}_{[t-T,t-1]}\right]^T$ of the nonlinear system (7.36) be given, where $\tilde{\mathbf{y}}_k = \mathbf{y}_k + \varepsilon_k$, and ε_k is a bounded noise or error.

The notion of an *artificial* steady state or equilibrium is introduced to steer the unknown system to a neighbourhood of an admissible and desired steady state. It is reminded that in model-based control strategies such as MPC, a desired equilibrium point of the system is stabilised and this equilibrium point is obtained via the available model of the system. However, in the data-driven control strategies, this desired equilibrium point must be defined in terms of the available input–output pairs. In this case, an input–output pair $(\mathbf{u}^s, \mathbf{y}^s)$ is defined as an equilibrium point of the system if the sequence containing $\mathbf{u}^s$ and $\mathbf{y}^s$ at all the required time instant is a trajectory of the system, and it is shown by $\mathbf{u}_n^s$ and $\mathbf{y}_n^s$ that are column vectors that contain $\mathbf{u}^s$ and $\mathbf{y}^s$ n times. Also, the addition of artificial reference can lead to an enlargement of the domain of attraction of the data-driven predictive controller. Also, by introducing an artificial set-point, the original set-point may not be initially reachable in a data-driven nonlinear predictive control and becomes reachable within the artificial input–output setup. It is shown that the introduced artificial variables will be employed as additional decision variables in the data-driven predictive control optimisation problem.

The closed-loop output constraints can be satisfied if the actual steady-state manifold lies strictly inside the constraints. For some convex polytope $\mathbb{U}^s \subseteq \mathbb{U}$, the artificial steady-state manifold and its projection on the output are defined as follows:

$$\mathcal{Z}^s = \{(\mathbf{x}^s, \mathbf{u}^s) \in \mathbb{R}^n \times \mathbb{U}^s \mid \mathbf{x}^s \ \textit{satisfies}\ (7.36)\}$$

$$\mathcal{Z}^s_{\mathbf{y}} = \{\mathbf{y}^s \in \mathbb{R}^p \mid \mathbf{y}^s \ \textit{satisfies}\ (7.36), (\mathbf{x}^s, \mathbf{u}^s) \in \mathcal{Z}^s\}.$$

The control goal is tracking a desired set-point $\mathbf{r} \in \mathbb{R}^p$. However, in general, $\mathbf{r} \notin \mathcal{Z}^s_{\mathbf{y}}$. Therefore, the goal is to stabilise the optimal reachable output equilibrium that corresponds to a given set-point, which may not lie on the output equilibrium manifold $\mathcal{Z}^s_{\mathbf{y}}$. This is depicted in Figure 7.6 from Ref. [20], where the output equilibrium manifold $\mathcal{Z}^s_{\mathbf{y}}$, the closed-loop output and artificial equilibrium at time 0, the closed-loop output and artificial equilibrium at time t, the past T_{ini} measurements used for prediction at time t, the optimal reachable equilibrium $\mathbf{y}^{sr}$ and the set-point $\mathbf{r}$ are shown.

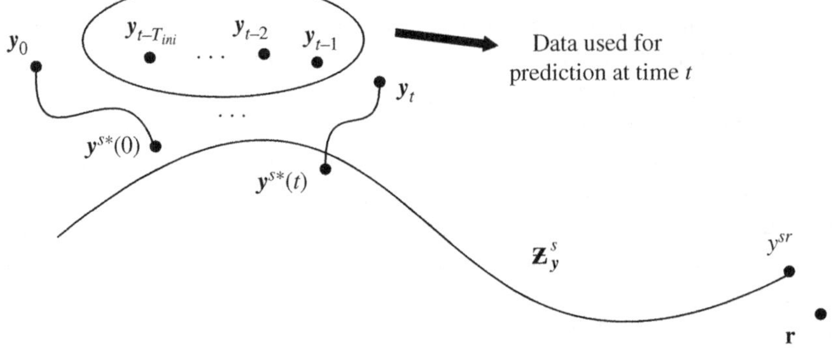

Figure 7.6 The schematic presentation of artificial steady state and artificial output.

Consider the following optimal control problem:

$$\min_{\mathbf{g},\sigma_{\mathbf{y}},\bar{\mathbf{u}},\bar{\mathbf{y}},\mathbf{u}^s,\mathbf{y}^s} \sum_{k=0}^{N-1} \|\bar{\mathbf{y}}_k(t) - \mathbf{y}^s(t)\|_{\mathbf{Q}}^2 + \|\bar{\mathbf{u}}_k(t) - \mathbf{u}^s(t)\|_{\mathbf{R}}^2 + \|\mathbf{y}^s(t) - \mathbf{r}_k(t)\|_{\mathbf{S}}^2$$

$$+ \lambda_{\mathbf{g}}\|\mathbf{g}(t)\|_2^2 + \lambda_\sigma\|\sigma_{\mathbf{y}}(t)\|_2^2 \tag{7.37a}$$

$$\text{s.t.} \quad \begin{bmatrix} \mathcal{H}_{N+T_{ini}}(\mathbf{u}_{[t-T,t-1]}) \\ \mathcal{H}_{N+T_{ini}}(\mathbf{y}_{[t-T,t-1]}) \end{bmatrix} \mathbf{g}(t) = \begin{bmatrix} \bar{\mathbf{u}}_{[-T_{ini},N-1]}(t) \\ \bar{\mathbf{y}}_{[-T_{ini},N-1]}(t) + \sigma_{\mathbf{y}}(t) \end{bmatrix} \tag{7.37b}$$

$$\begin{bmatrix} \bar{\mathbf{u}}_{[-T_{ini},-1]}(t) \\ \bar{\mathbf{y}}_{[-T_{ini},-1]}(t) \end{bmatrix} = \begin{bmatrix} \mathbf{u}_{[t-T_{ini},t-1]} \\ \mathbf{y}_{[t-T_{ini},t-1]} \end{bmatrix} \tag{7.37c}$$

$$\begin{bmatrix} \bar{\mathbf{u}}_{[N-T_{ini},N-1]}(t) \\ \bar{\mathbf{y}}_{[N-T_{ini},N-1]}(t) \end{bmatrix} = \begin{bmatrix} \mathbf{u}^s_{T_{ini}}(t) \\ \mathbf{y}^s_{T_{ini}}(t) \end{bmatrix} \tag{7.37d}$$

$$\sum_{i=0}^{T-T_{ini}-N} g_i(t) = 1, \bar{\mathbf{u}}_k(t) \in \mathbb{U}, \quad \mathbf{u}^s(t) \in \mathbb{U}^s, \qquad k = 0, \dots, N \tag{7.37e}$$

where the weighting matrices $\mathbf{Q}, \mathbf{R}, \mathbf{S} > \mathbf{0}$, the prediction horizon $N \geq 2n$ and the regularisation parameters $\lambda_{\mathbf{g}}, \lambda_\sigma > 0$. Also, $g_i(t)$ are the elements of the vector $\mathbf{g}(t)$. The optimal solution of (7.37) at time t is denoted by $\bar{\mathbf{u}}^*(t)$, $\bar{\mathbf{y}}^*(t)$, $\mathbf{g}^*(t)$, $\sigma_{\mathbf{y}}{}^*(t)$, $\mathbf{u}^{s*}(t)$, $\mathbf{y}^{s*}(t)$, and the closed-loop input, state and output at time t are $\mathbf{u}$, $\mathbf{x}$ and $\mathbf{y}$, respectively.

The key difference between (7.37) to the DeePC scheme considered in Section 7.6.1 is that the data used for prediction in the (7.37b) are updated online, thus providing a local linear approximation of the unknown nonlinear system (7.36).

Note that (7.37) contains regularisations of $\sigma_{\mathbf{y}}(t)$ and $\mathbf{g}(t)$, which are similar to the robust MPC problem (7.35) for LTI systems. This is because the error caused by the local linear approximation of (7.36) can also be viewed as output measurement noise similar to Section 7.6.2.

Also, (7.37) includes an artificial set-point $\mathbf{u}^s(t)$, $\mathbf{y}^s(t)$, which is optimised online and enters the terminal equality constraint (7.37d). The constraint (7.37d) is specified over T_{ini} steps such that $(\mathbf{u}^s(t), \mathbf{y}^s(t))$ is an (approximate) equilibrium of the system, and thus, the overall prediction horizon is of length N. At the same time, the distance of $\mathbf{y}^s(t)$ from the actual target set-point $\mathbf{r}(t)$ is penalised.

The idea of online optimisation of $\mathbf{u}^s(t)$, $\mathbf{y}^s(t)$ is inspired by the model-based tracking MPC [21], where artificial set-points can be used to increase the region of attraction or retain closed-loop properties despite online set-point changes. In the present problem setting, such an approach has the advantage that if $\mathbf{S}$ is sufficiently small, then the optimal artificial set-point $(\mathbf{u}^{s*}(t), \mathbf{y}^{s*}(t))$ appearing in the terminal equality constraint (7.37d) remains close to the optimal predicted input–output trajectory $(\overline{\mathbf{u}}^*(t), \overline{\mathbf{y}}^*(t))$. This means that the MPC first drives the system close to the steady-state manifold, where the linearity-based constraint (7.37b) is a good approximation of the nonlinear system dynamics (7.36), and therefore, the prediction error is small. Then, the artificial set-point is slowly shifted towards the target set-point $\mathbf{r}(t)$ along the steady-state manifold, and hence, the MPC also steers the closed-loop trajectory towards $\mathbf{r}(t)$. Finally, (7.37e) implies that the weighting vector $\mathbf{g}(t)$ sums up to 1. The explanation for this modification is that the linearisation of (7.36) at a point that is not a steady state of (7.36) generally leads to an affine system dynamic. To parametrise trajectories of an affine-based system on measured data, the constraint (7.37e) is necessary to ensure that the constant offset from the measured data is integrated into the predictions.

The data-driven predictive control algorithm for nonlinear systems based on local linearisation around the equilibrium points is presented in Algorithm 7.4. Note that Algorithm 7.4 allows controlling unknown nonlinear systems based only on measured data without any model knowledge, except for a potentially rough upper bound on the system order. It is worth noting that Algorithm 7.4 only requires solving the strictly convex quadratic programs (7.37) online, though the underlying control problem involves the nonlinear system (7.36). In Ref. [20], it is proved that under suitable assumptions on the design parameters and the initially collected data, and if the initial state is sufficiently close to the steady-state manifold, the closed-loop system under Algorithm 7.4 is practically exponentially stable. The concept of practical stability is useful for covering the cases where the performance is acceptable from a practical viewpoint, while it may show oscillations or lacks asymptotic stability in the strict mathematical sense. For an introduction to the subject refer to Ref. [22].

Algorithm 7.4 Nonlinear Data-driven Predictive Control Scheme

Input Select an upper bound on the system order n, prediction horizon N, cost matrices $\mathbf{Q}, \mathbf{R}, \mathbf{S} > \mathbf{0}$, regularisation parameters $\lambda_\mathbf{g}, \lambda_\sigma > 0$, constraint sets $\mathbb{U}, \mathbb{U}^s$, set-point $\mathbf{r}$ and generate data $\begin{bmatrix} \mathbf{u}_{[0,T-1]}^T & \mathbf{y}_{[0,T-1]}^T \end{bmatrix}^T$.

Step 1 Collect the past T_{ini} measurements $\begin{bmatrix} \mathbf{u}_{[t-T_{ini},t-1]}^T & \mathbf{y}_{[t-T_{ini},t-1]}^T \end{bmatrix}^T$.

Step 2 Solve (7.37).

Step 3 Apply the input sequence $\mathbf{u}_{[t,t+T_{ini}-1]}(t) = \overline{\mathbf{u}}^*_{[0,T_{ini}-1]}(t)$ over the next T_{ini} time steps.

Step 4 Set $t = t + T_{ini}$ and return to step 1.

Example 7.5 In this example, the proposed approach is applied for the control of a bi-linear model of a DC motor given in Ref. [23]:

$$\begin{aligned}
\dot{x}_1 &= -\left(\frac{R_a}{L_a}\right)x_1 - \left(\frac{k_m}{L_a}\right)x_2 u + \left(\frac{u_a}{L_a}\right) \\
\dot{x}_2 &= -\left(\frac{B}{J}\right)x_2 + \left(\frac{k_m}{J}\right)x_1 u - \left(\frac{\tau_l}{J}\right) \\
y &= x_2
\end{aligned} \tag{7.38}$$

where x_1 is the rotor current, x_2 the angular velocity and the control input u is the stator current and the output y is the angular velocity. The parameters are $L_a = 0.314$, $R_a = 12.345$, $k_m = 0.253$, $J = 0.00441$, $B = 0.00732$, $\tau_l = 1.47$ and $u_a = 60$. Notice in particular the bi-linearity between the state and the control input. The physical constraints on the control input are $\mathbb{U} \in [-1, 1]$. The goal is to design a nonlinear data-driven predictive controller based on Algorithm 7.4. The discrete-time nonlinear system is obtained by the Runge–Kutta four discretisation method with a discretisation period $T_s = 0.01s$. The control objective is to track a given angular velocity reference $\mathbf{r} = 0.4$, while satisfying the input constraints $\mathbb{U}, \mathbb{U}^s \in [-1, 1]$. To this end, an input sequence sampled uniformly from $[-1, 1]$ over the first $T = 100$ time steps is selected, where the system is initialised at $\mathbf{x}_0 = [0.887, 0.587]$. After that, for each $t \geq N$, we solve (7.37), apply the first component of the optimal predicted input, and update the data $\begin{bmatrix} \mathbf{u}_{[t-T_{ini},t-1]}^T & \mathbf{y}_{[t-T_{ini},t-1]}^T \end{bmatrix}^T$ used for prediction in (7.43b) in the next time step based on the current measurements. The design parameters are selected as follows:

$$\begin{aligned}
&\mathbf{Q} = 1e2\mathbf{I}_p, \quad \mathbf{S} = 1e4\mathbf{I}_p, \quad \mathbf{R} = 1\mathbf{I}_m, \lambda_\mathbf{g} = 1e-3, \lambda_\sigma = 1e5, \\
&T_{\text{ini}} = N, \quad N = 10 \text{ samples}
\end{aligned}$$

Note that updating the data used for prediction in (7.37b) is a critical component of the data-driven predictive control approach for nonlinear systems. In particular, if we do not update the data online but only use the first T input–output measurements $\begin{bmatrix} \mathbf{u}_{[0,T-1]}^T & \mathbf{y}_{[0,T-1]}^T \end{bmatrix}^T$ for prediction, the closed-loop does not

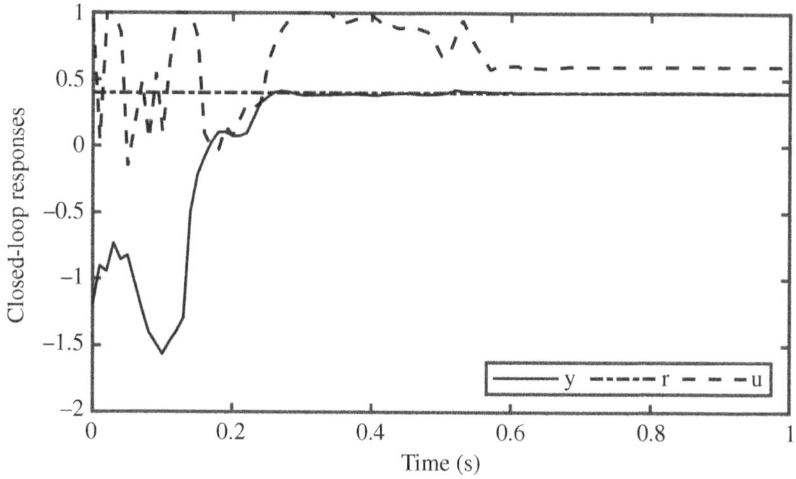

Figure 7.7 N-step data-enabled nonlinear predictive control responses.

converge to the desired set-point **r** and yields a significant permanent offset due to the model mismatch. The closed-loop response is shown in Figure 7.7.

7.7 Conclusion

Rooted in the behavioural system theory for discrete-time LTI systems, the Fundamental Lemma provides a framework for data-driven non-parametric system representation and control system design. In this chapter, the Fundamental Lemma is presented and proved. Based on this lemma, an equivalent data-based representation of LTI systems is derived, and different control systems designs, including state-feedback stabilisation, robust data-driven state-feedback stabilisation and data-driven predictive control systems entitled the DeePC algorithm, are derived. Extensions of data-driven predictive control for the cases of noisy measurements and nonlinear plants are provided. Avoiding detailed mathematical derivations, the basic principles are given with analogous algorithms to show the main steps of each design methodology. Finally, a data-driven predictive control for nonlinear systems is presented.

Problems

7.1 Discuss the minimality issue and the equivalence of controllability properties in the state-space and behavioural frameworks.
Hint: These issues are dealt with in Refs. [1, 2, 10].

7.2 Consider the state controllable LTI system $\mathbf{A}, \mathbf{B}$ with n states, and m inputs with the corresponding Hankel matrix given by

$$\begin{bmatrix} \mathcal{H}_1\left(\mathbf{x}_{[0,T-L]}\right) \\ \mathcal{H}_L\left(\mathbf{u}_{[0,T-1]}\right) \end{bmatrix} \tag{P7.2.1}$$

a) Show that the following equation holds for the Hankel matrix:

$$\mathcal{H}_1\left(\mathbf{x}_{[1,T-n-L+1]}\right) = [\mathbf{A}\,\mathbf{B}] \begin{bmatrix} \mathcal{H}_1\left(\mathbf{x}_{[0,T-n-L]}\right) \\ \mathcal{H}_1\left(\mathbf{u}_{[0,T-n-L]}\right) \end{bmatrix} \tag{P7.2.2}$$

b) Let $[\xi\ \boldsymbol{\eta}]$ be a vector in the left kernel of (P7.2.1), where $\boldsymbol{\eta}^T \in \mathbb{R}^{mL}$ and $\xi^T \in \mathbb{R}^n$. Show that for a *deeper* Hankel matrix of (P7.2.1) is given by

$$\begin{bmatrix} \mathcal{H}_1\left(\mathbf{x}_{[0,T-n-L]}\right) \\ \mathcal{H}_{L+n}\left(\mathbf{u}_{[0,T-1]}\right) \end{bmatrix} \tag{P7.2.3}$$

The following $n+1$ vectors derived from $[\xi\ \boldsymbol{\eta}]$ all belong to the left kernel of (P7.2.3):

$$\begin{aligned} \boldsymbol{w}_0 &\triangleq [\xi\ \boldsymbol{\eta}\ 0_{nm}] \\ \boldsymbol{w}_1 &\triangleq [\xi\mathbf{A}\ \ \xi\mathbf{B}\ \boldsymbol{\eta}\ 0_{(n-1)m}] \\ \boldsymbol{w}_2 &\triangleq [\xi\mathbf{A}^2\ \ \xi\mathbf{A}\mathbf{B}\ \xi\mathbf{B}\ \boldsymbol{\eta}\ 0_{(n-2)m}] \\ &\vdots \\ \boldsymbol{w}_n &\triangleq [\xi\mathbf{A}^n\ \ \xi\mathbf{A}^{n-1}\mathbf{B}\cdots\xi\mathbf{B}\ \boldsymbol{\eta}] \end{aligned} \tag{P7.2.4}$$

c) Show that for a persistently exciting input sequence of order $L+n$, $\mathbf{u}_{[0,T-1]}$, the vectors (P7.2.4) are linearly dependent.

d) Show that the linear dependence of the vectors (P7.2.4) implies that $\boldsymbol{\eta}$ must be the zero vector.

e) Prove that the Hankel matrix (P7.2.1) has full rank.
Hint: Employ the above results and the Cayley-Hamilton theorem that the matrix $\mathbf{A}$ satisfies its characteristic equation, that is $\sum_{i=0}^{n} \alpha_i \mathbf{A}^i = 0$, where α_i are the real coefficients of the characteristic equation of $\mathbf{A}$ with $\alpha_n = 1$. Let $\boldsymbol{v} = \sum_{i=0}^{n} \alpha_i\ \boldsymbol{w}_i$, and start the proof.

7.3 Consider an unstable discrete time reactor plant given by the following equations:

$$\mathbf{x}_{k+1} = \begin{bmatrix} 0.7792 & -2.1688 & -0.00372 \\ 0.0434 & 1.3901 & 0.0397 \\ 0.0037 & 0.1983 & 0.6738 \end{bmatrix} \mathbf{x}_k + \begin{bmatrix} 0.1786 & -0.0025 \\ 0.0042 & 0.0040 \\ 0.0003 & 0.1651 \end{bmatrix} \mathbf{u}_k$$

a) Generate a trajectory data $\mathbf{x}_{[1,T]}$ for zero initial conditions and a random input sequence of length $T = 20$, and identify a state-space realisation of the reactor. Compare the actual and identified state responses using simulation results.

b) Design a data-driven state feedback controller to stabilise the unstable reactor.

c) Determine the closed-loop eigenvalues and show the closed-loop responses.

7.4 The state-space equation of a plant is given as follows:

$$\mathbf{x}_{k+1} = \begin{bmatrix} 1 & 0.25 & 0.0036 \\ 0 & 1 & 0.029 \\ 0 & 0 & 1 \end{bmatrix} \mathbf{x}_k + \begin{bmatrix} 0.36 \\ 2.9 \\ 1 \end{bmatrix} \mathbf{u}_k$$

$$\mathbf{y}_k = \begin{bmatrix} 1.7 & 0 & 0 \end{bmatrix} \mathbf{x}_k$$

The initial conditions are assumed zero.

a) Apply the DeePC algorithm to stabilise the plant. The prediction horizon is $N = 30$, and the design parameters are selected as follows: $\mathbf{Q} = 10\mathbf{I}_p, \mathbf{R} = 1\mathbf{I}_m, T_{ini} = N$

b) Apply the DeePC algorithm in the presence of a uniformly distributed additive measurement noise with a bound of 0.002. The prediction horizon is $N = 30$, and the design parameters are selected as follows:

$$\mathbf{Q} = 10\mathbf{I}_p, \mathbf{R} = 1\mathbf{I}_m, \lambda_g = 50, \lambda_\sigma = 1000, T_{ini} = 10$$

Study the effect of different design parameters on the closed-loop response.

7.5 The following nonlinear state-space model of a drum boiler is given [24]:

$$\begin{bmatrix} \dot{x}_1(t) \\ \dot{x}_2(t) \end{bmatrix} = \begin{bmatrix} 0.014u_1(t) - 0.0033(x_1(t))^{1.125}u_2(t) \\ 0.014(x_1(t))^{1.125}u_2(t) - 0.064x_2(t) \end{bmatrix}$$

$$y(t) = 0.3(x_1(t))^{1.125}u_2(t) + 0.737x_2(t) + e(t)$$

where $x_1(t)$ is the drum pressure in kg/m^2, $x_2(t)$ is the reheater pressure in kg/m^2, $u_1(t)$ is the fuel flow in tons/h, $u_2(t)$ is a fractional control valve position and $y(t)$ is the output power in MW. The initial conditions are randomly selected.

a) Collect 1000 input–output data samples from the plant with the valve position fixed at 50%, and the fuel flow rate varied between 0 and 150 tons/h. The measurement noise is white with a standard deviation of 1 MW.

b) Design a nonlinear data-driven predictive controller for the boiler using Algorithm 7.4. Study the effect of the prediction horizon N, and the cost matrices $\mathbf{Q}, \mathbf{R}, \mathbf{S} > \mathbf{0}$ on the system closed-loop response.

References

1 I. Markovsky, J. C. Willems, S. Van Huffel, and B. De Moor, *Exact and Approximate Modeling of Linear Systems: A Behavioral Approach*. SIAM, 2006.

2 I. Markovsky and F. Dörfler, "Behavioral systems theory in data-driven analysis, signal processing, and control," *Annual Reviews in Control*, vol. 52, pp. 42–64, 2021.

3 J. C. Willems, P. Rapisarda, I. Markovsky, and B. L. De Moor, "A note on persistency of excitation," *Systems & Control Letters*, vol. 54, no. 4, pp. 325–329, 2005.

4 C. De Persis and P. Tesi, "Formulas for data-driven control: stabilization, optimality, and robustness," *IEEE Transactions on Automatic Control*, vol. 65, no. 3, pp. 909–924, 2019.

5 H. J. van Waarde, C. De Persis, M. K. Camlibel, and P. Tesi, "Willems' fundamental lemma for state-space systems and its extension to multiple datasets," *IEEE Control Systems Letters*, vol. 4, no. 3, pp. 602–607, 2020.

6 C. Verhoek, R. Tóth, S. Haesaert, and A. Koch, "Fundamental lemma for data-driven analysis of linear parameter-varying systems," in *2021 60th IEEE Conference on Decision and Control (CDC)*, 2021: IEEE, pp. 5040–5046.

7 J. Berberich, J. Köhler, M. A. Müller, and F. Allgöwer, "Data-driven model predictive control: closed-loop guarantees and experimental results," *at-Automatisierungstechnik*, vol. 69, no. 7, pp. 608–618, 2021.

8 M. Guo, C. De Persis, and P. Tesi, "Data-driven stabilization of nonlinear polynomial systems with noisy data," *IEEE Transactions on Automatic Control*, 2021. vol. 67, no. 8, pp. 4210–4217.

9 Y. Lian, R. Wang, and C.N. Jones, "Koopman based data-driven predictive control," 2021. *arXiv preprint arXiv:2102.05122*.

10 J. C. Willems, "Paradigms and puzzles in the theory of dynamical systems," *IEEE Transactions on Automatic Control*, vol. 36, no. 3, pp. 259–294, 1991.

11 P. Suetin, A. I. Kostrikin, and Y. I. Manin, *Linear Algebra and Geometry*. CRC Press, 1989.

12 K. Ogata, *Discrete-time Control Systems*. Prentice-Hall, Inc., 1995.

13 S. Boyd, L. El Ghaoui, E. Feron, and V. Balakrishnan, *Linear Matrix Inequalities in System and Control Theory*. SIAM, 1994.

14 J. Berberich, A. Koch, C. W. Scherer, and F. Allgöwer, "Robust data-driven state-feedback design," in *2020 American Control Conference (ACC)*, 2020: IEEE, pp. 1532–1538.

15 I. Khalil, J. Doyle, and K. Glover, *Robust and Optimal Control*. Prentice Hall, 1996.

16 C. W. Scherer, "Robust mixed control and linear parameter-varying control with full block scalings," in *Advances in Linear Matrix Inequality Methods in Control*, 2000: SIAM, pp. 187–207.

17 J. Coulson, J. Lygeros, and F. Dörfler, "Data-enabled predictive control: in the shallows of the DeePC," in *2019 18th European Control Conference (ECC)*, 2019: IEEE, pp. 307–312.

18 L. Huang, J. Coulson, J. Lygeros, and F. Dörfler, "Decentralized data-enabled predictive control for power system oscillation damping," *IEEE Transactions on Control Systems Technology*, vol. 30, no. 3, pp. 1065–1077, 2021.

19 J. Berberich, J. Köhler, M. A. Müller, and F. Allgöwer, "Data-driven model predictive control with stability and robustness guarantees," *IEEE Transactions on Automatic Control*, vol. 66, no. 4, pp. 1702–1717, 2020.

20 J. Berberich, J. Köhler, M. A. Muller, and F. Allgower, "Linear tracking MPC for nonlinear systems Part II: The data-driven case," *IEEE Transactions on Automatic Control*, vol. 67, no. 9, pp. 4406–4421, 2022.

21 D. Limón, I. Alvarado, T. Alamo, and E. F. Camacho, "MPC for tracking piecewise constant references for constrained linear systems," *Automatica*, vol. 44, no. 9, pp. 2382–2387, 2008.

22 V. Lakshmikantham, S. Leela, and A. A. Martynyuk, *Practical Stability of Nonlinear Systems*. World Scientific, 1990.

23 M. Korda and I. Mezić, "Linear predictors for nonlinear dynamical systems: Koopman operator meets model predictive control," *Automatica*, vol. 93, pp. 149–160, 2018.

24 T. Wigren, "Recursive identification of a nonlinear state space model," *International Journal of Adaptive Control and Signal Processing*, vol. 37, no. 2, pp. 447–473, 2021

8

Koopman Theory and Data-driven Control System Design of Nonlinear Systems

8.1 Introduction

This chapter presents a class of data-driven controllers based on Koopman theory and the Fundamental Lemma presented in Chapter 7. Koopman's theory provides an effective tool for handling nonlinear dynamical systems by *lifting* or *embedding* the nonlinear dynamics into a higher dimensional space where its evolution is approximately linear [1]. On the other hand, as discussed in Chapter 7, according to the Fundamental Lemma, persistently exciting data can represent the input–output behaviour of a linear system without needing to identify the linear system's matrices [2]. Hence, inspired by the Koopman theory and application of Willems' Fundamental Lemma, a class of data-driven control methods has been developed for unknown systems. Finally, motivated by these two ideas and the broad applicability of predictive controllers, a *data-driven Koopman-based predictive control* scheme is presented in this chapter. Koopman introduced his operator-theoretic perspective of dynamical systems in 1931, which provides a powerful tool to transform a nonlinear dynamical system into an infinite-dimensional, linear system [3]. However, Koopman's linearisation theory was not seriously noticed by engineers for more than eight decades. The main obstacle to applying the Koopman theory in practice was the fact that there did not exist an efficient method to compute the Koopman operators. However, with the development of the efficient numerical method of dynamic mode decomposition (DMD) algorithm, computation of the Koopman operator became readily available.

Koopman theory provides a platform for linear analysis and design to become feasible even for strongly nonlinear dynamical systems via the infinite-dimensional linear Koopman operator. The Koopman operator can completely characterise the nonlinear system's behaviour, and with effective

An Introduction to Data-Driven Control Systems, First Edition. Ali Khaki-Sedigh.

coordinate transformations, the nonlinear dynamics would appear linear [4]. Koopman's theory is treated in more detail in Ref. [5]. Following the application of Koopman's operator, the vast literature on linear control theory can be further employed to control strongly nonlinear systems. This practical application potential of Koopmans' approach was revealed in Ref. [6], where the usefulness of these linear representations in prediction and control applications is shown. As previously mentioned, interest in the Koopman approach was elevated after presenting the DMD algorithm in Ref. [7]. Theoretically, discrete Fourier transform (DFT) and the singular value decomposition (SVD) are the foundations of DMD. The DMD algorithm provides a finite-dimensional approximation of the Koopman operator and provides a platform for mathematical analysis of the Koopman approach. Hence, the spectral properties, including eigenvalues, eigenfunctions and Koopman modes, can be utilised for closed-loop system analysis [5, 8, 9]. However, obtaining Koopman eigenfunctions for even the simplest systems can be significantly difficult and often complex or uninterpretable. Also, the computation of Koopman linear state-space model matrices is an ill-conditioned problem because its solution is not unique [10]. However, trying to find the Koopman eigenfunctions is a one-time cost that leads to a global linear description. Obtaining proper Koopman eigenfunctions from data, or analytically, is a major challenge. An extensive presentation of different aspects of the Koopman theory for control is provided in Ref. [11].

8.2 Fundamentals of Koopman Theory for Data-driven Control System Design

8.2.1 Basic Concepts and Definitions

Consider the following discrete-time nonlinear system:

$$\mathbf{x}_{k+1} = \mathbf{F}(\mathbf{x}_k) \tag{8.1}$$

where $\mathbf{x}_k \in \boldsymbol{X} \subseteq \mathbb{R}^n$ is the system state, $\boldsymbol{X}$ is the state space and $\mathbf{F}$ is a vector field describing the nonlinear dynamics that can, in general, encompass discontinuous and hybrid systems. It is desired to find a new vector of coordinates $\mathbf{z}$ such that

$$\mathbf{z} = \boldsymbol{\varphi}(\mathbf{x}) \tag{8.2}$$

where the system dynamics are linearised as follows:

$$\mathbf{z}_{k+1} = \boldsymbol{K}\mathbf{z}_k \tag{8.3}$$

and the dynamics of $\mathbf{z}$ are entirely determined by the matrix $\boldsymbol{K}$.

Let $\psi: \boldsymbol{X} \rightarrow \mathbb{R}$ belong to the set of *measurement* or *output functions*, also called an *observable function*[1] in Koopman theory, where the observable functions $\psi(\mathbf{x})$ belong to the Hilbert function space $\mathcal{F}$. The infinite-dimensional discrete-time Koopman operator $\mathcal{K}_F$ associated with (8.1) for an *observable* function $\psi(\mathbf{x}_k)$ is defined through the following composition[2] that advances the measurement functions:

$$\mathcal{K}_F \psi(\mathbf{x}_k) := \psi(\mathbf{F}(\mathbf{x}_k)) = \psi(\mathbf{x}_{k+1}) \tag{8.4}$$

Remark 8.1 It can easily be shown that the Koopman operator is *linear* when the set of observable functions is a linear (vector) space of linear or nonlinear functions [4]. This linearity property holds regardless of the nonlinearity of the dynamical system under study.

Remark 8.2 The Koopman operator results in a dynamical system described by

$$\psi_{k+1} = \mathcal{K}_F \psi_k \tag{8.5}$$

analogous to (8.1), except that (8.5) is linear and infinite-dimensional.

Although the Koopman operator is linear, its infinite dimensionality makes it impractical for engineering and, in particular, control systems design purposes. To overcome this problem, a practical strategy is to identify certain key measurement functions that linearly evolve with the system's dynamics and capture the dominant behaviour of the system's dynamics. In fact, the subsequently defined Koopmans's eigenfunctions are a practical set of measurements that behave linearly in time and are suited for further control design purposes. The eigenfunctions and eigenvalues of the Koopman operator are called *Koopman eigenfunctions* and *Koopman eigenvalues*, and for the discrete-time system (8.1), the eigenfunction φ of $\mathcal{K}_F$ and the corresponding eigenvalue λ, are defined as follows:

$$\mathcal{K}_F \varphi(\mathbf{x}_k) = \lambda \varphi(\mathbf{x}_k) = \varphi(\mathbf{x}_{k+1}) \tag{8.6}$$

As the Koopman operator is infinite-dimensional, there will be an infinite number of Koopman eigenfunctions and eigenvalues for a system. However, to construct the Koopman finite dimensional linear state-space model, only a subset of the eigenfunctions φ_1 to φ_p with a finite p is sufficient.

For the continuous-time plants' analysis and the corresponding Koopman eigenfunctions, refer to [5]. A schematic diagram of the Koopman operator for nonlinear plants adopted from [5] is shown in Figure 8.1. Figure 8.1(a) represents the advancement of the state vector of the nonlinear plant in the state space governed

1 Note that this is not related to the observability concept from control theory and is borrowed initially from quantum mechanics.
2 Koopman operator is a composition operator. The composition operator is a linear operator denoted by $\circ$ and is defined by the rule $\mathcal{K}_F \psi = \psi \circ F$.

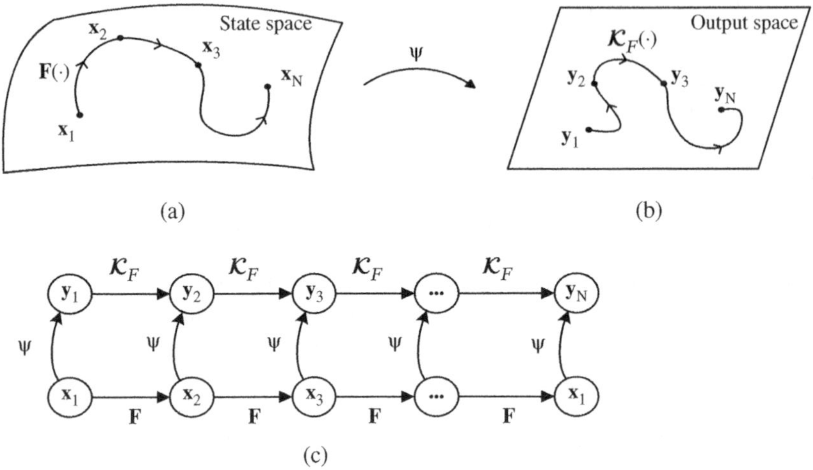

Figure 8.1 Schematic diagram of the Koopman operator in action.

by (8.1). Figure 8.1(b) shows the corresponding output vectors that are governed by (8.4) and $\mathbf{y}_k = \psi(\mathbf{x}_k)$. Figure 8.1(c) depicts the whole process of the state and output vector advancements by the plant dynamics and the Koopman operators, respectively, and the observables acting on the states to produce the outputs.

Remark 8.3 In the case of linear discrete-time systems,

$$\mathbf{x}_{k+1} = \mathbf{A}\mathbf{x}_k \tag{8.7}$$

where $\mathbf{x}_k \in \mathbb{R}^n$, let λ_i and $\mathbf{w}_i^T$ be the ith eigenvalue and the corresponding left eigenvector. Then, the spectrum of the Koopman operator contains these eigenvalues, and the corresponding eigenfunctions are given by

$$\mathcal{K}_F\varphi_{\lambda_i}(\mathbf{x}_k) = \varphi_{\lambda_i}(\mathbf{A}\mathbf{x}_k) = \mathbf{w}_i^T\mathbf{A}\mathbf{x}_k = \lambda_i\mathbf{w}_i^T\mathbf{x}_k = \lambda_i\varphi_{\lambda_i}(\mathbf{x}_k) \tag{8.8}$$

We have, $\varphi_{\lambda_i}(\mathbf{x}_k) = \mathbf{w}_i^T\mathbf{x}_k$ is the associated eigenfunctions.

The eigenfunctions and eigenvalues of the Koopman operator for continuous-time plants are similarly defined. Reference [11] presents the mathematical developments for both discrete-time and continuous-time plants. It is shown in Ref. [11] that for a sufficiently smooth continuous-time nonlinear plant given by

$$\dot{\mathbf{x}} = \mathbf{F}(\mathbf{x})$$

where $\mathbf{x} \in \boldsymbol{X} \subseteq \mathbb{R}^n$ is the system state, the equation governing the eigenfunction and eigenvalue of the corresponding Koopman operator is given as follows:

$$\mathbf{F} \cdot \nabla\varphi_{\lambda_i} = \lambda_i\varphi_{\lambda_i} \tag{8.9}$$

where $\cdot$ denotes the inner product and ∇ denotes the gradient operator.

Example 8.1 Consider the nonlinear system given by $\dot{x} = F(x) = -x - x^3, x \in \mathbb{R}$. It can be shown that $\varphi_\lambda = x/\sqrt{1+x^2}$ is a corresponding eigenfunction of the Koopman operator. Note that, $\nabla\varphi_\lambda = 1/\left((1+x^2)\sqrt{1+x^2}\right)$, and we have

$$-x(1+x^2)\frac{1}{(1+x^2)\sqrt{1+x^2}} = \lambda\frac{x}{\sqrt{1+x^2}}$$

which gives $\lambda = -1$.

Remark 8.4 Characterising the Koopman operator and approximating its eigenvalues and eigenfunctions, or spectral decomposition from measurement data, has been a major research interest in the Koopman theory [4]. There are several approaches to deriving the Koopman eigenfunctions. Generalised Laplace analysis (GLA) in Ref. [11] is presented for such derivation. However, GLA can lead to stability-related numerical problems. An alternative line of algorithms called the DMD algorithm, has also been advanced, enabling concurrent data-driven determination of approximate eigenvalues and eigenvectors of the underlying DMD operator. This approach is further discussed in Section 8.2.3.

Observables or measurement functions can be described as linear combinations of eigenfunctions, as the eigenfunctions span the space of observables. In the case of multiple measurements of a system, such as multiple outputs or when the system state variables are measured, we have

$$\boldsymbol{\Psi}(\mathbf{x}) = [\psi_1(\mathbf{x})\,\psi_2(\mathbf{x})\ldots\psi_p(\mathbf{x})]^T = \sum_{j=1}^{\infty}\varphi_j(\mathbf{x})\mathbf{v}_j \tag{8.10}$$

where $\mathbf{v}_j$ is called the jth *Koopman mode* associated with the eigenfunction $\varphi_j(\mathbf{x})$. Note that each of the individual measurements $\psi_i(\mathbf{x})$ can also be expanded in terms of the eigenfunctions $\varphi_j(\mathbf{x})$, that is $\psi_i(\mathbf{x}) = \sum_{j=1}^{\infty} v_{ij}\varphi_j(\mathbf{x})$ for appropriate constants v_{ij} and $i = 1, \ldots, p$. Hence, the dynamics of the measurements $\boldsymbol{\Psi}(\mathbf{x})$ can be written as follows:

$$\boldsymbol{\Psi}(\mathbf{x}_{k+1}) = \mathcal{K}_F\boldsymbol{\Psi}(\mathbf{x}_k) = \mathcal{K}_F\sum_{j=1}^{\infty}\varphi_j(\mathbf{x}_k)\mathbf{v}_j = \sum_{j=1}^{\infty}\mathcal{K}_F\varphi_j(\mathbf{x}_k)\mathbf{v}_j = \sum_{j=1}^{\infty}\lambda_j\varphi_j(\mathbf{x}_k)\mathbf{v}_j \tag{8.11}$$

The triple $\{\lambda_j, \varphi_j(\mathbf{x}), \mathbf{v}_j\}$ *for* $j = 1, \ldots, \infty$, is called the *Koopman mode decomposition* [12].

Remark 8.5 It should be noted that Koopman eigenfunctions and eigenvalues depend only on the dynamics of the system (8.1), and Koopman modes are specific to each observable of the nonlinear system.

Although the linearity of the Koopman operator is an attractive feature, its infinite dimensionality poses serious representational and computational challenges. It is, therefore, necessary to consider a finite-dimensional approximation of the Koopman operator for control design purposes.

8.2.2 Finite-dimensional Koopman Linear Model Approximation

Consider an p-dimensional linear subspace of the space of real-valued observables $\mathbf{\Psi}(\mathbf{x})$, spanned by the Koopman eigenfunctions $\{\varphi_1(\mathbf{x})\,\varphi_2(\mathbf{x})\ldots\varphi_p(\mathbf{x})\}$, where $p \gg n$. Let a finite-dimensional approximation of the Koopman operator $\mathcal{K}_F$ be denoted by $\mathcal{K}_{F_p}$. Then, it can be shown that there exists a matrix representation of $\mathcal{K}_{F_p}$, called the *Koopman Matrix* of the system, $\boldsymbol{K}$, that corresponds to the action of the Koopman operator on observables in the considered finite-dimensional subspace [11]. Let $\mathbf{\Phi}^T(\mathbf{x}) = \left[\varphi_1(\mathbf{x})\ \varphi_2(\mathbf{x})\ \ldots\ \varphi_p(\mathbf{x})\right]$, then the Koopman *lifted state* is defined as

$$\mathbf{z} = \mathbf{\Phi}(\mathbf{x}) = \begin{bmatrix} \varphi_1(\mathbf{x}) \\ \vdots \\ \varphi_p(\mathbf{x}) \end{bmatrix} \tag{8.12}$$

and the approximated Koopman *lifted dynamics* is

$$\mathbf{z}_{k+1} = \mathbf{A}\mathbf{z}_k \tag{8.13}$$

where, $\mathbf{A} = \boldsymbol{K}^T$ [11]. The system outputs are in the control design applications of the observables for some matrix $\mathbf{C} \in \mathbb{R}^{l \times p}$, as follows:

$$\mathbf{y}_k = \mathbf{C}\mathbf{z}_k \tag{8.14}$$

The results can be extended to systems with controlled inputs $\mathbf{u}_k$ and disturbances $\mathbf{w}_k$ as follows

$$\mathbf{x}_{k+1} = \mathbf{F}(\mathbf{x}_k, \mathbf{u}_k, \mathbf{w}_k) \tag{8.15}$$

In such cases, the Koopman operator can be approximated in a finite-dimensional subspace spanned by the eigenfunctions $\{\varphi_1(\mathbf{x}, \mathbf{u}, \mathbf{w})\ \varphi_2(\mathbf{x}, \mathbf{u}, \mathbf{w})\ldots \varphi_p(\mathbf{x}, \mathbf{u}, \mathbf{w})\}$. The lifted dynamics, in this case, is

$$\mathbf{z}_{k+1} = \mathbf{A}\mathbf{z}_k + \mathbf{B}\mathbf{u}_k + \mathbf{D}\mathbf{w}_k \tag{8.16}$$

where $\mathbf{z}$ is similarly defined and $\mathbf{A} \in \mathbb{R}^{p \times p}$, $\mathbf{B} \in \mathbb{R}^{p \times m}$, and $\mathbf{D} \in \mathbb{R}^{p \times d}$. The output vector can be similarly defined as (8.14). The Koopman model (8.16) incorporates the dominant features of the original nonlinear system (8.15). But the question that needs to be answered is how to obtain the Koopman linear model matrices (**A**, **B**, **D**) and the lifted state $\mathbf{\Phi}$ in (8.12), which is considered in the following example for a continuous time nonlinear system. The following example is from Ref. [11].

Example 8.2 Consider the following nonlinear system:

$$\dot{\mathbf{x}} = \begin{bmatrix} \rho x_1 \\ \mu\left(x_2 - x_1^2\right) \end{bmatrix}$$
$$y = \mathbf{h}(\mathbf{x}) = x_2 + x_1^2$$

Let's calculate a finite-dimensional Koopman linear model for the given nonlinear system. It can be shown that ρ, μ are Koopman eigenvalues with eigenfunctions $\varphi_\rho(\mathbf{x}) = x_1$, and $\varphi_\mu(\mathbf{x}) = x_2 - \alpha x_1^2$, respectively. Note that in this example, $\nabla\varphi_\rho(\mathbf{x}) = [1\ 0]^T$, and $\nabla\varphi_\mu(\mathbf{x}) = [-2\alpha x_1\ 1]^T$ and substituting these equations in (8.9) gives $\alpha = \frac{\mu}{\mu-2\rho}$. Also, note that 2ρ, $\rho+\mu$, etc. are the Koopman eigenvalues with the following corresponding eigenfunctions: φ_ρ^2, $\varphi_\rho\varphi_\mu$, etc. Let $\varphi_1 = \varphi_\rho$, $\varphi_2 = \varphi_\mu$ and $\varphi_3 = \varphi_\rho^2$. Then, it follows that

$$\mathbf{x} = \varphi_1 \times \begin{bmatrix} 1 \\ 0 \end{bmatrix} + \varphi_2 \times \begin{bmatrix} 0 \\ 1 \end{bmatrix} + \varphi_3 \times \begin{bmatrix} 0 \\ \alpha \end{bmatrix}$$
$$\mathbf{h}(\mathbf{x}) = \varphi_1 \times 0 + \varphi_2 \times 1 + \varphi_3 \times (1+\alpha)$$

then the Koopman *lifted state* is defined as

$$\mathbf{z} = \begin{bmatrix} z_1 \\ z_2 \\ z_3 \end{bmatrix} = \begin{bmatrix} \varphi_1 \\ \varphi_2 \\ \varphi_3 \end{bmatrix} = \begin{bmatrix} x_1 \\ x_2 - \alpha x_1^2 \\ x_1^2 \end{bmatrix}$$

and the Koopman linear model matrices are as follows:

$$\mathbf{A} = \begin{bmatrix} \rho & 0 & 0 \\ 0 & \mu & 0 \\ 0 & 0 & 2\rho \end{bmatrix}, \mathbf{C} = \begin{bmatrix} 0 & 1 & 1+\alpha \end{bmatrix}$$

The responses of the nonlinear system and the equivalent Koopman lifted state space model for $\rho = -0.6$, $\mu = -4$, and the initial condition $x_1(0) = -3$ and $x_2(0) = 1$ are shown in Figure 8.2.

However, obtaining Koopman eigenfunctions analytically for even the simplest systems can be significantly difficult. In practice, data-driven methods can be used to construct the Koopman eigenfunctions. DMD and extended dynamic mode decomposition (EDMD) are the two primarily used data-driven strategies to approximate the Koopman linear model for a nonlinear system.

8.2.3 Approximating the Koopman Linear Model from Measured Data: The DMD Approach

There are several model-based approaches to deriving the Koopman operator for a nonlinear system. The model-based approaches produce approximate

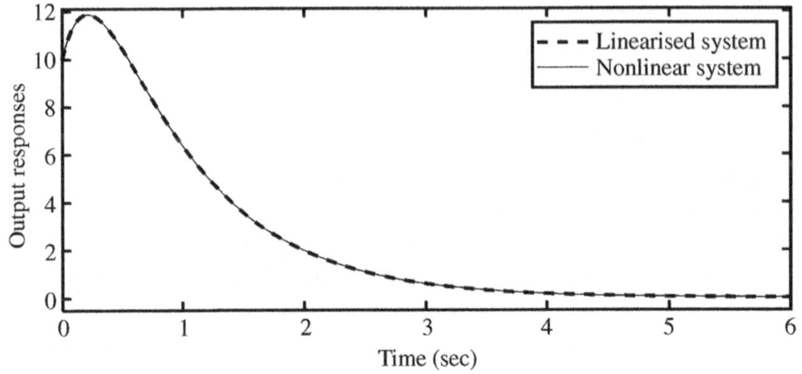

Figure 8.2 The output responses of the nonlinear system and its linear equivalent model.

finite-dimensional Koopman operators [4, 11, 12]. However, approximating the Koopman operator from the measured data is vital for data-driven control of unknown nonlinear systems.

The data-driven methods of the Koopman operator approximation are utilised to obtain the spectral properties of the unknown system dynamics, that is, the Koopman mode decomposition and the Koopman eigenfunctions. The acquired information is used to develop a finite-dimensional matrix approximation of the Koopman operator. DMD and EDMD are widely used standard algorithms to approximate the Koopman operator from measured data. The DMD technique was introduced in Koopman's theory in Ref. [13].

DMD employs the powerful and computationally efficient SVD technique. SVD is a well-known applied tool from linear algebra in control theory. It is effectively used for order reduction in high-dimensional data or mathematical models. The SVD orders modes in descending order of their role in a dynamic system and how much of the variance of the original data is captured by each mode. By truncating the SVD, large-energy content modes are retained, and the low-energy modes are discarded.

To briefly introduce SVD, let $\boldsymbol{A}$ be any $l \times m$ complex matrix. Then, it can be decomposed as $\boldsymbol{A} = \boldsymbol{U\Sigma V}^*$, called the SVD of $\boldsymbol{A}$. Where $\boldsymbol{U}$ and $\boldsymbol{V}$ are $l \times l$ and $m \times m$ unitary matrices and * denotes the conjugate transpose of a matrix. Hence, $\boldsymbol{U}^{-1} = \boldsymbol{U}^*$ and $\boldsymbol{V}^{-1} = \boldsymbol{V}^*$. The matrix $\boldsymbol{\Sigma}$ contains a diagonal matrix $\boldsymbol{\Sigma}_1$ of real, non-negative singular values of $\boldsymbol{A}$ arranged in descending order. If $k = min(l, m)$, then $\boldsymbol{\Sigma}_1 = \text{diag}(\sigma_1, \sigma_2, \cdots \sigma_k)$ and for $l \geq m$, we will have $\boldsymbol{\Sigma} = \begin{bmatrix} \boldsymbol{\Sigma}_1 \\ \boldsymbol{0} \end{bmatrix}$, and for $l \leq m$, we will have $\boldsymbol{\Sigma} = \begin{bmatrix} \boldsymbol{\Sigma}_1 & \boldsymbol{0} \end{bmatrix}$. The largest and smallest singular values of $\boldsymbol{A}$ are denoted as $\overline{\sigma} = \sigma_1$ and $\underline{\sigma} = \sigma_k$. It can be easily shown that singular values of $\boldsymbol{A}$ are the positive roots of the k largest eigenvalues of $\boldsymbol{A}\boldsymbol{A}^*$ and $\boldsymbol{A}^*\boldsymbol{A}$, and the columns of $\boldsymbol{U}$ and $\boldsymbol{V}$ are the eigenvectors of $\boldsymbol{A}\boldsymbol{A}^*$ and $\boldsymbol{A}^*\boldsymbol{A}$, respectively. For more information and applications of SVD in model-based control theory, refer to Ref. [14] for example.

The DMD modes are linear combinations of the selected SVD modes to extract structures with the same coherent linear behaviour over time.

The main attractive features of DMD are summarised as follows [4]:

- It approximates the best-fit linear matrix operator that advances high-dimensional measurements of a system forward in time.
- It approximates the Koopman operator restricted to the measurement subspace given by direct measurements of the state of a system.
- It is based entirely on measured data and does not require knowledge of the governing equations.
- It is naturally formulated in tractable linear algebra and can straightforwardly lead to practical results in control system theory and design.

Consider the discrete-time dynamical system (8.1). Assume that:

- A sequence of *snapshots* $\mathbf{x}_k$ of the dynamical system (8.1) is available.
- The system's dynamics $\mathbf{F}$ is unknown.
- The initial state is given, and the state variables are measurable.

The DMD technique finds the best-fit linear operator $\mathbf{A}$ that captures the evolution of the state vector $\mathbf{x}_k \in \mathbb{R}^n$ of the dynamical system (8.1). Assume that the available state vector measurements of the system are $\{\mathbf{x}_i, i = 1, \ldots, N_s\}$, where N_s is the number of samples. Assume for each snapshot $\mathbf{x}_i$ there is a corresponding snapshot $\mathbf{x}_i^+$ one-time step in the future, then form the two data matrices from the snapshots as follows:

$$\mathbf{X} = [\mathbf{x}_1\ \mathbf{x}_2 \ldots\ \mathbf{x}_{N_s}] \text{ and } \mathbf{X}^+ = \left[\mathbf{x}_1^+\ \mathbf{x}_2^+ \ldots \mathbf{x}_{N_s}^+\right]$$

The system given by (8.1) can be approximated in terms of the above data matrices as

$$\mathbf{X}^+ \approx \mathbf{A}\mathbf{X} \tag{8.17}$$

Finding the best-fit matrix $\mathbf{A}$ that approximates the subsequent sampling time measurement can be formulated as an optimisation problem

$$\mathbf{A} = \text{argmin}_{\mathbf{A}} \|\mathbf{X}^+ - \mathbf{A}\mathbf{X}\|_F = \mathbf{X}^+\mathbf{X}^\dagger \tag{8.18}$$

where $\|.\|_F$ is the Frobenius norm and $\dagger$ denotes the pseudo-inverse. The SVD of $\mathbf{X}$ gives $\mathbf{X} = \mathbf{U}\boldsymbol{\Sigma}\mathbf{V}^*$, where $\mathbf{U}$ and $\mathbf{V}$ are $n \times n$ unitary matrices, $\boldsymbol{\Sigma}$ is the matrix of singular values and * denotes the complex conjugate transpose. The pseudo-inverse of $\mathbf{X}$ is given by $\mathbf{X}^\dagger = \mathbf{V}\boldsymbol{\Sigma}^{-1}\mathbf{U}^*$.

$\mathbf{A}$ is an approximate representation of the Koopman operator that is derived from a finite-dimensional subspace of linear measurements, and its spectral decomposition is $\mathbf{A\Phi} = \mathbf{\Phi\Lambda}$, where $\mathbf{\Phi}$ is the matrix containing its eigenvector, and $\mathbf{\Lambda}$ is the diagonal matrix of eigenvalues. However, in practice, as p is a large number, computation of the spectral decomposition of matrix $\mathbf{A}$ is not feasible. Instead, the DMD algorithm seeks the leading spectral decomposition (i.e. eigenvalues and eigenvectors) of $\mathbf{A}$ without explicitly constructing it. Note that the data matrices $\mathbf{X} = [\mathbf{x}_1\ \mathbf{x}_2 \ldots\ \mathbf{x}_{N_s}]$ and $\mathbf{X}^+ = \left[\mathbf{x}_1^+\ \mathbf{x}_2^+ \ldots \mathbf{x}_{N_s}^+\right]$ have a much larger number of columns than rows, that is $n \ll N_s$ and therefore $\text{rank}(\mathbf{A}) = r \leq n$ and matrix $\mathbf{A}$ will have at most r nonzero eigenvalues and corresponding non-trivial eigenvectors. Hence, the following derived approximations can be used instead of directly handling (8.18). Let $\mathbf{U}_r$ denote the first r columns of $\mathbf{U}$, then $\mathbf{U} = \left[\mathbf{U}_r\ \mathbf{U}_{n-r}\right]$ and correspondingly $\mathbf{\Sigma} = diag\left(\mathbf{\Sigma}_r\ \mathbf{\Sigma}_{n-r}\right)$, $\mathbf{V} = \left[\mathbf{V}_r\ \mathbf{V}_{n-r}\right]$, where $\mathbf{\Sigma}_r$ contains the dominant singular values, and $\mathbf{V}_r$ denote the first r columns of $\mathbf{V}$. By direct substitution, we have $\mathbf{X} = \mathbf{U}_r\mathbf{\Sigma}_r\mathbf{V}_r^* + \mathbf{U}_{n-r}\mathbf{\Sigma}_{n-r}\mathbf{V}_{n-r}^*$ that gives the approximation $\mathbf{X} \approx \mathbf{U}_r\mathbf{\Sigma}_r\mathbf{V}_r^*$ for the dominant and most effective singular values. This concept is used in the model order reduction in the model-based analysis of dynamical systems. The $n \times n$ matrix $\mathbf{A}$ can now be reduced to the $r \times r$ matrix $\mathbf{A}_r$ by a projection of $\mathbf{A}$ and substitution of (8.18)

$$\widetilde{\mathbf{A}} = \mathbf{U}_r^*\mathbf{A}\mathbf{U}_r = \mathbf{U}_r^*\mathbf{X}^+\mathbf{V}_r\mathbf{\Sigma}_r^{-1}\mathbf{U}_r^*\mathbf{U}_r = \mathbf{U}_r^*\mathbf{X}^+\mathbf{V}_r\mathbf{\Sigma}_r^{-1} \tag{8.19}$$

Then, the leading spectral decomposition of $\mathbf{A}$ is approximated by the spectral decomposition of $\widetilde{\mathbf{A}}$, which can be efficiently computed as $\widetilde{\mathbf{A}}\mathbf{W} = \mathbf{W\Lambda}$, where the columns of $\mathbf{W}$ are the eigenvectors of $\widetilde{\mathbf{A}}$, and $\mathbf{\Lambda}$ is the diagonal matrix of DMD eigenvalues. It is important to note that the nonzero eigenvalues of the reduced matrix $\widetilde{\mathbf{A}}$ and the initial full high-dimensional matrix $\mathbf{A}$ are the same, and $\widetilde{\mathbf{A}}$ would provide the necessary information for data-driven control. Also, $\mathbf{U}$ can be employed to reconstruct the full state vector $\mathbf{x}$ from the reduced state $\mathbf{x}_r$ as $\mathbf{x} = \mathbf{U}_r\mathbf{x}_r$. Finally, the eigenvectors of the full matrix $\mathbf{A}$, that is $\mathbf{\Phi}$, can be reconstructed from the eigenvectors of the reduced matrix $\widetilde{\mathbf{A}}$, that is $\mathbf{W}$, by

$$\mathbf{\Phi} = \mathbf{X}^+\mathbf{V}_r\mathbf{\Sigma}_r^{-1}\mathbf{W} \tag{8.20}$$

8.2.4 System State Vector Response with DMD

The Koopman observables considered in the DMD approach are the states, that is $\mathbf{\Psi}(\mathbf{x}_k) = \mathbf{x}_k$. Then, we have from (8.11) that $\mathbf{x}_{k+1} = \sum_{j=1}^{\infty} \lambda_j \varphi_j(\mathbf{x}_k)\mathbf{v}_j$, and from the definition of an eigenfunction (8.6), we have $\varphi_j(\mathbf{x}_k) = \lambda_j \varphi_j(\mathbf{x}_{k-1})$, hence $\mathbf{x}_k = \sum_{j=1}^{\infty} \lambda_j^k \varphi_j(\mathbf{x}_0)\mathbf{v}_j$, which is the Koopman mode decomposition for the given observables. However, utilising the above DMD concept, this infinite series can be approximated with the following truncated series containing the dominant terms:

$$\mathbf{x}_k = \sum_{j=1}^{r} \lambda_j^k \varphi_j(\mathbf{x}_0)\mathbf{v}_j \tag{8.21}$$

and (8.21) can be rewritten in matrix form as

$$\mathbf{x}_k = \mathbf{\Phi}\mathbf{\Lambda}^k\mathbf{b} \tag{8.22}$$

where $\mathbf{\Phi}$ is the matrix of eigenvectors of $\mathbf{A}$ that approximate the Koopman modes, $\mathbf{\Lambda}$ is the diagonal matrix of the eigenvalues of $\mathbf{A}$ that approximate the corresponding Koopman eigenvalues and $\mathbf{b}$ is the vector containing the so-called *mode amplitudes* that approximate the Koopman eigenfunctions evaluated at the initial condition, $\mathbf{b} = [\varphi_1(\mathbf{x}_0)^T \ \varphi_2(\mathbf{x}_0)^T \dots \varphi_r(\mathbf{x}_0)^T]^T$. Note that (8.22) is directly derived from the Koopman mode decomposition (8.11). Hence, the DMD modes φ_j approximate the Koopman modes, the DMD eigenvalues approximate the corresponding Koopman eigenvalues, and the mode amplitudes approximate the Koopman eigenfunctions evaluated at the initial state vector [4].

Remark 8.6 In the linear model-based system analysis and control design methodologies, a nonlinear dynamical system is linearised at its operating points, and locally linear models of the system are derived. This linearisation approach enables the implementation of linear prediction, estimation and control techniques in the neighbourhood of the operating points. Alternatively, the Koopman operator seeks globally linear representations of nonlinear systems. In a data-driven context, this leads to finding a coordinate system or *embedding*, defined by nonlinear observable functions or measurements that span a Koopman-invariant subspace. However, the derivation, approximation and representation of these observables and embeddings are still a fundamental challenge in modern Koopman theory. The DMD technique introduced in Section 8.2.3 approximates the Koopman operator and is restricted to a space of linear measurements with a best-fit linear model. These utilised measurements are updated at each sampling time with the new information from the sensors. The main drawback of this tractable approach is that the linear DMD technique cannot capture many essential features of the nonlinear systems, such as the multiple fixed points and transients. Several methods are proposed in the literature to extend the DMD concept to overcome the associated linearity problems of DMD. The interested reader is referred to Refs. [4, 5] and the references therein for further discussions on this subject.

For nonlinear systems given by (8.15), considering system states as the observables, select N_s allowable control signal trajectories with length T, then apply them to the nonlinear system and collect the corresponding data. The collected data, with the applied input, form a complete data set as $\mathrm{X}^{+}, \mathrm{X}, \mathrm{U}, \mathrm{W} = \left\{\mathbf{x}_i^{+}, \mathbf{x}_i, \mathbf{u}_i, \mathbf{w}_i\right\}_{i=1}^{N_s}$. The following optimisation problem attempts to derive the matrices **A**, **B** and **D** of the DMD-based Koopman model:

$$\min_{\mathbf{A},\mathbf{B},\mathbf{D}} \sum_{i=1}^{N_s} \left\| \mathbf{x}_i^{+} - \mathbf{A}\mathbf{x}_i - \mathbf{B}\mathbf{u}_i - \mathbf{D}\mathbf{w}_i \right\|_2^2 \tag{8.23}$$

The analytical solution to (8.23) is

$$\left[\mathbf{A}\ \mathbf{B}\ \mathbf{D}\right]_{n\times(n+m+d)} = [\mathrm{X}^{+}][\mathrm{X}, \mathrm{U}, \mathrm{W}]^{\dagger} \tag{8.24}$$

In the following, the algorithm of DMD-based Koopman model identification is presented:

Algorithm 8.1 DMD-based Koopman Model Identification

Input Select the number of samples N_s, the data length T and the initial value for $\mathbf{x}_0$.
Step 1 Select N_s allowable control signal trajectories with length T, then apply them to the nonlinear system (8.15) to obtain $\mathrm{X}^{+}, \mathrm{X}, \mathrm{U}, \mathrm{W} = \left\{\mathbf{x}_i^{+}, \mathbf{x}_i, \mathbf{u}_i, \mathbf{w}_i\right\}_{i=1}^{N_s}$.
Step 2 Obtain the matrices **A**, **B** and **D** with the Eq. (8.24).

In the above algorithm, it is assumed that the disturbances W are measurable and available. Disturbances are defined as any system inputs that cannot be manipulated. For the disturbance-free cases, we have $\mathbf{D} = \mathbf{0}$. If the disturbances are unmeasurable, W must be estimated from the available data using the standard or Koopman operator-based nonlinear estimation techniques available in the literature, and in some cases, disturbance sequences can be obtained via simulation studies [11]. The assumption of measured disturbances is applicable in certain practical applications. For an example see Ref. [15], wherein for a solar collector field the direct irradiance, ambient temperature and the inlet temperature are measurable disturbances.

Example 8.3 Consider the plant given in Example 7.5 [11]. In this example, the Koopman model of the following bilinear DC motor system is obtained via the DMD approximation. We have

$$\dot{x}_1 = -\left(\frac{R_a}{L_a}\right)x_1 - \left(\frac{k_m}{L_a}\right)x_2 u + \left(\frac{u_a}{L_a}\right)$$

$$\dot{x}_2 = -\left(\frac{B}{J}\right)x_2 + \left(\frac{k_m}{J}\right)x_1 u - \left(\frac{\tau_l}{J}\right)$$

$$y = x_2$$

where x_1 is the rotor current, x_2 is the angular velocity, the control input u is the stator current and the output y is the angular velocity. The motor parameters are $L_a = 0.314$, $R_a = 12.345$, $k_m = 0.253$, $J = 0.00441$, $B = 0.00732$, $\tau_l = 1.47$ and $u_a = 60$.

There exists a bilinearity between the state and the control input. The discrete-time nonlinear system is obtained via the Runge–Kutta four discretisation method with a discretisation period $T_s = 0.01s$. The DMD-based Koopman model is approximated as follows:

$$\mathbf{z}_{k+1} = \begin{bmatrix} 0.999 & 0.0200 \\ -0.008 & 0.9973 \end{bmatrix} \mathbf{z}_k + \begin{bmatrix} -0.0001 \\ -0.0099 \end{bmatrix} u_k$$

$$\mathbf{x}_k = \begin{bmatrix} 1 & 0 \\ 0 & 1 \end{bmatrix} \mathbf{z}_k$$

$$y_k = x_2$$

In Figure 8.3, we compare the output of the DMD-based Koopman model and the output of the actual system. As it is shown, DMD provides an approximate equivalent linear model for the systems based on the plant measurements.

8.2.5 Approximating the Koopman Linear Model from Measured Data: The EDMD Approach

The DMD technique has become a standard numerical approach to approximate the Koopman operator. It is based on linear measurements of the system and cannot identify nonlinear changes of coordinates necessary to approximate the Koopman operator for strongly nonlinear systems. The EDMD is introduced to address this issue. The best-fit linear DMD regression with the extended approach is performed on an augmented vector containing nonlinear state measurements. Assume the formerly available state vector measurements of the system as $\{\mathbf{x}_i, i = 1, \ldots, N_s\}$ and assume for each snapshot $\mathbf{x}_i$, there is a corresponding snapshot $\mathbf{x}_i^+$ one-time step in the future. In EDMD, we seek a matrix $\mathbf{A}$ (the transpose of the finite-dimensional approximation of $\mathbf{K}$), minimising

$$\min_{\boldsymbol{\Psi},\mathbf{A}} \sum_{j=1}^{N_s} \left\| \boldsymbol{\Psi}\left(\mathbf{x}_i^+\right) - \mathbf{A}\boldsymbol{\Psi}(\mathbf{x}_i) \right\|_2^2 \tag{8.25a}$$

where

$$\boldsymbol{\Psi}(\mathbf{x}) = \begin{bmatrix} \psi_1^T(\mathbf{x}) \\ \vdots \\ \psi_p^T(\mathbf{x}) \end{bmatrix} \tag{8.25b}$$

is the vector of lifting functions and $\psi_j^T(\mathbf{x})\, j = 1, \ldots, p$ are the previously defined observables. If the vector $\boldsymbol{\Psi}(\mathbf{x})$ is known, the optimisation problem (8.25) will be easily solved. But the selection of $\boldsymbol{\Psi}(\mathbf{x})$ is a challenging issue. In the EDMD algorithm, canonical choices such as the radial basis functions are proposed for an initial $\boldsymbol{\Psi}(\mathbf{x})$ selection. Then, the Koopman linear model can be obtained as (8.13).

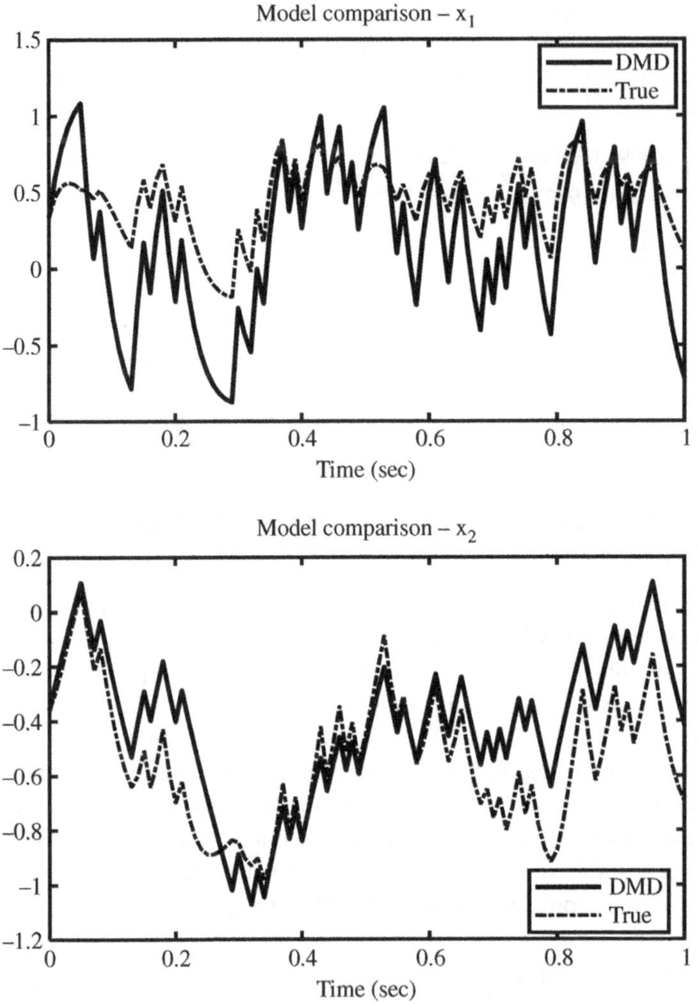

Figure 8.3 Comparison of the DMD-based Koopman model output with the actual output.

For nonlinear systems with inputs as (8.15), the vector of lifting functions in EDMD can be constructed as the vector $\left[\boldsymbol{\Psi}(\mathbf{x})^T \ \mathbf{u}^T \ \mathbf{w}^T\right]^T$ where $\boldsymbol{\Psi}(\mathbf{x})$ is defined in (8.25b) is the linear combination of Koopman eigenfunctions of the uncontrolled system. Therefore, the following optimisation problem attempts to derive the matrices **A**, **B** and **D** to minimise the following cost function:

$$\min_{\boldsymbol{\Psi},\mathbf{A},\mathbf{B},\mathbf{D}} \sum_{i=1}^{N_s} \left\| \boldsymbol{\Psi}\left(\mathbf{x}_i^+\right) - \mathbf{A}\boldsymbol{\Psi}(\mathbf{x}_i) - \mathbf{B}\mathbf{u}_i - \mathbf{D}\mathbf{w}_i \right\|_2^2 \tag{8.26}$$

To solve the optimisation problem (8.26), select N_s allowable control signal trajectories with length T, then apply them (along with an initial state vector trajectory with length T and N_s measured disturbance signal trajectories with length T) to the nonlinear system (8.15) to obtain the data set $X^+, X, U, W = \{\mathbf{x}_i^+, \mathbf{x}_i, \mathbf{u}_i, \mathbf{w}_i\}_{i=1}^{N_s}$. For $\mathbf{z}_k = [\mathbf{\Psi}^T(\mathbf{x}_k)\,\mathbf{x}_k^T]^T$ consists of n_{bf} radial basis functions and n nonlinear system state variables, in this case, $p = n_{bf} + n$, then obtain $X_{lift} = \{\mathbf{z}_i\}_{i=1}^{N_s}$ and $Y_{lift} = \{\mathbf{z}_i^+\}_{i=1}^{N_s}$. Then, the analytical solution to (8.26) is

$$[\mathbf{A}\ \mathbf{B}\ \mathbf{D}]_{p\times(p+m+d)} = Y_{lift}[X_{lift}, U, W]^{\dagger} \tag{8.27}$$

The analytical solution (8.27) is not the preferred method of solving the least-squares problem (8.26) in practice. In particular, for larger data sets with $N_s \gg p$ it is beneficial to solve the following alternative equation:

$$[\mathbf{A}\ \mathbf{B}\ \mathbf{D}]_{p\times(p+m+d)} = (Y_{lift}[X_{lift}, U, W])\left([X_{lift}, U, W]^T[X_{lift}, U, W]\right)^{\dagger} \tag{8.28}$$

The model (8.16) can be readily used to design controllers for the nonlinear dynamical system using linear controller design methodologies. However, the Koopman operator provides a global linearisation of the original dynamics provided that the set of selected basis functions is *rich enough*. The primary source of error stems from the basis functions selection and the finite size of the set of the basis functions used in the approximation of the Koopman operator. Recent studies, such as Refs. [16–18], have relied on machine learning methods to estimate the Koopman eigenfunctions automatically. Machine learning methods can effectively train and adapt dictionary elements to the existing system without selecting a pre-selected dictionary. Recently, in Ref. [19], a deep Long short-term memory (LSTM) autoencoder network has been designed to calculate the Koopman eigenfunctions of the solar collector field. The derived incremental Koopman linear model is used to design a linear MPC with terminal components to ensure closed-loop stability. The LSTM networks are known for learning complex nonlinear patterns and extracting features automatically. The deep networks can also help extract features from the data better. So, in Ref. [19], a deep LSTM neural network is used to calculate Koopman eigenfunctions of the complex nonlinear distributed model of the solar collector field.

However, as the Koopman finite-dimensional linear model is an approximated model, the computation of finite-dimensional Koopman linear state-space model matrices **A**, **B** and **D** for a nonlinear system is not unique. This section shows that without identifying a Koopman linear model, it can be possible to directly design a data-driven Koopman-based predictive controller from a finite-length dataset of a nonlinear system.

Remark 8.7 Radial basis functions are employed in this chapter for the lifting functions. For a practical introduction to radial basis functions and several candidate radial basis functions, refer to Ref. [20]. By definition, a function $\psi_i(\mathbf{x}) : \mathbb{R}^n \to \mathbb{R}$ is called radial if there exists a univariate function $\gamma : [0, \infty) \to \mathbb{R}$ such that $\psi_i(\mathbf{x}) = \gamma(r)$, where $r = ||\mathbf{x}||$. Note that the value of $\psi_i(\mathbf{x})$ for any two vectors of the same distance from the origin, that is, with equal Euclidean norms, is the same. Thus, the radial basis function is symmetric about its centre. As an example, let $\gamma(r) = e^{-(\varepsilon r)^2}$, $r \in \mathbb{R}$. Then, for any *fixed center* $\mathbf{c} \in \mathbb{R}^n$, the radial basis function can be defined as $\psi_i(\mathbf{x}) = e^{-\varepsilon^2 ||\mathbf{x}-\mathbf{c}||^2} = \gamma(||\mathbf{x} - \mathbf{c}||)$.

In the following, the algorithm of EDMD-based Koopman model identification is presented.

Algorithm 8.2 EDMD-based Koopman model identification

Input Select the number of radial basis functions p, the number of samples N_s, and the data length T. Select the initial values for $\mathbf{x}_0$.

Step 1 Select N_s allowable control signal trajectories with length T, then apply them to the nonlinear system (8.15) to obtain the data set $\mathrm{X}^+, \mathrm{X}, \mathrm{U}, \mathrm{W} = \{\mathbf{x}_i^+, \mathbf{x}_i, \mathbf{u}_i, \mathbf{w}_i\}_{i=1}^{N_s}$.

Step 2 Set $\mathbf{z}_k = [\boldsymbol{\Psi}^T(\mathbf{x}_k)\mathbf{x}_k^T]^T$ with n_{bf} radial basis functions and n nonlinear system state variables, then obtain $\mathrm{X}_{lift} = \{\mathbf{z}_i\}_{i=1}^{N_s}$ and $\mathrm{Y}_{lift} = \{\mathbf{z}_i^+\}_{i=1}^{N_s}$.

Step 3 Obtain the matrices $\mathbf{A}$, $\mathbf{B}$ and $\mathbf{D}$ from (8.28).

Example 8.4 In this example, the Koopman model of the bilinear DC motor in Example 8.3 is approximated via the EDMD algorithm. The lifting functions $\boldsymbol{\Psi}(\mathbf{x})$ are chosen to be the state vector $\mathbf{x}$ and 100 thin plate spline radial basis functions with centres selected randomly with uniform distribution over $[-1, 1]$. The thin plate spline radial basis function with a centre at $\mathbf{c}$ is defined by $\psi_i(\mathbf{x}) = ||\mathbf{x} - \mathbf{c}||^2\, log(||\mathbf{x} - \mathbf{c}||)$. The dimension of the lifted state-space is therefore $p = 102$. The EDMD-based Koopman model is approximated as follows:

$$\mathbf{z}_{k+1} = \begin{bmatrix} 1.0018 & 0.0177 & \dots & -0.0012 & -0.0008 \\ 0.1786 & 0.7776 & \dots & -0.1253 & -0.0786 \\ \vdots & \vdots & \ddots & \vdots & \vdots \\ 0.4508 & -0.4687 & \dots & 0.3584 & -1.1397 \\ -0.1470 & 0.3023 & \dots & 0.4407 & 0.1792 \end{bmatrix} \mathbf{z}_k + \begin{bmatrix} -0.0001 \\ -0.0099 \\ \vdots \\ -0.0153 \\ 0.0013 \end{bmatrix} u_k$$

$$\mathbf{x}_k = \begin{bmatrix} 1 & 0 & 0 & \dots \\ 0 & 1 & 0 & \dots \end{bmatrix} \mathbf{z}_k$$

$$y_k = x_2$$

In Figure 8.4, the outputs of the DMD and EDMD-based Koopman models and the output of the actual system are compared. As it is shown, the superiority of the EDMD approach with respect to the DMD algorithm is evident.

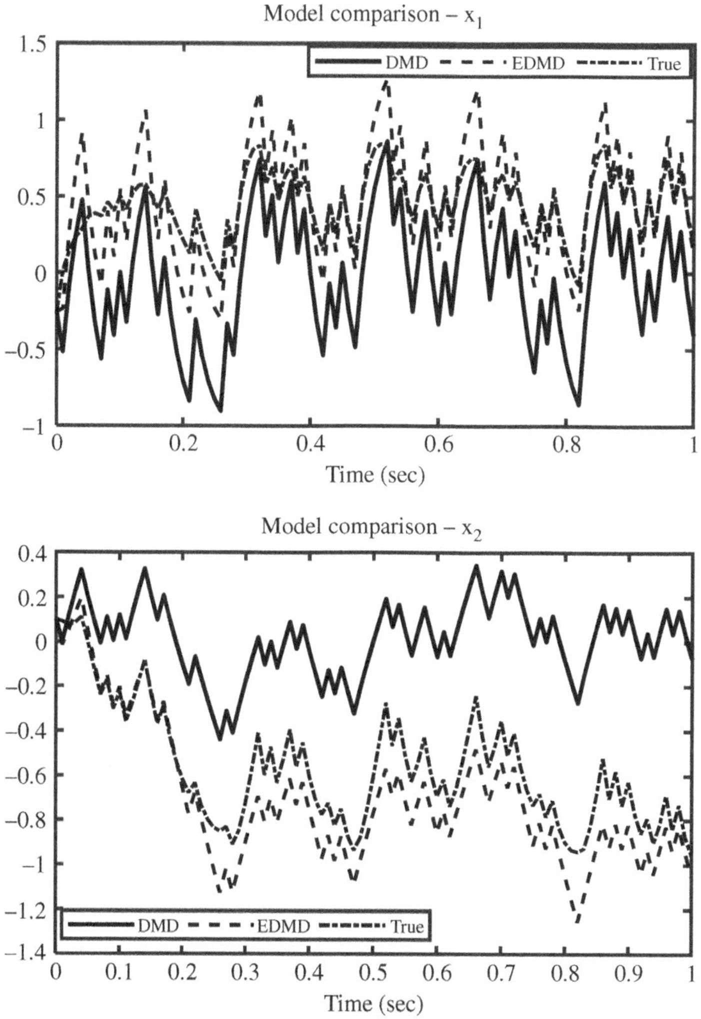

Figure 8.4 Comparison of the EDMD- and DMD-based Koopman model outputs with the actual output.

8.3 Koopman-based Data-driven Control of Nonlinear Systems

As indicated in Chapter 7, initiated by Willems' Fundamental Lemma [21], data-driven control methods have been developed for LTI systems. According to Willem's Lemma, persistently exciting data can be used to represent the input–output behaviour of a linear system without the need to identify the linear system's matrices. The application of Willems' Fundamental Lemma to nonlinear systems is investigated using the Koopman theory in Ref. [10]. Also, the application of the Koopman operator in Willems' Fundamental Lemma context is reported in Ref. [22]. Given that Willems' Fundamental Lemma applies to LTI systems, and the Koopman operator attempts to find a universal LTI model for a nonlinear system, the Koopman operator can be used to apply Willems' Fundamental Lemma to unknown nonlinear systems. However, closed-loop stability guarantees for the combined Koopman-Fundamental Lemma approaches is a challenging problem and providing non-conservative closed-loop stability guarantees using finite-length data, even for data generated from an LTI system, is still an open issue [23].

In this section, based on the results presented in Ref. [24], it is shown that without identifying a Koopman linear model, it can be possible to directly design a stable data-driven Koopman-based predictive controller from a finite-length dataset of a nonlinear system. In particular, for data provided from a nonlinear system, it is shown that a Koopman-based data-driven predictive controller can be designed, even though the Koopman model is not uniquely identified.

8.3.1 Koopman-Willems Lemma for Nonlinear Systems

In what follows, two lemmas adopted from Ref. [15] are presented. These lemmas are presented to provide a data-driven representation of nonlinear systems utilising Koopman and Willem's theories. In the first lemma, a linear Koopman model (8.16) is assumed and derived using an optimisation problem. The second lemma will exclude this limitation and is employed in the subsequent data-driven predictive control design in Section 8.3.2.[3]

Lemma 8.1 Consider the linear Koopman model (8.16) with m inputs and l outputs. If the sequences $\left[\mathbf{z}_{[0,T-1]}^T \ \mathbf{u}_{[0,T-1]}^T \ \mathbf{y}_{[0,T-1]}^T \ \mathbf{w}_{[0,T-1]}^T\right]^T$ are a trajectory set of the original system (8.15), where $\mathbf{u}_{[0,T-1]}$ is persistently exciting of order $p+N$, then $\left[\bar{\mathbf{z}}_{[0,N-1]}^T \ \bar{\mathbf{u}}_{[0,N-1]}^T \ \bar{\mathbf{y}}_{[0,N-1]}^T \ \bar{\mathbf{w}}_{[0,N-1]}^T\right]^T$ is also a trajectory of this system if there exists a vector $\mathbf{g} \in \mathbb{R}^{T-N+1}$ that satisfies

$$\begin{bmatrix} H_N(\mathbf{z}_{[0,T-1]}) \\ H_N(\mathbf{w}_{[0,T-1]}) \\ H_N(\mathbf{u}_{[0,T-1]}) \\ H_N(\mathbf{y}_{[0,T-1]}) \end{bmatrix} \mathbf{g} = \begin{bmatrix} \bar{\mathbf{z}}_{[0,N-1]}^T \\ \bar{\mathbf{w}}_{[0,N-1]}^T \\ \bar{\mathbf{u}}_{[0,N-1]}^T \\ \bar{\mathbf{y}}_{[0,N-1]}^T \end{bmatrix} \tag{8.29}$$

3 For Willems' Fundamental Lemma and the definition of the parameters and variables in this section refer to Chapter 7.

where $H_N(\mathbf{z}_{[0,T-1]}) \in \mathbb{R}^{Np\times(T-N+1)}$, $H_N(\mathbf{w}_{[0,T-1]}) \in \mathbb{R}^{Nd\times(T-N+1)}$, $H_N(\mathbf{u}_{[0,T-1]}) \in \mathbb{R}^{Nm\times(T-N+1)}$, and $H_N(\mathbf{y}_{[0,T-1]}) \in \mathbb{R}^{Nl\times(T-N+1)}$. Also, p is considered the dimension of the Koopman linear model (8.16).

In Lemma 8.1, using an available sequence of the Koopman state variables vector, all other sequences of the Koopman state variables vectors of the nonlinear system (8.1) can be obtained. According to (8.26) and (8.16), to compute the Koopman state variables vector of a system, the structure of the Koopman eigenfunctions must be pre-selected. If a linear Koopman model is assumed with $\mathbf{K} = \begin{bmatrix}\mathbf{A} & \mathbf{B} & \mathbf{D}\end{bmatrix}$, then the following formulation:

$$\mathbf{z}_{k+1} = \mathbf{K}^* \begin{bmatrix} \mathbf{z}_k \\ \mathbf{u}_k \\ \mathbf{w}_k \end{bmatrix}, \qquad k = 1, \dots, N \tag{8.30}$$

and

$$\mathbf{K}^* \in arg \min_{\mathbf{K}} \left\| \mathbf{z}_{k+1} - \mathbf{K} \begin{bmatrix} \mathbf{z}_k \\ \mathbf{u}_k \\ \mathbf{w}_k \end{bmatrix} \right\|_F^2, k = 1, \dots, T_{ini} \tag{8.31}$$

where F denotes the Frobenius norm and gives the equivalent Koopman linear state space matrices as a regression problem. To circumvent the need for the assumption of a Koopman linear model and the solution of the optimisation problem (8.31), the following lemma considers the application of the Koopman operator in Willems' Fundamental Lemma without the Koopman model (8.16).

Lemma 8.2 Consider the nonlinear system (8.15). If the sequences $[\mathbf{x}^T_{[0,T-1]}\ \mathbf{u}^T_{[0,T-1]}\ \mathbf{y}^T_{[0,T-1]}\ \mathbf{w}^T_{[0,T-1]}]^T$ are a trajectory of this nonlinear system, where $\mathbf{u}$ is persistently exciting of order $p+N$, then $\begin{bmatrix}\overline{\mathbf{x}}^T_{[-T_{ini},N-1]} & \overline{\mathbf{u}}^T_{[-T_{ini},N-1]} & \overline{\mathbf{y}}^T_{[-T_{ini},N-1]} & \overline{\mathbf{w}}^T_{[-T_{ini},N-1]}\end{bmatrix}^T$ is also a trajectory of this system if there exists a vector $\mathbf{g} \in \mathbb{R}^{T-N+1}$ that satisfies

$$\begin{bmatrix} H_{T_{ini}+N}(\mathbf{z}_{[0,T-1]}) \\ H_{T_{ini}+N}(\mathbf{w}_{[0,T-1]}) \\ H_{T_{ini}+N}(\mathbf{u}_{[0,T-1]}) \\ H_{T_{ini}+N}(\mathbf{y}_{[0,T-1]}) \end{bmatrix} \mathbf{g} = \begin{bmatrix} \overline{\mathbf{z}}^T_{[-T_{ini},N-1]} \\ \overline{\mathbf{w}}^T_{[-T_{ini},N-1]} \\ \overline{\mathbf{u}}^T_{[-T_{ini},N-1]} \\ \overline{\mathbf{y}}^T_{[-T_{ini},N-1]} \end{bmatrix} \tag{8.32}$$

where $\mathbf{z}_{[0,T-1]} = \mathbf{\Psi}(\mathbf{x}_{[0,T-1]})$ is a trajectory of the Koopman eigenfunctions for the nonlinear system (8.15). The Koopman lifting functions are defined using known basic functions such as the radial basis functions, and the T_{ini} index shows the number of most recent measured system data.

Lemma 8.2 indicates that instead of deriving the linear matrices of the Koopman model based on the latest data measured from the system, we can derive the vector $\mathbf{g}$, which satisfies Willems' Fundamental Lemma relation (8.32).

The following points are to be noted regarding Lemma 8.2:

- A sufficient volume of data (large enough T) is required in Lemma 8.2 so that the other input–output sequences of the nonlinear system (8.15) can be generated from the dataset $\begin{bmatrix} \mathbf{x}^T_{[0,T-1]} & \mathbf{u}^T_{[0,T-1]} & \mathbf{y}^T_{[0,T-1]} & \mathbf{w}^T_{[0,T-1]} \end{bmatrix}^T$.
- Unlike the existing data-driven modelling methods to extract the Koopman model, such as Ref. [1, 25], in which matrices $\mathbf{A}$, $\mathbf{B}$ and $\mathbf{D}$ are calculated and used in the controller design, the proposed Lemma 8.2 does not require the calculation of these matrices.
- The Koopman linear state and output predictions are simultaneously constructed in this scheme.

8.3.2 Data-driven Koopman Predictive Control

The Data-enabled Predictive Control (DeePC) presented in Section 7.6 uses non-parametric and data-based Hankel matrix time series to replace linear system state space matrices in the predictive control optimisation problem via Willems' Fundamental Lemma. Accordingly, the data-driven Koopman predictive control problem for a dataset $\begin{bmatrix} \mathbf{x}^T_{[0,T-1]} & \mathbf{u}^T_{[0,T-1]} & \mathbf{y}^T_{[0,T-1]} & \mathbf{w}^T_{[0,T-1]} \end{bmatrix}^T$ of the nonlinear system (8.15) can be defined as the following convex optimisation problem:

$$
\begin{aligned}
&\min_{\mathbf{u}_k,\mathbf{y}_k} \sum_{k=0}^{N-1} \|\mathbf{y}_k(t) - \mathbf{y}_s\|^2_{\mathbf{Q}} + \|\mathbf{u}_k(t)\|^2_{\mathbf{R}} \\
&\text{s.t.} \quad \begin{bmatrix} H_{T_{ini}+N}\left(\mathbf{z}_{[0,T-1]}\right) \\ H_{T_{ini}+N}\left(\mathbf{u}_{[0,T-1]}\right) \\ H_{T_{ini}+N}\left(\mathbf{w}_{[0,T-1]}\right) \\ H_{T_{ini}+N}\left(\mathbf{y}_{[0,T-1]}\right) \end{bmatrix} \mathbf{g} = \begin{bmatrix} \bar{\mathbf{z}}^T_{[-T_{ini},N-1]}(t) \\ \bar{\mathbf{w}}^T_{[-T_{ini},N-1]}(t) \\ \bar{\mathbf{u}}^T_{[-T_{ini},N-1]}(t) \\ \bar{\mathbf{y}}^T_{[-T_{ini},N-1]}(t) \end{bmatrix} \\
&\mathbf{z}_k(t) = \boldsymbol{\Psi}(\mathbf{x}_k(t)) \\
&\mathbf{u}_k(t) \in \mathbb{U}, k = 0, \dots, N-1 \\
&\mathbf{y}_k(t) \in \mathbb{Y}, k = 0, \dots, N-1
\end{aligned} \tag{8.33}
$$

where $\mathbb{U}$ and $\mathbb{Y}$ are the m and l dimensional input and output spaces, respectively, t is the discrete working time of the system starting from t_0, T_{ini} index represents the number of recently measured or available system data, $\mathbf{y}_s$ is the set-point reference signal, N is the prediction horizon assumed to be equal to the control horizon. $\mathbf{Q}$ and $\mathbf{R}$ are the positive semi-definite and positive definite weighting matrices to penalise the tracking error and control effort, respectively. Finally, $\Psi(\cdot)$ is the Koopman lifting functions vector of the nonlinear system (8.15) defined using known basic functions such as the radial basis functions.

Algorithm 8.3 presents the workflow of the proposed data-driven Koopman predictive controller for nonlinear systems. The entire control scheme for the model-based Koopman predictive control and the data-driven Koopman predictive control are presented in Figures 8.5 and 8.6, respectively. A comparative analysis of Figures 8.5 and 8.6 shows that the main difference between the two schemes is in the underlying constraints employed in the optimisation problems of Koopman model predictive control (MPC) and data-driven Koopman predictive

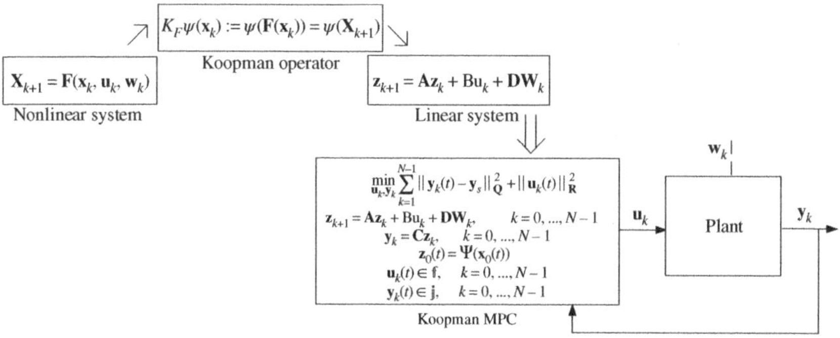

Figure 8.5 Block diagram of the model-based Koopman predictive control system.

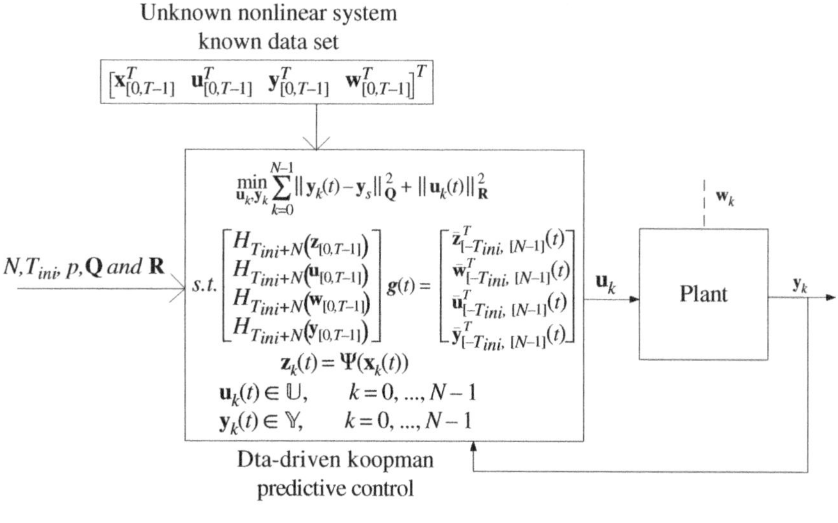

Figure 8.6 Block diagram of the data-driven Koopman predictive control system.

control (DDKPC). In Koopman MPC, the underlying constraints are the plant's linear state-space model found in typical MPC problem formulations. In this case, the linear state-space model is directly derived from the Koopman operator representation of the original nonlinear system. On the other hand, in DDKPC, the underlying constraints are the equations derived from the Fundamental Lemma that is directly written from the measured plant data set. An important point to note is that the relation $\mathbf{z}_k(t) = \boldsymbol{\Psi}(\mathbf{x}_k(t))$ gives the Koopman state vector $\mathbf{z}_k(t)$ from the measured states $\mathbf{x}_k(t)$.

Algorithm 8.3 Data-Driven Koopman Predictive Control

Input A dataset $\begin{bmatrix} \mathbf{x}^T_{[0,T-1]} & \mathbf{u}^T_{[0,T-1]} & \mathbf{y}^T_{[0,T-1]} & \mathbf{w}^T_{[0,T-1]} \end{bmatrix}^T$ from the unknown nonlinear system (8.15), and the parameters N, T_{ini}, p, $\mathbf{Q}$, and $\mathbf{R}$.
For $t = t_0, \ldots$ repeat:
Step 1 Set $\mathbf{z}_k(t) = \boldsymbol{\Psi}(\mathbf{x}_k(t))$ where $\boldsymbol{\Psi}$ consists of p radial basis functions and n nonlinear system state variables.
Step 2 Solve the optimisation problem (8.33) to obtain an optimal input sequence $\mathbf{u}^*_k(t), k = 0,1, \ldots, N-1$.
Step 3 Apply $\mathbf{u}_0(t) = \mathbf{u}^*_{k=0}(t)$ to the nonlinear system.

Example 8.5 Consider the bilinear DC motor in Example 7.5. The physical constraints on the control input are $u \in [-4, 4]$, which we scale to $[-1, 1]$. The goal is to design a Koopman data-driven predictive controller based on Section 8.3.2, i.e. assuming only that measured data are available and no explicit knowledge of the model is utilised. We discretise the scaled dynamics using the Runge–Kutta four method with a discretisation period $T_s = 0.01$s and simulate 1 trajectory over 1000 sampling periods. The control input for each trajectory is a random signal uniformly distributed on $[-1, 1]$. The lifting function $\boldsymbol{\Psi}(\mathbf{x})$ is chosen to be the state vector $\mathbf{x}$ and 100 thin plate spline radial basis functions with centres selected randomly with uniform distribution over $[-1, 1]$. The thin plate spline radial basis function with a centre at $\mathbf{c}$ is defined by $\psi_i(\mathbf{x}) = \|\mathbf{x} - \mathbf{c}\|^2 \, log\,(\|\mathbf{x} - \mathbf{c}\|)$. The dimension of the lifted state-space is, therefore $p = 102$. The control objective is to track a given angular velocity reference $\mathbf{y}_s$ according to (8.33). The cost function matrices were chosen as $\mathbf{Q} = 100$, $\mathbf{R} = 1$, $T_{ini} = 1$ and $N = 0.1$ which results in $N = 10$. We do not impose any constraints on the output and track a piecewise constant reference. The simulation result is shown in Figure 8.7, which shows appropriate tracking. Note that in this example, the magnitude of the vector $\mathbf{g}$ must be minimised in the DDKPC objective function with the coefficient $\lambda_g = 2$ (see (8.40a) in Section 8.3.4), which is an important tuning parameter. More detail is provided in the next section.

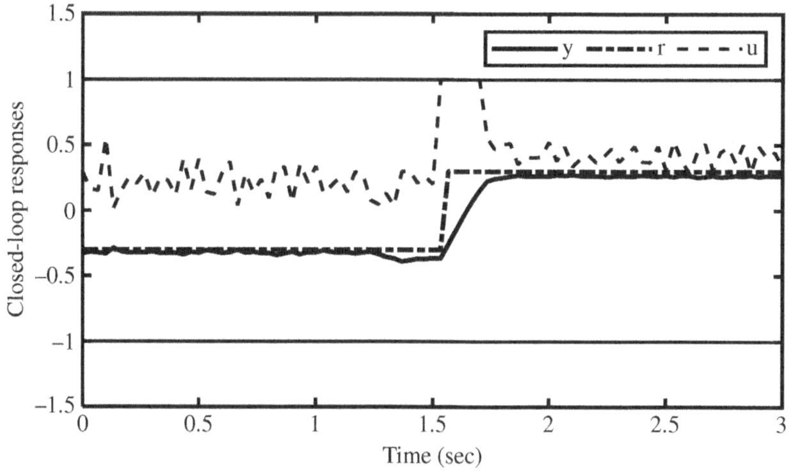

Figure 8.7 Output of the bilinear DC motor by applying the proposed data-driven Koopman predictive control.

8.3.3 Robust Stability Analysis of the Data-driven Koopman Predictive Control

The previous section presented the application of Willems' Fundamental Lemma in the Koopman MPC of a general nonlinear system. In the proposed approach, basis functions are employed as the Koopman lifting functions. However, the selected basis functions are finite and may not be an exact choice for the Koopman lifting functions; hence, the DDKPC (8.33) would involve unstructured uncertainties. Accordingly, this section analyses the optimisation problem given by (8.33) in the face of unstructured uncertainties in the Koopman state variables.

Robust Koopman–Willems Lemma. The two main strategies to ensure closed-loop stability of the model-based predictive controllers are providing bounds on the minimal required prediction horizon and including terminal components such as terminal cost functions or terminal region constraints. These approaches are model-based and rely on the model information. They cannot, therefore, be used directly in a data-driven framework [26]. In this section, the terminal components approach is employed to prove the closed-loop stability of the data-driven Koopman predictive control using only the following T-length measured and available trajectories of the unknown nonlinear system (8.15):

$$\begin{aligned}
\mathbf{u}_{[0,T-1]} &= \{\mathbf{u}_k, k = 0, \dots, T-1\},\\
\mathbf{w}_{[0,T-1]} &= \{\mathbf{w}_k, k = 0, \dots, T-1\},\\
\mathbf{y}_{[0,T-1]} &= \{\mathbf{y}_k, k = 0, \dots, T-1\},
\end{aligned}$$

$$\mathbf{x}_{[0,T-1]} = \{\mathbf{x}_k, k = 0, \dots, T-1\},$$
$$\mathbf{x}^+_{[1,T]} = \{\mathbf{x}_k, k = 1, \dots, T\}. \tag{8.34}$$

From the above data trajectories, the Koopman state trajectory $\mathbf{z}_{[0,\,T-1]}$ is defined as follows:

$$\mathbf{z}_{[0,T-1]} = \left\{ \begin{bmatrix} \psi_1(\mathbf{x}_{[0,T-1]}) \\ \vdots \\ \psi_p(\mathbf{x}_{[0,T-1]}) \\ \mathbf{x}_{[0,T-1]} \end{bmatrix}, k = 0, \dots, T-1 \right\},$$

$$\mathbf{z}^+_{[1,T]} = \left\{ \begin{bmatrix} \psi_1(\mathbf{x}_{[1,T]}) \\ \vdots \\ \psi_p(\mathbf{x}_{[1,T]}) \\ \mathbf{x}_{[1,T]} \end{bmatrix}, k = 1, \dots, T \right\}, \tag{8.35}$$

where $\psi_j(\mathbf{x}), j = 1, \dots, p$ are the basis functions. Let the actual Koopman linear model be as follows:

$$\mathbf{z}^*_{k+1} = \mathbf{A}^*\mathbf{z}^*_k + \mathbf{B}^*\mathbf{u}_k + \mathbf{D}^*\mathbf{w}_k,$$
$$\mathbf{y}_k = \mathbf{C}\mathbf{z}^*_k, \tag{8.36}$$

where $\mathbf{z}^*$, $\mathbf{A}^*$, $\mathbf{B}^*$ and $\mathbf{D}^*$ minimise the optimisation problem (8.26). In the DDKPC problem (8.33), by selecting the Koopman state vector as the basis functions, the Koopman linear model matrices $\mathbf{A}$, $\mathbf{B}$ and $\mathbf{D}$ in (8.16) will not necessarily be equal to the actual Koopman's model matrices $\mathbf{A}^*$, $\mathbf{B}^*$ and $\mathbf{D}^*$ for a nonlinear system satisfying (8.16). Accordingly, the Koopman actual model is defined based on the Koopman approximated model (8.16) matrices as follows:

$$\mathbf{z}^*_{k+1} = \underbrace{\mathbf{A}\mathbf{z}_k + \mathbf{B}\mathbf{u}_k + \mathbf{D}\mathbf{w}_k}_{\mathbf{z}_{k+1}} + \boldsymbol{\delta}_{\mathbf{z},k}, \tag{8.37}$$

where

$$\boldsymbol{\delta}_{\mathbf{z},k} = \Delta\mathbf{A}\mathbf{z}^*_k + \Delta\mathbf{B}\mathbf{u}_k + \Delta\mathbf{D}\mathbf{w}_k,$$

and $\boldsymbol{\delta}_{\mathbf{z},\,k}$ represents the uncertainties in the Koopman state variables.

Lemma 8.3 Consider the Koopman actual model (8.36) of the nonlinear system (8.15) and the available measured trajectory of the Koopman state variables as defined in (8.34) and (8.35). If the sequences $\begin{bmatrix}\mathbf{x}^T_{[0,T-1]} & \mathbf{u}^T_{[0,T-1]} & \mathbf{w}^T_{[0,T-1]} & \mathbf{y}^T_{[0,T-1]}\end{bmatrix}^T$ are a trajectory of the nonlinear system (8.15), where $\mathbf{u}_{[0,\,T-1]}$ is persistently exciting of order $p+N$, then $\begin{bmatrix}\overline{\mathbf{z}}^T_{[-T_{ini},N-1]} & \overline{\mathbf{u}}^T_{[-T_{ini},N-1]} & \overline{\mathbf{w}}^T_{[-T_{ini},N-1]} & \overline{\mathbf{y}}^T_{[-T_{ini},N-1]}\end{bmatrix}^T$

(where $-T_{ini}$ shows the number of samples measured before working time t)[4] is also a trajectory of (8.15) if there exists a vector $\mathbf{g} \in \mathbb{R}^{T-N+1}$ that satisfies:

$$\begin{aligned} &\min_{\mathbf{g}} \ \|H_k\left(\mathbf{z}_{[0,T-1]}\right)\mathbf{g} - \overline{\mathbf{z}}_{[-T_{ini},N-1]}\| + \beta\sqrt{\|\mathbf{g}\|_2^2 + 1}, \\ &\text{s.t.} \begin{bmatrix} H_{T_{ini}+N}\left(\mathbf{u}_{[0,T-1]}\right) \\ H_{T_{ini}+N}\left(\mathbf{w}_{[0,T-1]}\right) \\ H_{T_{ini}+N}\left(\mathbf{y}_{[0,T-1]}\right) \end{bmatrix} \mathbf{g} = \begin{bmatrix} \overline{\mathbf{u}}_{[-T_{ini},N-1]} \\ \overline{\mathbf{w}}_{[-T_{ini},N-1]} \\ \overline{\mathbf{y}}_{[-T_{ini},N-1]} \end{bmatrix}, \end{aligned} \tag{8.38}$$

where β is the uncertainty bound in the vector of Koopman actual state variables $\mathbf{z}^*$.

Proof: Refer to [15].

Remark 8.8 In a similar problem presented in Ref. [27], by selecting $\beta = \lambda_g\sqrt{\|\mathbf{g}\|_2^2 + 1}$, it can be shown that the optimal solution of $\mathbf{g}$ in (8.38) also minimises the following optimisation problem:

$$\begin{aligned} &\min_{\mathbf{g}} \ \|H_k\left(\mathbf{z}_{[0,T-1]}\right)\mathbf{g} - \overline{\mathbf{z}}_{[-T_{ini},N-1]}\|^2 + \lambda_{\mathbf{g}}\|\mathbf{g}\|_2^2, \\ &\text{s.t.} \begin{bmatrix} H_{T_{ini}+N}\left(\mathbf{u}_{[0,T-1]}\right) \\ H_{T_{ini}+N}\left(\mathbf{w}_{[0,T-1]}\right) \\ H_{T_{ini}+N}\left(\mathbf{y}_{[0,T-1]}\right) \end{bmatrix} \mathbf{g} = \begin{bmatrix} \overline{\mathbf{u}}_{[-T_{ini},N-1]} \\ \overline{\mathbf{w}}_{[-T_{ini},N-1]} \\ \overline{\mathbf{y}}_{[-T_{ini},N-1]} \end{bmatrix}. \end{aligned} \tag{8.39}$$

The quadratic regularisation[5] of the vector $\mathbf{g}$ in (8.39) is equivalent to the robust formulation (8.38) with an explicit bound for the uncertainty term. Therefore, by selecting a large enough value for λ_g in (8.39), a large enough bound for β can be obtained.

8.3.4 Robust Data-driven Koopman Predictive Control

According to Lemma 8.3 and Remark 8.8, for the dataset (8.34) and (8.35) of the unknown nonlinear system (8.15), the DDKPC problem (8.33) with terminal components and uncertainty in the Koopman state variables according to (8.37) can be defined as the following convex optimisation problem:

$$J_N^* = \min_{\mathbf{u}_k,\mathbf{y}_k,\mathbf{g}} \sum_{k=0}^{N-1} \|\mathbf{y}_k(t) - \mathbf{y}_s\|_{\mathbf{Q}}^2 + \|\mathbf{u}_k(t) - \mathbf{u}_s\|_{\mathbf{R}}^2 + \lambda_{\mathbf{g}}\|\mathbf{g}\|_2^2 + l_f(\mathbf{z}_N), \tag{8.40a}$$

4 Note that the time indices start at $k = -T_{ini}$, since the last T_{ini} inputs and outputs will be used to invoke a unique initial state at working time t.
5 Regularisation is also used in Section 7.6.2.

$$\text{s.t.} \quad \begin{bmatrix} H_{T_{ini}+N}\left(\mathbf{z}_{[0,T-1]}\right) \\ H_{T_{ini}+N}\left(\mathbf{u}_{[0,T-1]}\right) \\ H_{T_{ini}+N}\left(\mathbf{w}_{[0,T-1]}\right) \\ H_{T_{ini}+N}\left(\mathbf{y}_{[0,T-1]}\right) \end{bmatrix} \mathbf{g} = \begin{bmatrix} \overline{\mathbf{z}}^{T}_{[-T_{ini},N-1]}(t) \\ \overline{\mathbf{w}}^{T}_{[-T_{ini},N-1]}(t) \\ \overline{\mathbf{u}}^{T}_{[-T_{ini},N-1]}(t) \\ \overline{\mathbf{y}}^{T}_{[-T_{ini},N-1]}(t) \end{bmatrix}, \tag{8.40b}$$

$$\mathbf{z}_N(t) \in Z_f, \tag{8.40c}$$

$$\mathbf{u}_k(t) \in \mathbb{U}, \qquad k = 0, \ldots, N-1 \tag{8.40d}$$

$$\mathbf{y}_k(t) \in \mathbb{Y}, \qquad k = 0, \ldots, N-1 \tag{8.40e}$$

where $\mathbf{z}$ is the vector of the Koopman state variables selected by basis functions, $\mathbf{y}_s$ is the set-point reference signal and $\mathbf{u}_s$ is the control input that sustains the output at the desired reference. Constraints (8.40d) and (8.40e) show the pointwise-in-time constraints on the predicted input and output, which can be defined as follows for some real numbers c and d:

$$\widetilde{Z} = \left\{ \begin{bmatrix} \mathbf{y}_k \\ \mathbf{u}_k \end{bmatrix} \in \mathbb{R}^{p+m} \; : \; |c(\mathbf{y}_k - \mathbf{y}_s) + d(\mathbf{u}_k - \mathbf{u}_s)| \leq 1 \right\}, \; k = 0, \ldots, N-1. \tag{8.41}$$

Finally, $l_f(\mathbf{z}_N)$ specifies the terminal cost and $\mathbf{z}_N \in Z_f$ specifies the terminal region constraint on the N-step predicted Koopman states.

Consider the DDKPC given by (8.33). In the case of uncertainties presented in Lemma 8.3, inspired by (8.39), the robust data-driven Koopman predictive control can be formulated as (8.40). This shows that by embedding quadratic regularisation, the problem of robust data-driven predictive control with unstructured uncertainty can be converted to the problem of data-driven predictive control with quadratic regularisation. This shows the importance of regularisation from the perspective of robust optimisation [28].

At an initial time t_0 and an initial state $\mathbf{z}(t_0)$ in the feasibility region, minimisation of the cost function (8.40a) under the system trajectories (8.40b), the input and output constraints (8.40d) and (8.40e), and the terminal state constraint (8.40c), $\mathbf{z}_N(t_0) \in Z_f$ where Z_f is a defined terminal set, yields the optimal control signal $\mathbf{u}_k(t_0)$, $0 \leq k \leq N-1$ in the moving horizon time frame. The terminal state $\mathbf{z}_N(t_0)$ generated under the control of $\mathbf{u}_k(t_0)$, $0 \leq k \leq N-1$ reaches the terminal region Z_f. But only the first value of the control signal $\mathbf{u}_k(t_0)$, $0 \leq k \leq N-1$, i.e. the control signal $\mathbf{u}_{k=0}(t_0)$ is implemented, and then this process is repeated. This is the *receding horizon strategy* employed in predictive control systems.

Remark 8.9 According to Willems' Fundamental Lemma and the results presented in Refs. [22, 26], the order of persistence of excitation of the input signal

$\mathbf{u}_{[0,T-1]}$ in the data-driven predictive control problem (8.40a) to (8.40e) must be $N + T_{ini} + p$. This is due to the fact that the reconstructed trajectories in (8.40b) are of length $N + T_{ini}$. Persistence of excitation order for the Koopman linear model is $N + T_{ini} + p$, which will often be a very large number since usually $p \gg n$. According to the results presented in Refs. [23, 24, 29], this level of persistence of excitation is a sufficient but not necessary condition for data-driven control. Also, due to the uncertainty in the number of Koopman state variables of a nonlinear system and as obtaining the exact Koopman matrices is generally infeasible in practice, there is no need to have a persistence of excitation order of the input signal trajectory that can be used to form the exact Koopman model. Inspired by the concept of *data informativity* introduced in Ref. [24], it is sufficient to have a persistently exciting input signal such that all the Koopman models identified from the dataset (8.34) and (8.35) can solve the optimisation problem (8.40a) while satisfying the constraints (8.40b), (8.40d) and (8.40e).

In the following, the asymptotic stability of the closed-loop nonlinear system controlled by the DDKPC (8.40a)–(8.40e) is proved using the trajectories (8.35) of the nonlinear system (8.15). The terminal components are required to prove the stability and recursive feasibility of the closed-loop system with the DDKPC (8.40a)–(8.40e).

Assumption 8.1 The matrices $\mathbf{E} \in \mathbb{R}^{m \times p}$, $\mathbf{P} = \mathbf{P}^T \in \mathbb{R}^{p \times p} > 0$ and the set Z_f exist such that by applying $\mathbf{u} - \mathbf{u}_s = \mathbf{E}(\mathbf{z} - \mathbf{z}_s)$ for all $\mathbf{z} \in Z_f$, we have:

$$\mathbf{Az} + \mathbf{Bu} + \mathbf{Dw} + \boldsymbol{\delta}_\mathbf{z} \in Z_f$$

(1) $\mathbf{u}_k \in \mathbb{U}$ and $\mathbf{y}_k \in \mathbb{Y}$
(2) If the terminal cost function l_f is a Lyapunov function as

$$Z_f = \left\{ \mathbf{z} \in \mathbb{R}^p : l_f(\mathbf{z}) = (\mathbf{z} - \mathbf{z}_s)^T \mathbf{P}(\mathbf{z} - \mathbf{z}_s) \le \alpha, \alpha > 0 \right\}, \tag{8.42}$$

then for the terminal region Z_f of the DDKPC (8.40a)–(8.40e), the following inequality holds:

$$\|\mathbf{z}_{k+1} - \mathbf{z}_s\|_\mathbf{P}^2 - \|\mathbf{z}_k - \mathbf{z}_s\|_\mathbf{P}^2 \le -\|\mathbf{y}_k - \mathbf{y}_s\|_\mathbf{Q}^2 - \|\mathbf{u}_k - \mathbf{u}_s\|_\mathbf{R}^2 \tag{8.43}$$

where $\mathbf{z}_s$ is constructed through $\mathbf{x}_s$ as (8.35), and $\mathbf{x}_s$ is the system desired reference point corresponding to $\mathbf{u}_s$ and $\mathbf{y}_s$.

Lemma 8.4 For the data trajectory in (8.34) and (8.35), assume that there exists a matrix $\mathbf{G}$ such that

$$\mathbf{z}_{[0,T-1]}\mathbf{G} = \mathbf{I} \tag{8.44}$$

Then, there exist matrices $\mathbf{E} \in \mathbb{R}^{m \times p}$, $\mathbf{P} = \mathbf{P}^T \in \mathbb{R}^{p \times p} > 0$, a terminal set Z_f for the DDKPC (8.40a)–(8.40e), and a positive constant ε, such that for all $\mathbf{z} \in Z_f$, we have

$$\left\| \mathbf{z}^+_{[1,T]} \mathbf{G}(\mathbf{z}_k - \mathbf{z}_s) \right\|_{\mathbf{P}}^2 - \|\mathbf{z}_k - \mathbf{z}_s\|_{\mathbf{P}}^2 \leq -\|\mathbf{y}_k - \mathbf{y}_s\|_{\mathbf{Q}}^2 - \|\mathbf{u}_k - \mathbf{u}_s\|_{\mathbf{R}}^2 - \varepsilon \|\mathbf{z}_k - \mathbf{z}_s\|^2 \tag{8.45}$$

Proof: Please refer to Ref. [15].

Assumption 8.1 is a standard condition in MPC to ensure closed-loop stability. Under Assumption 8.1 and Lemma 8.4, which are conditions for stability of the DDKPC (8.40a)–(8.40e), the following Theorem for the closed-loop stability of the predictive controller is presented:

Theorem 8.1 Let Assumption 8.1 and Lemma 8.4 hold. If the problem (8.40a)–(8.40e) is feasible for $t = t_0$, then

(1) It is recursively feasible at any time $t \geq t_0$.
(2) The closed-loop trajectories satisfy the constraints $\mathbf{u}_k(t) \in \mathbb{U}$ and $\mathbf{y}_k(t) \in \mathbb{Y}$ for all $t \geq t_0$.
(3) The equilibrium $\mathbf{u}_s$, $\mathbf{y}_s$ is asymptotically stable for the closed-loop system.

Proof: Please refer to Ref. [15].

As is observed from Assumption 8.1 and Theorem 8.1, for the data-driven control, it is desired to derive the matrix $\mathbf{P}$ and the set Z_f using the data trajectory $(\mathbf{z}_{[0,T-1]}, \mathbf{u}_{[0,T-1]}, \mathbf{w}_{[0,T-1]})$ with no prior plant information. If $\mathbb{U}$, $\mathbb{Y}$ and Z_f are convex polytopic, then the problem (8.40a)–(8.40e) is a quadratic programming problem that can be solved effectively.

Theorem 8.2 Assume that the dataset $(\mathbf{z}_{[0,T-1]}, \mathbf{u}_{[0,T-1]}, \mathbf{w}_{[0,T-1]})$ from the unknown nonlinear system (8.15) is available and $\mathbf{z}_{[0,T-1]}$ has full row rank. Then, there exist matrices $\mathbf{X}^{-1} = \mathbf{P} > 0$, $\mathbf{Y} = \mathbf{G}\mathbf{X}$ and a constant $\alpha > 0$, such that the following optimisation problem is feasible:

$$\min_{\alpha,\mathbf{X},\mathbf{Y}} - \log\det(\alpha\,\mathbf{X}), \tag{8.46a}$$

$$s.t. \begin{bmatrix} \boldsymbol{X} & \left(\mathbf{z}^+_{[1,T]}\boldsymbol{Y}\right)^T & (\boldsymbol{C}\boldsymbol{X})^T & \boldsymbol{u}_{[0,T-1]}\boldsymbol{Y} & \sqrt{\varepsilon}\boldsymbol{X} \\ * & \boldsymbol{X} & \mathbf{0} & \mathbf{0} & \mathbf{0} \\ * & * & \boldsymbol{Q}^{-1} & \mathbf{0} & \mathbf{0} \\ * & * & * & \boldsymbol{R}^{-1} & \mathbf{0} \\ * & * & * & * & \boldsymbol{I} \end{bmatrix} \geq \mathbf{0}, \tag{8.46b}$$

$$\begin{bmatrix} \frac{1}{\alpha} & c\mathbf{C} + d\mathbf{u}_{[0,T-1]}\mathbf{z}^{\dagger}_{[0,T-1]} \\ \left(c\mathbf{C} + d\mathbf{u}_{[0,T-1]}\mathbf{z}^{\dagger}_{[0,T-1]}\right)^{T} & \mathbf{X} \end{bmatrix} \geq 0. \tag{8.46c}$$

where * denote matrix entries, which can be inferred from symmetry. In this case:

(1) The convex set $Z_f \subset Z$ with the radius $\alpha \, det \, \mathbf{X}$ is the terminal region of the data-driven predictive control problem (8.40a)–(8.40e).
(2) In the Z_f region, by selecting $\mathbf{E} = \mathbf{u}_{[0,T-1]}\mathbf{z}^{\dagger}_{[0,T-1]}$, the closed-loop system with $\mathbf{u} - \mathbf{u}_s = \mathbf{E}(\mathbf{z} - \mathbf{z}_s)$ is asymptotically stable and satisfies the input–output constraints (8.40d) and (8.40e).
(3) The closed-loop system under the data-driven predictive controller (8.40a)–(8.40e) is asymptotically stable if it is feasible at the time t_0.

Proof: Please refer to Ref. [15].

Remark 8.10 For the uncertainty-free data-driven Koopman predictive controller proposed in Section 8.3.2, similar results to Theorem 8.2 are derived for $\varepsilon = 0$ in (8.46b).

Algorithm 8.4 Robust Data-driven Koopman predictive control

Input

- A dataset $\left[\mathbf{x}^{T}_{[0,T-1]} \; \mathbf{u}^{T}_{[0,T-1]} \; \mathbf{y}^{T}_{[0,T-1]} \; \mathbf{w}^{T}_{[0,T-1]}\right]^{T}$ from the unknown nonlinear system (8.15).
- The parameters N, T_{ini}, p, λ_g, $\mathbf{Q}$ and $\mathbf{R}$.
- Calculating the matrix $\mathbf{P}$ and the constant α from (8.46).

For $t = t_0, \ldots$ repeat:
Step 1 Set $\mathbf{z}_k(t) = \boldsymbol{\Psi}(\mathbf{x}_k(t))$ where $\boldsymbol{\Psi}$ consists of p radial basis functions and n nonlinear system state variables.
Step 2 Solve the optimisation problem (8.40) to obtain an optimal input sequence $\mathbf{u}^{*}_{k}(t), k = 0, \ldots, N-1$.
Step 3 Apply $\mathbf{u}_0(t) = \mathbf{u}^{*}_{k=0}(t)$ to the nonlinear system.

8.4 A Case Study: Data-driven Koopman Predictive Control of the ACUREX Parabolic Solar Collector Field

In this section, the proposed DDKPC (8.33) and the proposed robust data-driven Koopman predictive control with terminal components in (8.40a)–(8.40e) are applied to the ACUREX solar collector field. ACUREX is a solar plant located in the Tabernass desert in Spain. ACUREX is an available experimental plant at Plataforma Solar de Almería and its data is widely used for solar systems studies.

ACUREX parabolic solar collector field mathematical modelling and control

There is a growing interest in using renewable energy due to global warming and the need to reduce the environmental impact of fossil fuels. Sun is the most abundant source of renewable energy. Solar energy is mainly used to produce electricity in solar thermal plants and photovoltaic plants [30]. A concentrated solar power (CSP) plant is composed of a solar field which heats a heat-transfer fluid (usually synthetic oil) to the desired temperature, a steam generator which uses the heated fluid to produce electricity, a storage system which provides energy and auxiliary elements such as valves and pipes [31]. The main objective of a control system in solar trough plants is to maintain the outlet temperature of the field around the desired set-point. The MPC methodologies have been most effective for CSP systems in practical implementations. Although nonlinear MPC is the most suitable control approach due to the highly nonlinear behaviour of the distributed solar collector field [32], it is not effective in practice due to the complexity and time-consuming solution of non-convex optimisation problems and the lack of globally optimal solutions. The proposed linear MPC approaches are either based on the gain scheduling generalised predictive control (GPC) [33] or based on the observer-based state-space MPC presented in Refs. [34, 35]. The linear MPC designs rely on the local linearisation of the solar collector field model at several operating points and may lead to incorrect results if changes in the model or system inputs are rapid and frequent (for example, under severe changes in solar radiation). In fact, due to the slow dynamics of the solar collector field system and the rapid changes in operating conditions and system inputs, the local linearisation of the solar collector field system model will never reach the optimal accuracy for the model over the entire system time-interval [36].

The stable DDKPC can be a suitable candidate for controlling complex nonlinear solar fields. It was shown in the previous section that the proposed approach transforms the nonlinear system into an approximately global linear state-space model without the need for local linearisation, and the controller is designed solely based on a finite-length dataset of a nonlinear system.

System model. The ACUREX solar collector field comprises 480 single-axis parabolic trough collectors arranged in 10 parallel loops with 48 collectors in each loop, and each loop with a length of $L_{loop} = 174$ m, which comprises active parts (142 m), and passive parts, i.e. joints and other parts not reached by concentrated radiation (30 m). It can provide about 1.2 MW of peak thermal power with solar radiation of 900 W/m^2 [35]. Solar collectors are made up of linear parabolic mirrors that reflect and concentrate sunlight onto a pipe located at the focal axis of the collectors. To control the solar collector field, it is essential to describe the outlet oil temperature as a function of the heat transfer fluid (HTF) flow. Such a model can be formed by applying the energy conservation law to a volume dx of the collector pipe around the period dt as follows:

$$\rho_m C_m A_m \frac{\partial T_m}{\partial t} = K_{opt}\eta_0 IG - H_l \pi D_m (T_m - T_a) - H_t \pi D_f (T_m - T_f),$$
$$\rho_f C_f A_f \frac{\partial T_f}{\partial t} + \rho_f C_f q \frac{\partial T_f}{\partial x} = H_t \pi D_f (T_m - T_f) \tag{8.47}$$

The subscripts m and f denote the metal pipe and the fluid, respectively. Other parameters and variables are defined in Table 8.1. Table 8.2 presents the measured parameters of the ACUREX distributed solar collector field, and a schematic diagram of the ACUREX solar collector field is shown in Figure 8.8. In Figure 8.8, the heated fluid is synthetic oil Therminol 55, which can be heated up to 300 °C without straying its thermal properties. The oil is stored in a thermal storage tank of 140 m^3 capacity. The oil at low inlet temperature is pumped through a pipe from the bottom of the tank to the field and, after heating by the collectors, is deposited via a return pipe at the top of the tank. The oil properties permit stratified energy storage according to their density. The HTF in the ACUREX field is the synthetic oil Therminol 55. Following the technical datasheet, its density and specific heat capacity are as follows [37]:

$$\rho_f = 903 - 0.672T_f, \quad C_f = 1820 + 3.478T_f. \tag{8.48}$$

Table 8.1 Model parameters and variables of the solar collector field.

Unit	Description	Symbol	Unit	Description	Symbol
kg/m^3	Density	ρ	W/m^2 K	Coefficient of thermal losses	H_l
J/kg K	Specific heat capacity	C	W/m^2 K	Coefficient of transmission metal-fluid	H_t
m^2	Pipe cross-section	A	m^3/s	Fluid volume flow	q
Without unit	Optical efficiency of collectors	K_{opt}	m	Aperture of collectors	G
Without unit	Geometric efficiency of collectors	η_0	m	The outer diameter of the pipe	D_m
W/m^2	Solar radiation	I	m	The inner diameter of the pipe	D_f
°C	The temperature of the fluid leaving the loop	T	°C	Ambient temperature	T_a

Table 8.2 Parameters measured of the ACUREX parabolic collector field.

Value	Parameter	Value	Parameter
$G = 1.82$ m	Collectors Aperture	$\rho_m = 7800\ \mathrm{kg/m^3}$	Metal pipe density
$n_0 = K_{opt}\eta_0 = 0.56$	Collectors' efficiency mean value	$C_m = 550\ \mathrm{J/kg\,{}^\circ C}$	Metal pipe specific heat capacity
$A_f = 7.55 \times 10^{-4}\ \mathrm{m^2}$	Fluid cross-section	$A_m = 2.48 \times 10^{-4}\ \mathrm{m^2}$	Pipe cross-section
$D_f = 2.758 \times 10^{-2}$ m	Pipe inner diameter	$D_m = 3.180 \times 10^{-2}$ m	Pipe outer diameter

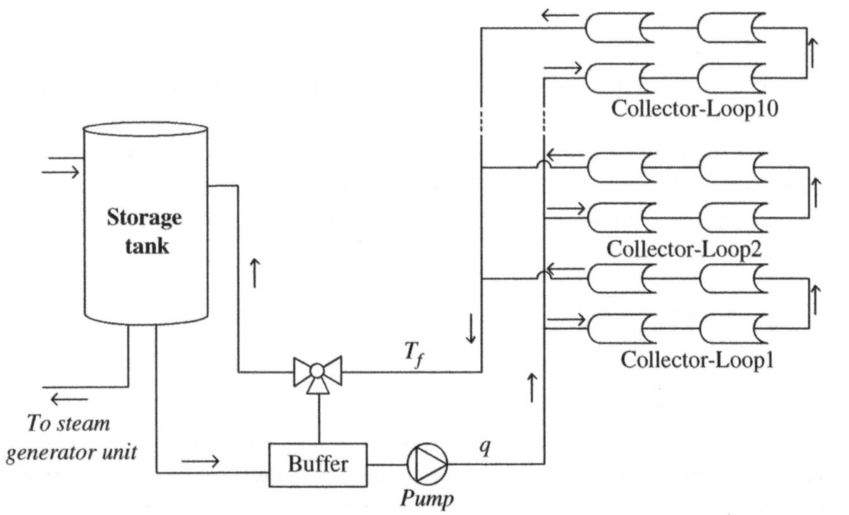

Figure 8.8 A schematic diagram of the ACUREX solar collector field.

Also, the metal-fluid heat transmission coefficient H_t, and the global coefficient of thermal loss H_l for the ACUREX solar collector field were determined experimentally as follows [37]:

$$H_t = \left(2.17 \times 10^6 - 5.01 \times 10^4 T_f + 4.53 \times 10^2 T_f^2 - 1.64 T_f^3 + 2.10 \times 10^{-3} T_f^4\right) q^{0.8},$$
$$H_l = 0.00249(T_f - T_a) - 0.06133 \tag{8.49}$$

Therefore, the properties of HTF (ρ_f and C_f), the metal-fluid heat transmission coefficient H_t and the global coefficient of thermal loss H_l are dependent on the system variables T_f, T_a and q. The above equations show the highly nonlinear dynamics of the process. Measured data can be used to directly control the plant instead of using the complex nonlinear distributed-parameter model (8.47).

In order to use a linear controller over the entire working interval, the measured data can be used to form the Koopman linear state-space model of the nonlinear system.

8.4.1 Data-driven Koopman Predictive Control of the ACUREX Solar Collector Field

This section applies the proposed DDKPC (8.33) to the ACUREX solar collector field. To achieve a reasonable trade-off between computational requirements and accuracy, the 174 m length loop path of the ACUREX solar collector field is divided into 10 parts ($n_p = 10$) and therefore, the number of system state variables will be $n = 21$ ($n = 2n_p + 1$), $2n_p$ states represent the outlet temperature of the n_p parts of the metal pipe and fluid and the other state represents the inlet temperature of the field. So, according to Algorithm 8.1, the Koopman state vector $\mathbf{z}$ consists of 21 state variables of the distributed-parameter model (8.47) and 650 Gaussian radial basis functions are selected as

$$\psi_i(\mathbf{x}) = e^{-\frac{\|\mathbf{x}-\mathbf{c}\|^2}{2}} \tag{8.50}$$

where c is randomly selected and $\|\cdot\|$ is the 2-norm. To construct the Hankel matrices in (8.32), an allowable input signal $\mathbf{u_d}$ with length $T = 2000$ is constructed as shown in Figure 8.9 and applied to the nonlinear model (8.47). A controller sampling time of 120 seconds has been chosen in this example. The prediction and control horizons are considered 3 seconds ($N = 6$), $\mathbf{Q}$ and $\mathbf{R}$ are selected as

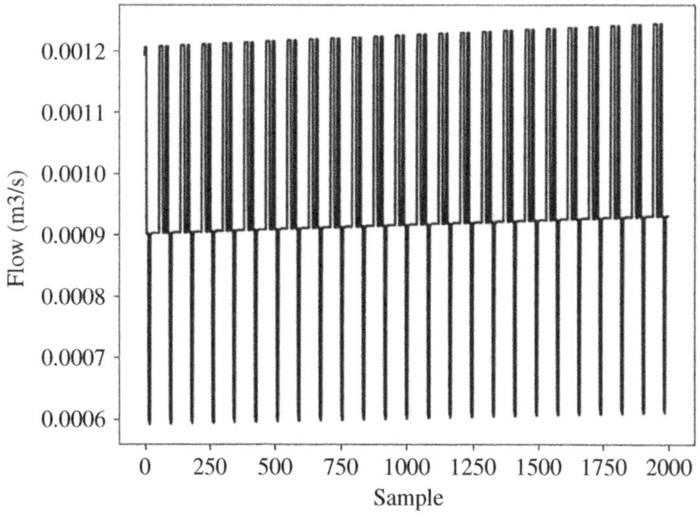

Figure 8.9 Persistently excitation signal $\mathbf{u_d}$.

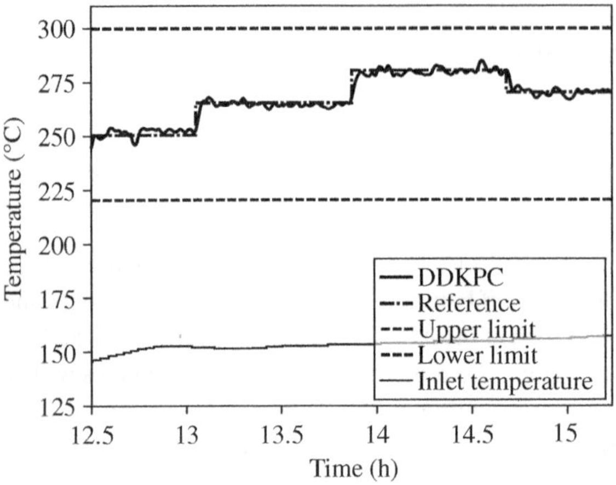

Figure 8.10 Output temperature of the ACUREX solar collector field distributed-parameter model by applying the proposed data-driven Koopman predictive control.

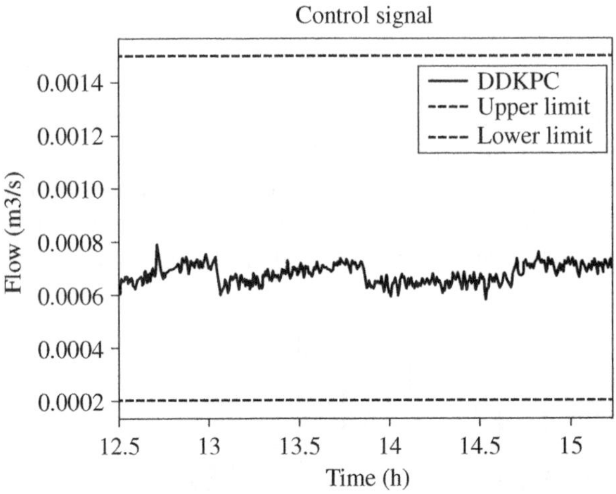

Figure 8.11 Irradiance from 12 : 30 to 15 : 18.

diagonal matrices with diagonal elements equal to 2 and 1, respectively, and the system sampling time is considered equal to 0.5 seconds. The simulation data are acquired from [37]. The optimisation problem (8.33) was solved using the SCS solver from the CVXPY library in the Python environment and a 4 GHz Intel Corei7 with 16 GB RAM. The results are shown in Figures 8.10 and 8.11 for $T_{ini} = 1$. The results indicate that the tracking characteristics are appropriate, the control signal is smooth and it is within an admissible range. Also, the time needed for calculating the controller is around 0.29 s, whereas using the nonlinear MPC, it is around 3 seconds. As shown in Figure 8.11, the input signal $\mathbf{u}$ with

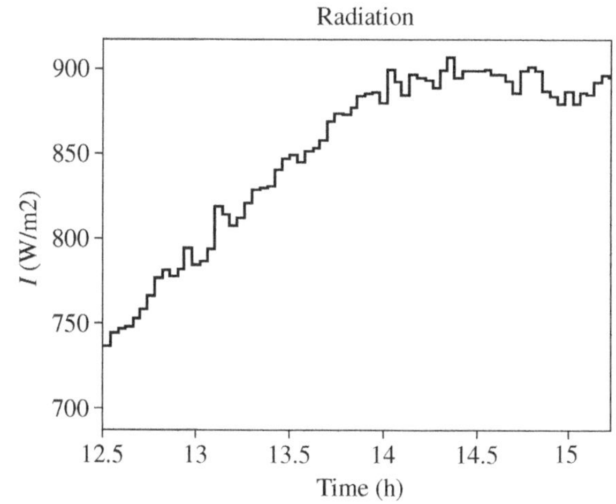

Figure 8.12 HTF flow of the ACUREX solar collector field model by applying the proposed data-driven Koopman predictive control.

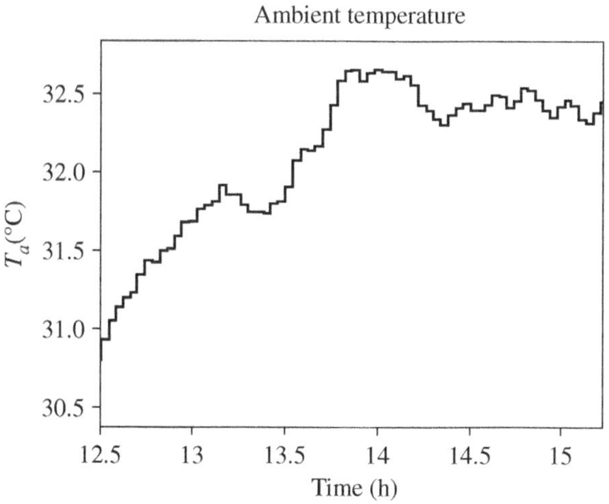

Figure 8.13 Ambient Temperature from 12 : 30 to 15 : 18.

persistently excitation order of $N+2T_{ini} = 8$ and selecting $T_{ini} = 1$, which is much less than order $p+T_{ini}+N = 657$ in Remark 8.2, is sufficient to have a data-driven predictive control with satisfactory performance. It is also necessary to have $N \geq 2$ in order to limit $l_f(z_N)$ in (8.40a). Also, Figures 8.12 and 8.13 and inlet temperature in Figure 8.10 show the measured disturbance signals that are acquired from Ref. [37].

Remark 8.11 Since the dimension of the Koopman linear state-space model is usually a large number, to reduce the dimension of the equations set (8.33), the equation $H_N(\mathbf{z}_{[0,T-1]})\mathbf{g} = \bar{\mathbf{z}}^T_{[0,N-1]}(t)$, that is the prediction part, is omitted. The simulation results show that this simplification does not interfere with the control of the system and leads to a considerable reduction in the control signal computation time.

8.5 Conclusion

Koopman's theory provides an effective tool for linearising nonlinear systems in the large. The Koopman operator yields an infinite-dimensional linear representation of nonlinear systems. To apply standard model-based or data-driven control systems analysis and design techniques, these infinite-dimensional representations can be reduced to finite-dimensional linear approximations through a selected set of basis functions. To achieve this, by appropriately selecting a set of observables, the DMD approach provides a finite-dimensional spectral approximation of the Koopman operator associated with the nonlinear system. The Koopman operator framework is similarly extended to controlled dynamical systems by applying the (EDMD). Also, in this chapter, a data-driven predictive controller based on the Koopman operator for unknown nonlinear systems is proposed. A major challenge in applying Koopman theory in control system design is identifying an equivalent Koopman linear model for the nonlinear system. Inspired by Willems' Fundamental Lemma, without identifying linear system's matrices, a Koopman-based data-driven predictive control scheme for nonlinear systems is proposed. Also, a data-driven Koopman predictive controller with uncertainty inclusion in the Koopman state variables, with closed-loop stability guarantees, is presented. The proposed data-driven Koopman predictive controller is employed for the control of the ACUREX solar collector field distributed-parameter model.

Problems

8.1 Consider the linear discrete-time system $\mathbf{x}_{k+1} = \boldsymbol{A}\mathbf{x}_k$ where $\boldsymbol{A} \in \mathbb{R}^{n\times n}$. Assume that the matrix $\boldsymbol{A}$ has distinct real eigenvalues. Then, $\boldsymbol{A}$ can be diagonalised as $\boldsymbol{A} = \boldsymbol{V\Lambda W}$, where the matrix of eigenvalues is $\Lambda = \text{Diag}\left[\lambda_1 \ \cdots \ \lambda_n\right]$ and the columns of the similarity transformation matrix $\boldsymbol{V}$ and $\boldsymbol{W}^T$ are the right and left eigenvectors of $\boldsymbol{A}$, respectively.

(a) Set $\mathbf{x}_k = \boldsymbol{A}^k\mathbf{x}_0$ where $\mathbf{x}_0$ is the initial state vector. Show that $\mathbf{x}_k = \sum_{i=1}^{n} \lambda_i^k \left(\boldsymbol{w}_i^T\mathbf{x}_0\right) \boldsymbol{v}_i$ where $\boldsymbol{v}_i$ and $\boldsymbol{w}_i^T$ are the column and row vectors of $\boldsymbol{V}$ and $\boldsymbol{W}$, respectively.

(b) Compare the results of part a with the Eqs. (8.21) and (8.22).

(c) What are the interpretations of $\mathbf{v}_i$ in the state-space and $\mathbf{v}_j$ from the Koopman space?
(d) What are the interpretations of $\mathbf{w}_i^T \mathbf{x}_0$ in the state-space and $\varphi_j(\mathbf{x}_0)$ from the Koopman space?
(e) Prove Eq. (8.8).

8.2 (a) Prove the linearity property of the Koopman operator for the observable functions $\psi_i(\mathrm{x}_k)$.
(b) Show that if $\varphi_1(\mathrm{x})$, $\varphi_2(\mathrm{x})$, … , and $\varphi_r(\mathrm{x})$ are the Koopman eigenfunctions of the system (8.1), then $\varphi_{r+1}(\mathrm{x}) = \prod_{i=1}^{r} \varphi_i(\mathrm{x})$ is also a Koopman eigenfunction and compute the corresponding eigenvalue of $\varphi_r(\mathrm{x})$.
(c) Using the result of part b explain why the dimension of the Koopman operator is infinite.

8.3 The eigenfunctions of a sufficiently smooth dynamical system can be derived from Eq. (8.9) utilising the Taylor or Laurent series for the eigenfunction.
(a) Consider the system $dx(t)/dt = x(t)$ and let the Taylor series expansion for an eigenfunction be given as $\varphi(x) = \sum_{i=0}^{\infty} c_i x^i$. Solve Eq. (8.9) to find the possible values for c_i and λ_i.
(b) Determine the eigenfunctions corresponding to $\lambda_1 = 1, \lambda_2 = 2, \ldots, \lambda_n = n$.
(c) Suggest two observable functions from these eigenfunctions.
(d) Consider the system $dx(t)/dt = ax^n(t)$ and $\varphi(x) = \sum_{i=-\infty}^{\infty} c_i x^i$ as the Laurent series expansion. Show that $\varphi(x) = e^{\frac{\lambda}{(1-n)a} x^{1-n}}$ is an eigenfunction for λ.

8.4 The equations governing a duffing oscillator are given as follows:

$$\dot{x}_1 = x_2$$

$$\dot{x}_2 = x_1 - x_1^3$$

(a) Determine the fixed points of the duffing oscillator.
(b) Is it possible to have a linear Koopman model representation for a system with multiple fixed points?
(c) Consider the Hamiltonian energy function as $H = x_2^2 - \frac{x_1^2}{2} + \frac{x_1^4}{4}$, use the Eq. (8.9) and verify if H is an eigenfunction and find the corresponding eigenvalue.

8.5 Consider the following Van der Pol oscillator continuous-time model:

$$\begin{bmatrix} \dot{x}_1 \\ \dot{x}_2 \end{bmatrix} = \begin{bmatrix} x_2 \\ 2x_2 - 10x_1^2 x_2 - 0.8x_1 - u \end{bmatrix}$$

where $\mathbf{x} = [x_1, x_2]^T$ is the state vector that represents the position and speed, respectively, and u is the control input. The constraints to be enforced are $-[2.5\,\text{m}, 2.5\,\text{m/s}]^T \leq \mathbf{x} \leq [2.5\,\text{m}, 2.5\,\text{m/s}]^T$, $-10\,\text{m}^2/\text{s} \leq u \leq 10\,\text{m}^2/\text{s}$.

(a) Construct DMD- and EDMD-based approximated linear Koopman models and compare their performances with the actual nonlinear model.

(b) Design a data-driven Koopman predictive controller for the Van der Pol oscillator using the results of part **a**.

8.6 Design a robust data-driven Koopman predictive controller for a non-affine system whose continuous-time model is

$$\begin{bmatrix} \dot{x}_1 \\ \dot{x}_2 \end{bmatrix} = \begin{bmatrix} x_2 \\ x_1^2 + 0.15u^3 + 0.1\left(1 + x_2^2\right) u + \sin(0.1u) \end{bmatrix}$$

where $\mathbf{x} = (x_1, x_2)$ are the state variables that represent the angle and its rate, respectively, while u is the control input. The constraints to be enforced are $-[2.5, 2.5]^T \leq \mathbf{x} \leq [2.5, 2.5]^T$, $-25 \leq u \leq 25$.

References

1 M. Korda and I. Mezić, "Linear predictors for nonlinear dynamical systems: Koopman operator meets model predictive control," *Automatica*, vol. 93, pp. 149–160, 2018.

2 I. Markovsky, J. C. Willems, S. Van Huffel, and B. De Moor, *Exact and Approximate Modeling of Linear Systems: A Behavioral Approach*. SIAM, 2006.

3 B. O. Koopman, "Hamiltonian systems and transformation in Hilbert space," *Proceedings of the National Academy of Sciences*, vol. 17, no. 5, pp. 315–318, 1931.

4 S. L. Brunton, M. Budišić, E. Kaiser, and J. N. Kutz. "Modern Koopman theory for dynamical systems," *arXiv preprint arXiv:2102.12086*, 2021.

5 S. L. Brunton and J. N. Kutz, *Data-driven Science and Engineering: Machine Learning, Dynamical Systems, and Control*. Cambridge University Press, 2022.

6 I. Mezić and A. Banaszuk, "Comparison of systems with complex behavior," *Physica D: Nonlinear Phenomena*, vol. 197, no. 1–2, pp. 101–133, 2004.

7 M. O. Williams, I. G. Kevrekidis, and C. W. Rowley, "A data–driven approximation of the Koopman operator: extending dynamic mode decomposition," *Journal of Nonlinear Science*, vol. 25, no. 6, pp. 1307–1346, 2015.

8 Y. Lian and C. N. Jones, "Learning feature maps of the Koopman operator: a subspace viewpoint," in *2019 IEEE 58th Conference on Decision and Control (CDC)*, 2019: IEEE, pp. 860–866.

9 M. O. Williams, M. S. Hemati, S. T. Dawson, I. G. Kevrekidis, and C. W. Rowley, "Extending data-driven Koopman analysis to actuated systems," *IFAC-PapersOnLine*, vol. 49, no. 18, pp. 704–709, 2016.

10 Y. Lian, R. Wang, and C. N. Jones. "Koopman based data-driven predictive control,". *arXiv preprint arXiv:2102.05122*, 2021.

11 A. Mauroy, Y. Susuki, and I. Mezić, *Koopman Operator in Systems and Control.* Springer, 2020.

12 I. Mezić, "Spectral properties of dynamical systems, model reduction and decompositions," *Nonlinear Dynamics*, vol. 41, no. 1, pp. 309–325, 2005.

13 P. J. Schmid, "Dynamic mode decomposition of numerical and experimental data," *Journal of fluid mechanics*, vol. 656, pp. 5–28, 2010.

14 S. Skogestad and I. Postlethwaite, *Multivariable Feedback Control: Analysis and Design*. Wiley, 2005.

15 T. Gholaminejad and A. Khaki-Sedigh, "Stable data-driven Koopman predictive control: concentrated solar collector field case study," *IET Control Theory & Applications*, vol. 17, no. 9, pp. 1116–1131, 2023.

16 Q. Li, F. Dietrich, E. M. Bollt, and I. G. Kevrekidis, "Extended dynamic mode decomposition with dictionary learning: a data-driven adaptive spectral decomposition of the Koopman operator," *Chaos: An Interdisciplinary Journal of Nonlinear Science*, vol. 27, no. 10, p. 103111, 2017.

17 E. Yeung, S. Kundu, and N. Hodas, "Learning deep neural network representations for Koopman operators of nonlinear dynamical systems," in *2019 American Control Conference (ACC)*, 2019: IEEE, pp. 4832–4839.

18 B. Lusch, J. N. Kutz, and S. L. Brunton, "Deep learning for universal linear embeddings of nonlinear dynamics," *Nature Communications*, vol. 9, no. 1, pp. 1–10, 2018.

19 T. Gholaminejad and A. Khaki-Sedigh, "Stable deep Koopman model predictive control for solar parabolic-trough collector field," *Renewable Energy*, vol. 198, pp. 492–504, 2022.

20 G. E. Fasshauer, *Meshfree Approximation Methods with MATLAB.* World Scientific, 2007.

21 J. C. Willems, P. Rapisarda, I. Markovsky, and B. L. De Moor, "A note on persistency of excitation," *Systems & Control Letters*, vol. 54, no. 4, pp. 325–329, 2005.

22 J. Berberich and F. Allgöwer, "A trajectory-based framework for data-driven system analysis and control," in *2020 European Control Conference (ECC)*, 2020: IEEE, pp. 1365–1370.

23 J. Berberich, A. Koch, C. W. Scherer, and F. Allgöwer, "Robust data-driven state-feedback design," in *2020 American Control Conference (ACC)*, 2020: IEEE, pp. 1532–1538.

24 H. J. Van Waarde, J. Eising, H. L. Trentelman, and M. K. Camlibel, "Data informativity: a new perspective on data-driven analysis and control," *IEEE Transactions on Automatic Control*, vol. 65, no. 11, pp. 4753–4768, 2020.

25 J. L. Proctor, S. L. Brunton, and J. N. Kutz, "Generalizing Koopman theory to allow for inputs and control," *SIAM Journal on Applied Dynamical Systems*, vol. 17, no. 1, pp. 909–930, 2018.

26 J. Berberich, J. Köhler, M. A. Müller, and F. Allgöwer, "Data-driven model predictive control with stability and robustness guarantees," *IEEE Transactions on Automatic Control*, vol. 66, no. 4, pp. 1702–1717, 2020.

27 L. Huang, J. Zhen, J. Lygeros, and F. Dörfler. "Robust data-enabled predictive control: Tractable formulations and performance guarantees," *arXiv preprint arXiv:2105.07199*, 2021.

28 L. Huang, J. Coulson, J. Lygeros, and F. Dörfler, "Decentralized data-enabled predictive control for power system oscillation damping," *IEEE Transactions on Control Systems Technology*, vol. 30, no. 3, pp. 1065–1077, 2021.

29 Y. Yu, S. Talebi, H. J. van Waarde, U. Topcu, M. Mesbahi, and B. Açıkmeşe, "On controllability and persistency of excitation in data-driven control: extensions of willems' fundamental lemma," in *2021 60th IEEE Conference on Decision and Control (CDC)*, 2021: IEEE, pp. 6485–6490.

30 E. F. Camacho, A. J. Gallego, A. J. Sanchez, and M. Berenguel, "Incremental state-space model predictive control of a Fresnel solar collector field," *Energies*, vol. 12, no. 1, p. 3, 2018.

31 E. F. Camacho, A. J. Gallego, J. M. Escaño, and A. J. Sánchez, "Hybrid nonlinear MPC of a solar cooling plant," *Energies*, vol. 12, no. 14, p. 2723, 2019.

32 E. Masero, J. R. D. Frejo, J. M. Maestre, and E. F. Camacho, "A light clustering model predictive control approach to maximize thermal power in solar parabolic-trough plants," *Solar Energy*, vol. 214, pp. 531–541, 2021.

33 A. Sánchez, A. Gallego, J. Escaño, and E. Camacho, "Temperature homogenization of a solar trough field for performance improvement," *Solar Energy*, vol. 165, pp. 1–9, 2018.

34 A. J. Gallego and E. F. Camacho, "Adaptative state-space model predictive control of a parabolic-trough field," *Control Engineering Practice*, vol. 20, no. 9, pp. 904–911, 2012.

35 A. Gallego, F. Fele, E. F. Camacho, and L. Yebra, "Observer-based model predictive control of a parabolic-trough field," *Solar Energy*, vol. 97, pp. 426–435, 2013.

36 A. Alsharkawi and J. A. Rossiter, "Dual mode mpc for a concentrated solar thermal power plant," *IFAC-PapersOnLine*, vol. 49, no. 7, pp. 260–265, 2016.

37 A. Alsharkawi, "*Automatic Control of a Parabolic Trough Solar Thermal Power Plant*," Master Thesis, University of Sheffield, 2017.

9

Model-free Adaptive Control Design

9.1 Introduction

Model-free adaptive control (MFAC) is a popular methodology in the class of data-driven control systems. The term MFAC is used in different contexts. According to the bibliographic database of Scopus, the first paper that simultaneously addressed *model-free* and *adaptive control* appeared in an International Federation of Automatic Control (IFAC) 1991 symposium on Intelligent Tuning and Adaptive Control [1]. This paper employed expert systems and pattern recognition techniques for proportional integral (PI) controller tuning. It can therefore be categorised in the class of model-free adaptive PI controllers. The subsequent two papers, according to Scopus, appeared in 1994, and both papers used neural networks for their underlying design strategy. Neural networks became the dominant design methodology in the MFAC literature. Following the PhD thesis [2] in 1994, the first paper appeared in 1997 that proposed MFAC from a different perspective to neural networks [3]. The proposed MFAC applicable to a class of nonlinear systems was based on the so-called *pseudo-partial-derivative* (PPD) concept. The PPD approach to MFAC design is presented in this chapter. The MFAC systems have the following general features:

- It is assumed that no precise quantitative information about the plant is available.
- In the design process, no system identification techniques are utilised.
- Only minimum plant assumptions are necessary.
- There are no complicated tuning procedures.
- Mathematical proofs for closed-loop stability are available.

In the identification and adaptive control methodologies, plants can be categorised as white boxes, grey boxes or black boxes, as shown in Figure 9.1. Model-based control design techniques are most suitable for white box plants, where quantitative information on the plant is available. In many practical

An Introduction to Data-Driven Control Systems, First Edition. Ali Khaki-Sedigh.

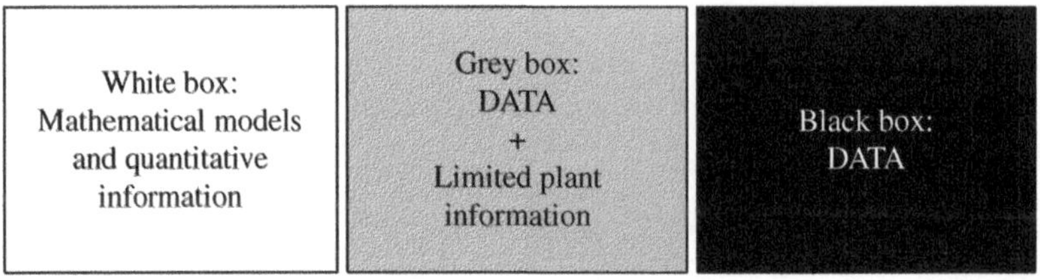

Figure 9.1 Different modelling boxes in control systems identification, modelling and design.

cases, there is limited information about the process, and the accuracy of the information may be questionable.

In these cases, the designer is dubious about the plant information and is encountering a grey box problem. On the other hand, the only available information about the plant in the black box problems is the measured input–output data. It is interesting to note that most present industrial processes can be categorised as grey box problems. MFAC can deal with black, grey and white box problems. However, it is most suitable to employ MFAC to deal with the grey and black box problems, and the designer should be more vigilant in the case of black box problems. An attractive feature of the MFAC systems is that they do not require a system identification mechanism as in the classical adaptive control approaches. Hence, a general system stability criterion can be developed, and the closed-loop MFAC system stability is guaranteed for linear, nonlinear, time-invariant or time-varying plants.

Linearisation is a widely used approach to handle the complexities of nonlinear systems and adaptive control of real plants. An overview of dynamic linearisation techniques and their application in control is given in Ref. [4]. The commonly applied linearisation techniques are feedback linearisation, Taylor's linearisation, piecewise linearisation, polynomial functions approximation and the more general Koopman approach presented in Chapter 8. However, apart from Koopman, which leads to very high-order linear models, the other techniques would require a mathematical model of the plant. In the case of feedback linearisation, the accuracy of the model is essential, and in the Taylor approach, unmodelled dynamics would be evident due to deleting higher-order terms. The main problems associated with the piecewise linearisation techniques are computational issues and the breakpoints selection and switching. A linearised model to be employed in an adaptive data-driven control application should have a *simple structure*, *few adjustable parameters* and be *readily applicable* to measured input–output data. The different forms of dynamic linearisation discussed in this chapter and presented in Ref. [5] have such features and are employed in the MFAC architecture. These methodologies capture the dynamical system's behaviour by the variations of the output signal resulting from the changes in the current input signal. However, it must be emphasised that the dynamic linearisation data models are most suited for controller design and are control-design-oriented, and cannot be used for other system analysis purposes. They provide a *virtual* description of the closed-loop input–output data. Controller structures in MFAC are derived from these dynamically linearised data models and are, therefore, totally data-driven schemes. This is in contrast to the classical adaptive control methodologies, where the controller structure is derived from the plant's model.

MFAC and its applications are presented in many research papers; in particular, the textbook [5] deals explicitly with MFAC; also Ref. [6] presents MFAC principles and more advanced variants of the MFAC approach.

9.2 The Dynamic Linearisation Methodologies

The dynamic linearisation concept provides an equivalent linearised data model of an unknown nonlinear plant based on the closed-loop input–output data. The linearised data model is indeed a virtual equivalent dynamic linearisation data model and is fundamentally different from the standard mathematical analytically derived or identified models available in the classical adaptive control literature. The dynamic linearisation methodologies provide an alternative equivalent model in the strict input–output sense. That is, a dynamic linearised data model does not provide other plant information, such as the poles and zeros locations and cannot be used for objectives other than control systems design, such as estimation and fault diagnosis.

Consider the nonlinear discrete-time plants described by

$$y(t+1) = f(y(t), \dots, y(t-n_y), u(t), \dots, u(t-n_u)) \tag{9.1}$$

where $u(t) \in \mathbb{R}$ and $y(t) \in \mathbb{R}$ are the plant's input and output at the time instant t, respectively. The nonlinear function $f(\cdot)$, n_y and n_u are assumed unknown. But the measured input–output sequences are assumed to be available at $t = 0,1,2, \dots$. The dynamic linearisation approach provides at every sampling time instant, a pseudo local dynamic data model of (9.1) with unknown parameters. The main dynamic linearisation techniques are as follows:

- The compact form dynamic linearisation (CFDL). Where in the case of single input single output (SISO) plants, a scalar parameter called the *PPD* is obtained.
- The partial form dynamic linearisation (PFDL).
- The full form dynamic linearisation (FFDL).

Time-varying model parameters in the above three models are initially unknown and are estimated and recursively updated at each sampling time once a new set of measured input–output data are received.

9.2.1 The Compact Form Dynamic Linearisation

The CFDL of the general nonlinear plant described by (9.1) is presented in this chapter based on the materials from [5]. The following two assumptions regarding (9.1) are the minimum necessary assumptions regarding the nonlinear plant:

Assumption 1. The partial derivatives of (9.1) are continuous with respect to the control input $u(t)$.

Assumption 2. The nonlinear plant given by (9.1) satisfies the *generalised Lipschitz condition*. That is

$$|y(t_1+1)-y(t_2+1)| \leq b|u(t_1)-u(t_2)| \text{ for any } t_1 \neq t_2, t_1, t_2 \geq 0 \text{ and } u(t_1) \neq u(t_2)$$

where b is a positive constant.

Let Δ be subsequently defined as the change operator, that is $\Delta\zeta(t)=\zeta(t)-\zeta(t-1)$ for any variable $\zeta(t)$. Then from (9.1), we have

$$\Delta y(t+1) = f(y(t), \dots, y(t-n_y), u(t), \dots, u(t-n_u)) - f(y(t-1), \dots, y(t-n_y-1), u(t-1), \dots, u(t-n_u-1)) \quad (9.2)$$

Adding and subtracting $f(y(t), \dots, y(t-n_y), u(t-1), u(t-1), \dots, u(t-n_u))$ to (9.2), and defining

$$\mathcal{Y}(t) \triangleq f(y(t), \dots, y(t-n_y), u(t-1), u(t-1), \dots, u(t-n_u)) - f(y(t-1), \dots, y(t-n_y-1), u(t-1), \dots, u(t-n_u-1))$$

Gives

$$\Delta y(t+1) - \mathcal{Y}(t) = f(y(t), \dots, y(t-n_y), u(t), u(t-1), \dots, u(t-n_u)) - f(y(t), \dots, y(t-n_y), u(t-1), u(t-1), \dots, u(t-n_u))$$

Let $a = (y(t), \dots, y(t-n_y), u(t-1), u(t-1), \dots, u(t-n_u))$ and $b = (y(t), \dots, y(t-n_y), u(t), u(t-1) \dots, u(t-n_u))$, then according to Cauchy's mean-value theorem also known as the extended mean-value theorem [7], $\exists c \in (a, b)$ such that

$$\left.\frac{\partial f}{\partial u}\right|_c = \frac{f(b)-f(a)}{u(t)-u(t-1)} = \frac{\Delta y(t+1)-\mathcal{Y}(t)}{\Delta u(t)}$$

Hence,

$$\Delta y(t+1) = \left.\frac{\partial f}{\partial u}\right|_c \Delta u(t) + \mathcal{Y}(t) \quad (9.3)$$

At each time instant t, let

$$\mathcal{Y}(t) = \eta(t)\Delta u(t) \quad (9.4)$$

Since $\Delta u(t) \neq 0$, (9.4) has a unique solution $\overline{\eta}(t)$, and (9.3) can be written as

$$\Delta y(t+1) = \phi_c(t)\Delta u(t) \tag{9.5}$$

where $\phi_c(t) = \overline{\eta}(t) + \frac{\partial f}{\partial u}\Big|_c \in \mathbb{R}$. $\phi_c(t)$ is called the PPD, and Assumption 2 directly implies that that $\phi_c(t)$ is bounded for any time instant t. Equation (9.5) is called the *CFDL data model* of (9.1).

Remark 9.1 Note that (9.5) is a linear time-varying data model of the nonlinear plant given by (9.1). $\phi_c(t)$ cannot, in general, be written in a closed-form analytical equation. This is because the input–output data up to t are used to calculate the PPD, and PPD is determined by the mean value of the partial derivative at some point within an interval and an added nonlinear term. The mean value in Cauchy's mean-value theorem cannot be formulated analytically even for a simple known nonlinear function.

Remark 9.2 The data model (9.5) is purely derived from the measured input–output data. Apart from the two stated assumptions that hold in most practical nonlinear systems, no other assumptions or information regarding the plant structures, parameters or dynamics is required. Hence, it is a genuinely data-driven model.

Remark 9.3 In some applications, we may have $\Delta u(t) = 0$ for some t. In this case, there will exist a time instant t_0 such that

$$\Delta u(j) \begin{cases} = 0, & if j = 1, \dots, t_{0-1} \\ \neq 0, & if j = t_0 \end{cases}$$

Then for any $t \geq t_0$, $\exists \sigma_t$ such that

$$\Delta u(t-j) \begin{cases} = 0, & if j = 0, \dots, \sigma_t - 2 \\ \neq 0, & if j = \sigma_t - 1 \end{cases}$$

This ensures the existence of a time instant after which the non-zero condition holds. Then, the equivalent data model of (9.1) is given as follows:

$$y(t+1) - y(t - \sigma_k + 1) = \phi_c(t)(u(t) - u(t - \sigma_k)) \tag{9.6}$$

9.2.2 The Partial Form Dynamic Linearisation

The scalar time-varying parameter $\phi_c(t)$ of the CFDL approach introduced in Section 9.2.1 has the following features:

- In providing an equivalent linear representation of the plant given by (9.1), it should capsulise all the characteristics of an unknown nonlinear time-varying

plant into $\phi_c(t)$. This obligation makes $\phi_c(t)$ highly complicated and analytically unpresentable.
- In deriving $\phi_c(t)$ the change of $y(t)$ at the next sampling time and the change of $u(t)$ at the current sampling time are utilised.

However, it is well-known that in dynamical systems, the effect of many past inputs (not just the change in the present input) is evident in the formation of $y(t)$. Hence, in this section, a modification of the CFDL approach is presented that considers the effect of past inputs on the next plant output. This dynamic linearisation approach is called the *PFDL* methodology. In PFDL, multiple degrees of freedom are added to capture the nonlinear time-varying plant characteristics of the plant given by (9.1).

Let us define a time length L called the input *linearisation length constant* (LLC) that determines the number of past inputs the designer wishes to participate in the proposed dynamically linearised data model. A moving time window $[t-L+1,t]$ and a past control input's vector is subsequently defined as follows:

$$U_L(t) = [u(t), \ldots, u(t-L+1)]^T \tag{9.7}$$

and $U_L(t)=\mathbf{0}_L$, where $\mathbf{0}_L$ is a zero vector of dimension L for $t<0$.

Assumption 1′. Assumption 1 is extended to accommodate the continuity of the partial derivatives of the function $f(\cdot)$ with respect to the inputs $u(t), \ldots, u(t-L+1)$.

Assumption 2′. The generalised Lipschitz condition in Assumption 2 is modified to hold for any two different time instants t_1 and t_2 as follows: $|y(t_1+1)-y(t_2+1)| \le b\|U_L(t_1)-U_L(t_2)\|$ for $U_L(t_1)\ne U_L(t_2)$, where $\|\cdot\|$ denotes the Euclidean norm, and b is a positive constant.

Similar to the derivation in Section 9.2.1 for the CFDL approach, assuming that the above two assumptions hold, the following result can be established [5].

Consider the nonlinear plant given by (9.1). Then, for any selected L, provided that $\|\Delta U_L(t)\|\ne 0$, where $\Delta U_L(t)=U_L(t)-U_L(t-1)$, there exists a time-varying vector $\boldsymbol{\phi}_{p,L}(t)\in\mathbb{R}^L$, called the *pseudo-gradient* (PG) such that (9.1) can be presented in the following PFDL data model:

$$\Delta y(t+1) = \boldsymbol{\phi}_{p,L}^T(t)\Delta U_L(t) \tag{9.8}$$

where $\boldsymbol{\phi}_{p,L}(t)=[\phi_1(t), \ldots, \phi_L(t)]^T$ is bounded for all t. To further exploit $\boldsymbol{\phi}_{p,L}(t)$ in more detail, (9.1) gives the output at the last sampling time as

$$y(t) = f(y(t-1), \ldots, y(t-n_y-1), u(t-1), \ldots, u(t-n_u-1)) \tag{9.9}$$

and define

$$\mathcal{Y}_1(y(t-1), \ldots, y(t-n_y-1), u(t-1), u(t-2), \ldots, u(t-n_u-1))$$

$$\triangleq f(f(y(t-1),\ldots,y(t-n_y-1),u(t-1),\ldots,u(t-n_u-1)),$$
$$y(t-1),\ldots,y(t-n_y),u(t-1),u(t-1),\ldots,u(t-n_u))$$
$$-f(y(t-1),\ldots,y(t-n_y-1),u(t-1),\ldots,u(t-n_u-1)) \quad (9.10)$$

and

$$\mathcal{Y}_2(y(t-2),\ldots,y(t-n_y-2),u(t-2),u(t-3),\ldots,u(t-n_u-2))$$
$$\triangleq \mathcal{Y}_1(f(y(t-2),\ldots,y(t-n_y-2),u(t-2),u(t-3),\ldots,u(t-n_u-2)),$$
$$y(t-2),\ldots,y(t-n_y-1),u(t-2),u(t-2),\ldots,u(t-n_u-1)) \quad (9.11)$$

and in general for $i=2,\ldots,L$,

$$\mathcal{Y}_i(y(t-i),\ldots,y(t-n_y-i),u(t-i),u(t-i-1),\ldots,u(t-n_u-i))$$
$$\triangleq \mathcal{Y}_{i-1}(f(y(t-i),\ldots,y(t-n_y-i),u(t-i),u(t-i-1),\ldots,u(t-n_u-i)),$$
$$y(t-i),\ldots,y(t-n_y-i+1),u(t-i),u(t-i),\ldots,u(t-n_u-i+1)) \quad (9.12)$$

Similar to the arguments that led to (9.3), by substitution of (9.10), we have for some c in the corresponding Cauchy's mean-value theorem:

$$\Delta y(t+1) = \left.\frac{\partial f}{\partial u(t)}\right|_c \Delta u(t) + \mathcal{Y}_1(y(t-1),\ldots,y(t-n_y-1),u(t-1),$$
$$u(t-2)\ldots,u(t-n_u-1)) - \mathcal{Y}_1(y(t-1),\ldots,y(t-n_y-1),$$
$$u(t-2),u(t-2)\ldots,u(t-n_u-1)) + \mathcal{Y}_1(y(t-1),\ldots,$$
$$y(t-n_y-1),u(t-2),u(t-2)\ldots,u(t-n_u-1))$$
$$= \left.\frac{\partial f}{\partial u(t)}\right|_c \Delta u(t) + \left.\frac{\partial \mathcal{Y}_1}{\partial u(t-1)}\right|_c \Delta u(t-1) + \mathcal{Y}_2(y(t-2),\ldots,$$
$$y(t-n_y-2),u(t-2),\ldots,u(t-n_u-2)) \quad (9.13)$$

Repeatedly applying a similar argument, the general equation for a given L is derived as follows:

$$\Delta y(t+1) = \left.\frac{\partial f}{\partial u(t)}\right|_c \Delta u(t) + \left.\frac{\partial \mathcal{Y}_1}{\partial u(t-1)}\right|_c \Delta u(t-1) + \ldots + \left.\frac{\partial \mathcal{Y}_{L-1}}{\partial u(t-L+1)}\right|_c$$
$$\times \Delta u(t-L+1) + \mathcal{Y}_L(y(t-L),\ldots,y(t-n_y-L),$$
$$u(t-L),\ldots,u(t-n_u-L)) \quad (9.14)$$

Similarly, as in (9.4), for a given L, we have

$$\mathcal{Y}_L(y(t-L),\ldots,y(t-n_y-L),u(t-L),\ldots,u(t-n_u-L)) = \boldsymbol{\eta}^T(t)\Delta U_L(t) \quad (9.15)$$

Assuming $\|\Delta U_L(t)\| \neq 0$, for a solution to (9.15) $\overline{\eta}(t)$, from comparing (9.14) and (9.8), the PG can be represented as in the following equation:

$$\boldsymbol{\phi}_{p,L}(t) = \overline{\eta}(t) + \left[\left.\frac{\partial f}{\partial u(t)}\right|_c, \left.\frac{\partial \mathcal{Y}_1}{\partial u(t-1)}\right|_c, \ldots, \left.\frac{\partial \mathcal{Y}_{L-1}}{\partial u(t-L+1)}\right|_c \right]^T \tag{9.16}$$

Remark 9.4 In calculating the PG given by (9.8), satisfying the minimum plant assumptions, only measured input–output data up to time t are required. The PFDL data model is a data-driven linear time-varying model of (9.1) with an adjustable length L selected by the designer. The adjustable length L is a design parameter, and different PFDL data models are obtained by selecting different LLC L. Also, the PFDL data model takes into account the effect of control input changes within the fixed length time window $[t-L+1,t]$ on the plant output change at $t+1$. This provides more degrees of freedom for representing the unknown plant characteristics than the CFDL model with one parameter.

9.2.3 The Full Form Dynamic Linearisation

The final dynamic linearisation method presented in this section is the FFDL approach. In the PFDL data modelling approach, the effect of past inputs in a specified LLC was considered to form the PG in the equivalent time-varying linear model of the nonlinear plant (9.1). However, a better time-varying linear picture of the unknown nonlinear plant is attainable if the past plant outputs are considered in obtaining the next sampling time output in addition to the previous inputs and the current output. In this framework, degrees of freedom are increased in comparison with the PFDL data model to capture the main dynamical characteristics of the unknown plant more effectively.

Define two moving time windows $[t-L_u+1,t]$ and $[t-L_y+1,t]$ corresponding to the inputs and the outputs of the plant, respectively. The positive integers L_u and L_y are called the plant *pseudo orders* or the *control input* LLC and the *controlled output* LLC, respectively. Then, a vector is defined to entail the respective outputs and inputs in these time windows as follows:

$$H_{L_y,L_u}(t) = [y(t), \ldots, y(t-L_y+1), u(t), \ldots, u(t-L_u+1)]^T \tag{9.17}$$

and $H_{L_y,L_u}(t) = \mathbf{0}_{L_y+L_u}$ for $t<0$.

Assumption 1″. Assumption 1 is extended to accommodate the continuity of the partial derivatives of the function $f(\cdot)$ with respect to all the variables.

Assumption 2″. The generalised Lipschitz condition in Assumption 2 is modified to hold for any two different time instants t_1 and t_2 as follows: $|y(t_1+1) - y(t_2+1)| \leq b\,\|H_{L_y,L_u}(t_1) - H_{L_y,L_u}(t_2)\|$ for $H_{L_y,L_u}(t_1) \neq H_{L_y,L_u}(t_2)$, and b is a positive constant.

Similar to the derivation in Section 9.2.2 for the PFDL approach and avoiding any details, assuming that the above two assumptions hold, the following result can be established [5]:

Consider the nonlinear plant given by (9.1). Then, for any selected $0 \leq L_y \leq n_y$ and $0 \leq L_u \leq n_u$, provided that $\|\Delta H_{L_y,L_u}(t)\| \neq 0$, there exists a time-varying vector $\boldsymbol{\phi}_{f,L_y,L_u}(t) \in \mathbb{R}^{L_y+L_u}$, called the PG such that (9.1) can be presented in the following FFDL data model:

$$\Delta y(t+1) = \boldsymbol{\phi}^T_{f,L_y,L_u}(t)\Delta H_{L_y,L_u}(t) \tag{9.18}$$

where $\boldsymbol{\phi}_{f,L_y,L_u}(t) = [\phi_1(t), \ldots, \phi_{L_y}(t), \ldots, \phi_{L_y+L_u}(t)]^T$ is bounded for all t. Where it can be shown that for some c in the corresponding Cauchy's mean-value theorem, we have

$$\boldsymbol{\phi}_{f,L_y,L_u}(t) = \overline{\eta}(t) + \left[\left.\frac{\partial f}{\partial y(t)}\right|_c, \ldots, \left.\frac{\partial f}{\partial y(t-L_y)}\right|_c, \left.\frac{\partial f}{\partial u(t)}\right|_c, \ldots, \left.\frac{\partial f}{\partial u(t-L_u)}\right|_c \right]^T \tag{9.19}$$

And $\overline{\eta}(t)$ is derived as the solution to the following equation:

$$\mathcal{Y}(t) = \boldsymbol{\eta}^T(t)\Delta H_{L_y,L_u}(t) \tag{9.20}$$

In the derivation of (9.19), after deriving $\Delta y(t+1)$, the following term:

$$f(y(k-1), \ldots, y(k-L_y), y(k-L_y), \ldots, y(k-n_y), u(k-1), \ldots,$$
$$u(k-L_u), u(k-L_u), \ldots, u(k-n_u))$$

is added and subtracted to $\Delta y(t+1)$. Similar to the previous derivations $\mathcal{Y}(t)$ is defined as

$$\begin{aligned}\mathcal{Y}(t) \triangleq\, & f(y(k-1), y(k-2), \ldots, y(k-L_y), y(k-L_y), \ldots, y(k-n_y),\\ & u(k-1), u(k-2), \ldots, u(k-L_u), u(k-L_u), \ldots, u(k-n_u))\\ & - f(y(k-1), y(k-2), \ldots, y(k-L_y), y(k-L_y-1), \ldots, y(k-n_y-1),\\ & u(k-1), u(k-2), \ldots, u(k-L_u), u(k-L_u-1), \ldots, u(k-n_u-1))\end{aligned}$$

And Cauchy's mean-value theorem is then applied to obtain the desired equation.

Remark 9.5 CFDL and PFDL are two special cases of FFDL. For $L_y = 0, L_u = 1$ and $L_y = 0, L_u = L$, CFDL and PFDL are attained, respectively.

Remark 9.6 The pseudo orders can be considered as design parameters, and there is a trade-off between computational burden and the dynamically linearised model accuracy for complicated and simple plants. In the case of a known plant, the pseudo orders would be naturally selected as their exact values.

Example 9.1 Consider the following linear deterministic autoregressive moving average (DARMA) model that is widely used in classical adaptive control techniques [8].

$$y(t+1) = \alpha(q^{-1})y(t) + \beta(q^{-1})u(t)$$

where q is the shift operator and

$$\alpha(q^{-1}) = \alpha_0 + \alpha_1 q^{-1} + \ldots + \alpha_{n-1} q^{-(n-1)}$$
$$\beta(q^{-1}) = \beta_0 + \beta_1 q^{-1} + \ldots + \beta_{m-1} q^{-(m-1)}$$

Let $\Delta y(t+1) = y(t+1) - y(t)$, then we have $\Delta y(t+1) = \alpha(q^{-1})y(t) + \beta(q^{-1})u(t) - (\alpha(q^{-1})y(t-1) + \beta(q^{-1})u(t-1))$, which gives

$$\Delta y(t+1) = \alpha(q^{-1})\Delta y(t) + \beta(q^{-1})\Delta u(t)$$

and

$$\Delta y(t+1) = \alpha_0 \Delta y(t) + \alpha_1 \Delta y(t-1) + \ldots + \alpha_{n-1}\Delta y(t-n+1) + \beta_0 \Delta u(t)$$
$$+ \beta_1 \Delta u(t-1) + \ldots + \beta_{m-1}\Delta u(t-m+1)$$

where α_i, $i = 0, \ldots, n-1, \beta_j, j = 0, \ldots, m-1$ are constant coefficients, and n, m are assumed to be known in the model-based adaptive control techniques. In the case of model-based techniques, $L_y = n$, $L_u = m$, and the FFDL data model can be written as (9.18) with,

$$\boldsymbol{\phi}_{f,n,m}(t) = [\alpha_0, \alpha_1, \ldots, \alpha_{n-1}, \beta_0, \beta_1, \ldots, \beta_{m-1}]$$
$$H_{n,m}(t) = [y(t), \ldots, y(t-n+1), u(t), \ldots u(t-m+1)]$$

Note that $\boldsymbol{\phi}_{f,\,n,\,m}(t)$ in this case, is a time-invariant vector. However, in the case of the linear unknown DARMA model, (9.18) is written as

$$\boldsymbol{\phi}_{f,n,m}(t) = [\phi_1(t), \ldots, \phi_{L_y-1}(t), \phi_{L_y}(t), \ldots, \phi_{L_y+L_u}(t)]$$
$$H_{L_y,L_u}(t) = [y(t), \ldots, y(t-L_y+1), u(t), \ldots u(t-L_u+1)]$$

where L_y, L_u are selected by the designer, and the PG vector is time-varying, while the plant is originally time-invariant. It is clear that if the plant order and relative degree are known then, $L_y = n$ and $L_u = m$.

9.3 Extensions of the Dynamic Linearisation Methodologies to Multivariable Plants

An appealing feature of the dynamic linearisation methodologies presented in Section 9.2 is the relatively straightforward extension of the SISO results to the multivariable plants. Consider the following nonlinear multivariable plant:

$$\mathbf{y}(t+1) = \boldsymbol{f}(\mathbf{y}(t), \ldots, \mathbf{y}(t-n_y), \mathbf{u}(t), \ldots, \mathbf{u}(t-n_u)) \tag{9.21}$$

where $\mathbf{u}(t) \in \mathbb{R}^m$ and $\mathbf{y}(t) \in \mathbb{R}^m$ are the plant's input and output vectors at the time instant t, respectively. The results in this section are presented for square multivariable plants. However, these results can be readily extended to non-square multi-input single-output multivariable plants, as given in [5]. The nonlinear m-dimensional vector function $\boldsymbol{f}(\cdot) = (f_1(\cdot), \ldots f_m(\cdot))^T$, n_y and n_u are assumed unknown. But the measured input–output vector sequences are assumed to be available at $t = 0,1,2, \ldots$. The minimum necessary assumptions regarding the nonlinear multivariable plant (9.21) are as follows:

Assumption 3. The partial derivatives of (9.21), that is $f_j(\cdot)$, $j = 1, \ldots, m$, are continuous with respect to all the control inputs $u_j(t)$, $j = 1, \ldots, m$.

Assumption 4. The nonlinear plant given by (9.21) satisfies the *generalised Lipschitz condition*. That is

$$\|\mathbf{y}(t_1 + 1) - \mathbf{y}(t_2 + 1)\| \leq b\|\mathbf{u}(t_1) - \mathbf{u}(t_2)\| \text{ for any } t_1 \neq t_2, t_1, t_2 \geq 0 \text{ and } \mathbf{u}(t_1) \neq \mathbf{u}(t_2)$$

where b is a positive constant.

9.3.1 CFDL Data Model for Nonlinear Multivariable Plants

Consider the nonlinear multivariable plant represented by (9.21) satisfying assumptions 3 and 4. Then for $\|\Delta\mathbf{u}(t)\| \neq 0$, there exists an $m \times m$ time-varying matrix $\boldsymbol{\Phi}_c(t)$, called the *pseudo-Jacobian matrix* (PJM), such that

$$\Delta\mathbf{y}(t + 1) = \boldsymbol{\Phi}_c(t)\Delta\mathbf{u}(t) \tag{9.22}$$

Equation (9.22) is the linear time-varying multivariable CFDL data model of (9.21), and $\boldsymbol{\Phi}_c(t)$ is bounded for all t. In the multivariable plants, PJM and the corresponding dynamic linearisation data model given by (9.22) are not unique.

9.3.2 PFDL Data Model for Nonlinear Multivariable Plants

Consider the nonlinear multivariable plant represented by (9.21) satisfying the following assumptions:

Assumption 3′. The partial derivatives of (9.21), that is $f_j(\cdot)$, $j = 1, \ldots, m$, are continuous with respect to all the control inputs in $\mathbf{u}(t), \ldots, \mathbf{u}(t - L + 1)$.

Assumption 4′. The nonlinear plant given by (9.21) satisfies the *generalised Lipschitz condition*. That is $\|\mathbf{y}(t_1 + 1) - \mathbf{y}(t_2 + 1)\| \leq b\,\|\overline{\boldsymbol{U}}_L(t_1) - \overline{\boldsymbol{U}}_L(t_2)\|$ for any $t_1 \neq t_2$, $t_1, t_2 \geq 0$ and $\overline{\boldsymbol{U}}_L(t_1) \neq \overline{\boldsymbol{U}}_L(t_2)$

where $\overline{\boldsymbol{U}}_L(t) = [\mathbf{u}^T(t), \ldots, \mathbf{u}^T(t - L + 1)]^T$ is an mL vector, and L is the LLC as defined in Section 9.2.2, the corresponding moving time window is $[t - L + 1, t]$, and b is a positive constant. Also, $\overline{\boldsymbol{U}}_L(t) = \mathbf{0}_{mL}$, for $t < 0$.

Then, for a given designer specified L, there exists a time-varying $m \times mL$ matrix $\mathbf{\Phi}_{p,L}(t)$, called the *pseudo-partitioned-Jacobian matrix* (PPJM), such that

$$\Delta \mathbf{y}(t+1) = \mathbf{\Phi}_{p,L}(t)\Delta \overline{\boldsymbol{U}}_L(t) \tag{9.23}$$

where $\Delta \overline{\boldsymbol{U}}_L(t) = [\Delta \mathbf{u}^T(t), \ldots, \Delta \mathbf{u}^T(t-L+1)]^T$. Equation (9.23) is the linear time-varying multivariable PFDL data model of (9.21), and $\mathbf{\Phi}_{p,L}(t) = [\mathbf{\Phi}_1(t), \ldots, \mathbf{\Phi}_L(t)]$ is bounded for all t with $m \times m$ matrices $\mathbf{\Phi}_j(t), j = 1, \ldots, L$.

9.3.3 FFDL Data Model for Nonlinear Multivariable Plants

Consider the nonlinear multivariable plant represented by (9.21) satisfying the following assumptions:

Assumption 3″. The partial derivatives of (9.21), that is $f_j(\cdot)$, $j = 1, \ldots, m$, are continuous with respect to all the input and output variables with correspondingly specified lengths.

Assumption 4″. The nonlinear plant given by (9.21) satisfies the *generalised Lipschitz condition*. That is

$$\|\mathbf{y}(t_1+1) - \mathbf{y}(t_2+1)\| \leq b\, \|\overline{\boldsymbol{H}}_{L_y,L_u}(t_1) - \overline{\boldsymbol{H}}_{L_y,L_u}(t_2)\| \text{ for any}$$
$$t_1 \neq t_2, t_1, t_2 \geq 0 \text{ and } \overline{\boldsymbol{H}}_{L_y,L_u}(t_1) \neq \overline{\boldsymbol{H}}_{L_y,L_u}(t_2)$$

where $\overline{\boldsymbol{H}}_{L_y,L_u}(t) = [\mathbf{y}^T(t), \ldots, \mathbf{y}^T(t-L_y+1)\ \mathbf{u}^T(t), \ldots, \mathbf{u}^T(t-L_u+1)]^T$ is an $m(L_u + L_y)$ vector, and L_u, L_y are previously defined pseudo orders in Section 9.2.3, the corresponding moving time windows are $[t-L_u+1, t]$, $[t-L_y+1, t]$, and b is a positive constant. Also, $\overline{\boldsymbol{H}}_{L_y,L_u}(t) = \mathbf{0}_{m(L_y+L_u)}$ for $t < 0$.

Then, for a given designer specified $0 \leq L_u \leq n_u$ and $0 \leq L_y \leq n_y$, there exists a time-varying $m \times m(L_u + L_y)$ matrix $\mathbf{\Phi}_{f,L_y,L_u}(t)$, called the PPJM, such that

$$\Delta \mathbf{y}(t+1) = \mathbf{\Phi}_{f,L_y,L_u}(t)\Delta \overline{\boldsymbol{H}}_{L_y,L_u}(t) \tag{9.24}$$

Equation (9.24) is the linear time-varying multivariable FFDL data model of (9.21), and $\mathbf{\Phi}_{f,L_y,L_u}(t) = [\mathbf{\Phi}_1(t), \ldots, \mathbf{\Phi}_{L_y+L_u}(t)]$ is bounded for all t with $m \times m$ matrices $\mathbf{\Phi}_j(t), j = 1, \ldots, L_y + L_u$.

9.4 Design of Model-free Adaptive Control Systems for Unknown Nonlinear Plants

The dynamic linearisation data model approaches presented in Section 9.3 can be readily employed in control system design techniques. This section proposes MFAC designs based on the three data models, CFDL, PFDL and FFDL.

Exploiting these data models in a control design framework will result in a data-driven control system with minimum plant assumptions. The following general assumption on controllability is necessary for the design of dynamic linearisation-based MFAC schemes.

Assumption 5. The plant to be controlled must be *output controllable*. For a given feasible reference input, control signal sequences exist that derive the tracking error to zero within a finite time.

The model-based control systems design has standard tests for output controllability of linear and nonlinear plants. These tests are employed before a controller design process starts. However, for unknown nonlinear plants (9.1), it is not possible to verify the plant's output controllability prior to the design. In such cases, the PPD tool can be utilised to study the output controllability property. It can be observed that the plant (9.1) is output controllable if the corresponding PPD is finite and non-zero for all t.

9.4.1 Model-free Adaptive Control Based on the CFDL Data Model

Consider the unknown nonlinear plant given by (9.1) satisfying Assumptions 1, 2 and 5. Satisfying Assumptions 1 and 2, the plant given by (9.1) can be equivalently represented in the CFDL data model given by (9.5). Hence, we have for $\Delta u(t) \neq 0$,

$$y(t+1) = y(t) + \phi_c(t)\Delta u(t) \tag{9.25}$$

Introduce the *weighted one-step-ahead prediction cost function* of the following form:

$$J(u(t)) = |y_d(t+1) - y(t+1)|^2 + \lambda |u(t) - u(t-1)|^2 \tag{9.26}$$

where $y_d(t)$ is the desired output, and $\lambda > 0$ is the controller weighting factor to restrict the changing rate of the control input for practical implementations regarding the actuator's limitations in real plants[1].

Inserting (9.25) in (9.26) and minimising (9.26) by $\frac{\partial (J(u(t))}{\partial u(t)} = 0$, gives

$$u(t) = u(t-1) + \frac{\phi_c(t)}{\lambda + \phi_c(t)^2}(y_d(t+1) - y(t)) \tag{9.27}$$

Remark 9.7 Note that in (9.27), the term $\frac{\phi_c(t)}{\lambda + \phi_c(t)^2}(y_d(t+1) - y(t))$ can be considered as a correcting term for updating $u(t)$ in each iteration. Hence, a step factor

1 $\lambda = 0$ corresponds to the *one-step-ahead* control, that in model-based linear control systems design leads to the dead-beat control. It is the best achievable control in terms of output response, but requires excessive control signals, and the minimum phase assumption.

$\rho \in (0,1]$ is added in (9.27) to provide a degree of freedom for controlling the effect of the correcting term. This gives the control signal as follows:

$$u(t) = u(t-1) + \frac{\rho \phi_c(t)}{\lambda + \phi_c(t)^2}(y_d(t+1) - y(t)) \tag{9.28}$$

To implement the control law (9.28), $y_d(t+1)$ is provided by the designer or the actual control problem, λ is a tuning parameter that must be selected by the designer. The tuning parameter affects the closed-loop performance and stability and is an essential parameter in the control law that requires careful consideration. The input–output plant data are also assumed available. The only variable that is unknown in (9.28) is the PPD, $\phi_c(t)$. PPD cannot be analytically derived and should be identified at each sampling time to calculate the control action. It is desired to determine an equation to recursively estimate $\phi_c(t)$. To derive such an equation, an appropriate cost function should be introduced. Let $\hat{\phi}_c(t)$ be the estimate of $\phi_c(t)$, and consider the following cost function:

$$J(\phi_c(t)) = |y(t) - y(t-1) - \phi_c(t)\Delta u(t-1)|^2 + \mu|\phi_c(t) - \hat{\phi}_c(t-1)|^2 \tag{9.29}$$

where $\mu > 0$ is a weighting factor. Minimisation of (9.29) with respect to $\phi_c(t)$, gives

$$\hat{\phi}_c(t) = \hat{\phi}_c(t-1) + \frac{\eta \Delta u(t-1)}{\mu + \Delta u(t-1)^2}[\Delta y(t) - \hat{\phi}_c(t-1)\Delta u(t-1)] \tag{9.30}$$

Similar to the arguments in Remark 9.7, the step factor $\eta \in (0,2]$ is added as a free parameter to control the correcting term. The following assumption is necessary for the CFDL-MFAC structure that will be presented.

Assumption 6. The sign of PPD must remain unchanged for all t, and $\Delta u(t)$ must be non-zero.

This assumption implies that the output and input changes must always be in the same direction. In linear systems, this indicates the *high-frequency* or the *instantaneous gain* must remain unchanged, which is a critical condition for the stability of classical adaptive control systems [9].

Equations (9.28) and (9.30) comprise the CFDL-based MFAC algorithm. The following three situations can deteriorate the identification quality in (9.30), especially in the face of time-varying parameters or sudden parameter changes: $\hat{\phi}_c(t) \approx 0$, $\Delta u(t-1) \approx 0$, or sign change in $\hat{\phi}_c(t)$. Hence, a reset mechanism is defined to avoid such circumstances.[2] That is,

$$\hat{\phi}_c(t) = \hat{\phi}_c(0)$$

2 In the conventional recursive least squares estimation or projection type algorithms, *covariance resetting* is a common technique to deal with parameter changes when the estimation error becomes small.

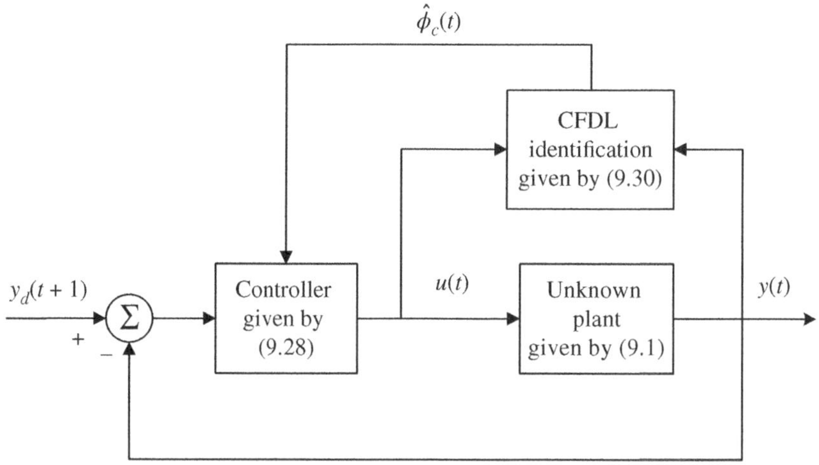

Figure 9.2 The MFAC closed-loop plant based on the PPD identification.

where $\hat{\phi}_c(0)$ is the initial value of the PDD in the case of

$$|\hat{\phi}_c(t)| \leq \varepsilon, \text{or } |\Delta u(t-1)| \leq \varepsilon, \text{or } sign(\hat{\phi}_c(t)) \neq sign(\hat{\phi}_c(0))$$

where ε is a designer-specified small positive constant. The block diagram of the closed-loop MFAC based on the CFDL dynamic linearisation is shown in Figure 9.2.

Algorithm 9.1 CFDL–MFAC

Input Select μ, λ, ε, ρ, η, initialise $\hat{\phi}_c(t)$ and collect the measured system input–output data.

Step 1 Estimate $\hat{\phi}_c(t)$ using (9.30).

Step 2 Reset $\hat{\phi}_c(t)$ if the resetting condition occurred.

Step 3 Compute $u(t)$ by utilising $\hat{\phi}_c(t)$ in (9.28).

Update $y(t)$ and go to step 1.

The following theorem proved in [5] provides the steady-state constant set-point tracking and closed-loop stability guarantee of the closed-loop system shown in Figure 9.2.

Theorem 9.1 Consider the unknown nonlinear plant given by (9.1), satisfying Assumptions 1, 2, 5 and 6. Then, there exists a constant $\lambda_{min} > 0$, such that for all $\lambda > \lambda_{min}$, the closed-loop control system shown in Figure 9.2 has the following properties for a constant set-point y_d:

1. $\lim_{t\to\infty} |y_d - y(t)| \to 0$.
2. The closed-loop system is BIBO stable.

9.4.2 Model-free Adaptive Control Based on the PFDL Data Model

As previously discussed in Section 9.2.2, considering the effects of past inputs on the next plant output provides more degrees of freedom in the dynamically linearised model. This modified the CFDL data model and resulted in the PFDL data model. In this section, the PFDL approach is employed for the MFAC design. Consider the nonlinear plant given by (9.1) satisfying the Assumption 1′, and 2′, as conferred in Section 9.2.2, and for any selected L with $\|\Delta U_L(t)\| \neq 0$ for all t, we have

$$y(t+1) = y(t) + \boldsymbol{\phi}_{p,L}^T(t)\Delta U_L(t) \tag{9.31}$$

where all the parameters and variables are as previously defined. Considering the cost function given by (9.26) and upon the substitution of (9.31) in (9.26), differentiating with respect to $u(t)$, and noting that $\boldsymbol{\phi}_{p,L}^T(t) = [\phi_1(t), \dots, \phi_L(t)]^T$, the derived control law is given by

$$u(t) = u(t-1) + \frac{\rho_1\phi_1(t)(y_d(t+1) - y(t))}{\lambda + \phi_1(t)^2} - \frac{\phi_1(t)\sum_{i=2}^{L}\rho_i\phi_i(t)\Delta u(t-i+1)}{\lambda + \phi_1(t)^2} \tag{9.32}$$

where $\rho_i \in (0,1]$, $i = 1,2, \dots, L$, and it is inserted in (9.32) in accordance with the arguments in Remark 9.7. The only unknown variables in (9.32) are the elements of the vector $\boldsymbol{\phi}_{p,L}(t)$. Hence, similar to the arguments in Section 9.4.1, an equation to recursively estimate the vector $\boldsymbol{\phi}_{p,L}(t)$ is obtained by minimising the following cost function:

$$J(\boldsymbol{\phi}_{p,L}(t)) = |y(t) - y(t-1) - \boldsymbol{\phi}_{p,L}^T(t)\Delta U_L(t-1)|^2 + \mu\|\boldsymbol{\phi}_{p,L}(t) - \hat{\boldsymbol{\phi}}_{p,L}(t-1)\|^2 \tag{9.33}$$

where $\mu > 0$ is a weighting factor. Minimisation of (9.33) with respect to $\boldsymbol{\phi}_{p,L}(t)$ and using the matrix inversion lemma[3] gives

$$\hat{\boldsymbol{\phi}}_{p,L}(t) = \hat{\boldsymbol{\phi}}_{p,L}(t-1) + \frac{\eta\Delta U_L(t-1)}{\mu + \|\Delta U_L(t-1)\|^2}\left[\Delta y(t) - \hat{\boldsymbol{\phi}}_{p,L}^T(t-1)\Delta U_L(t-1)\right] \tag{9.34}$$

Similar to the arguments in Remark 9.7, the step factor $\eta \in (0,2]$ is added as a free parameter to control the correcting term. As in Assumption 6, the modified Assumption 6′ is necessary for the stability of the PFDL–MFAC scheme.

---

3 The matrix inversion lemma is a widely used lemma in deriving recursive estimation equations to reduce the matrix inversion sizes. This lemma states that for any given set of real matrices A, B, C and D, of appropriate dimensions, we have $[A + BCD]^{-1} = A^{-1} - A^{-1}B[DA^{-1}B + C^{-1}]^{-1}DA^{-1}$, provided that the respective inverses exist.

Assumption 6′. The sign of $\phi_1(t)$ in $\boldsymbol{\phi}_{p,L}(t)$ is unchanged for all t and $\|\Delta U_L(t)\| \neq 0$ for all t. Note that the sign may be positive or negative, but it should not vary with time.

Equations (9.32) and (9.34) comprise the PFDL-based MFAC algorithm. Similar to Section 9.4.1, a reset mechanism is defined as $\hat{\boldsymbol{\phi}}_{p,L}(t) = \hat{\boldsymbol{\phi}}_{p,L}(0)$ to improve the estimation process for time-varying plants, in the case of

$$\|\hat{\boldsymbol{\phi}}_{p,L}(t)\| \leq \varepsilon, \text{or } \|\Delta U_L(t-1)\| \leq \varepsilon, \text{or } \operatorname{sign}(\hat{\phi}_1(t)) \neq \operatorname{sign}(\hat{\phi}_1(0))$$

Algorithm 9.2 PFDL–MFAC

Input Select $\mu, \lambda, \varepsilon, \rho, \eta$, initialise $\hat{\boldsymbol{\phi}}_{p,L}(t)$ and collect the measured system input–output data.

Step 1 Estimate $\hat{\boldsymbol{\phi}}_{p,L}(t)$ using (9.34).

Step 2 Reset $\hat{\boldsymbol{\phi}}_{p,L}(t)$ if the resetting condition occurred.

Step 3 Compute $u(t)$ by utilising $\hat{\boldsymbol{\phi}}_{p,L}(t)$ in (9.32).

Update $y(t)$ and go to step 1.

The following theorem proved in Ref. [5] provides the steady-state constant set-point tracking and closed-loop stability guarantee of the closed-loop system shown in Figure 9.3.

Theorem 9.2 Consider the unknown nonlinear plant given by (9.1), satisfying Assumptions 1′, 2′, 5 and 6′. Then, there exists a constant $\lambda_{min} > 0$, such that for

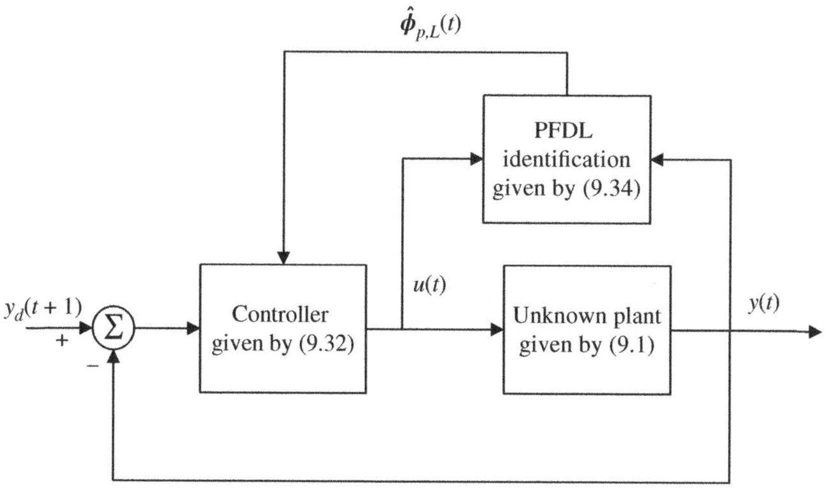

Figure 9.3 The MFAC closed-loop plant based on the PFDL data model identification.

all $\lambda > \lambda_{\min}$, the closed-loop control system shown in Figure 9.3 has the following properties for a constant set-point y_d:

1. $\lim_{t\to\infty} |y_d - y(t)| \to 0$.
2. The closed-loop system is BIBO stable.

9.4.3 Model-free Adaptive Control Based on the FFDL Data Model

As discussed in Section 9.2.3, the FFDL data modelling approach considers the effect of past plant inputs and outputs in obtaining the next sampling time. Hence, all possible degrees of freedom are utilised to capture the main input–output dynamical characteristics of the unknown plant more effectively. In this case, Assumption 1 is extended to accommodate the continuity of the partial derivative of the function $f(\cdot)$ with respect to all variables in Assumption 1″. Also, the *generalised Lipschitz* condition in Assumption 2 is modified to Assumption 2″, as presented in Section 9.2.3. Then, we have for $\|\Delta H_{L_y,L_u}(t)\| \neq 0$ for all t,

$$y(t+1) = y(t) + \boldsymbol{\phi}^T_{f,L_y,L_u}(t)\Delta H_{L_y,L_u}(t) \tag{9.35}$$

where all the parameters and variables are as previously defined. Considering the cost function given by (9.26) and upon the substitution of (9.35) in (9.26) and differentiating with respect to $u(t)$, and noting that $\boldsymbol{\phi}_{f,L_y,L_u}(t) = [\phi_1(t), \dots, \phi_{L_y}(t), \dots, \phi_{L_y+L_u}(t)]^T$, the derived control law is given by

$$\begin{aligned} u(t) = u(t-1) &+ \frac{\rho_{L_y+1}\phi_{L_y+1}(t)(y_d(t+1) - y(t))}{\lambda + \phi_{L_y+1}(t)^2} \\ &- \frac{\phi_{L_y+1}\sum_{i=1}^{L_y}\rho_i\phi_i(t)\Delta y(t-i+1)}{\lambda + \phi_{L_y+1}(t)^2} \\ &- \frac{\phi_{L_y+1}\sum_{i=L_y+2}^{L_y+L_u}\rho_i\phi_i(t)\Delta u(t+L_y-i+1)}{\lambda + \phi_{L_y+1}(t)^2} \end{aligned} \tag{9.36}$$

where $\rho_i \in (0,1]$, $i = 1,2, \dots, L_y + L_u$, and it is inserted in (9.36) in accordance with the arguments in Remark 9.7. The only unknown variables in (9.36) are the elements of the vector $\boldsymbol{\phi}_{f,L_y,L_u}(t)$. Hence, similar to the arguments in 9.4.2, an equation to recursively estimate the vector $\boldsymbol{\phi}_{f,L_y,L_u}(t)$ is obtained by minimising the following cost function:

$$\begin{aligned} J(\boldsymbol{\phi}_{f,L_y,L_u}(t)) = &\left| y(t) - y(t-1) - \boldsymbol{\phi}^T_{f,L_y,L_u}(t)\Delta H_{L_y,L_u}(t-1) \right|^2 \\ &+ \mu \|\boldsymbol{\phi}_{f,L_y,L_u}(t) - \widehat{\boldsymbol{\phi}}_{f,L_y,L_u}(t-1)\|^2 \end{aligned} \tag{9.37}$$

where $\mu > 0$ is a weighting factor. Minimisation of (9.37) with respect to $\boldsymbol{\phi}_{f,L_y,L_u}(t)$, and using the matrix inversion lemma gives

$$\hat{\boldsymbol{\phi}}_{f,L_y,L_u}(t) = \hat{\boldsymbol{\phi}}_{f,L_y,L_u}(t-1) + \frac{\eta \Delta H_{L_y,L_u}(t-1)}{\mu + \|\Delta H_{L_y,L_u}(t-1)\|^2} \times \left[\Delta y(t) - \hat{\boldsymbol{\phi}}^T_{f,L_y,L_u}(t-1)\Delta H_{L_y,L_u}(t-1)\right] \tag{9.38}$$

Similar to the arguments in Remark 9.7, the step factor $\eta \in (0,2]$ is added as a free parameter to control the correcting term.

As in Assumption 6, the modified Assumption 6″ is necessary for the stability of the FFDL–MFAC scheme.

Assumption 6″. The sign of $\phi_{L_y+1}(t)$ is known and unchanged for all t. Note that the sign may be positive or negative, but it should not vary with time.

Equations (9.36) and (9.38) comprise the FFDL-based MFAC algorithm. Following Section 9.4.2, a reset mechanism is defined as $\hat{\boldsymbol{\phi}}_{f,L_y,L_u}(t) = \hat{\boldsymbol{\phi}}_{f,L_y,L_u}(0)$, to improve the estimation process for time-varying plants, in the case of

$$\|\hat{\boldsymbol{\phi}}_{f,L_y,L_u}\| \leq \varepsilon, \text{or } \|\Delta H_{L_y,L_u}(t-1)\| \leq \varepsilon, \text{or } \operatorname{sign}(\hat{\phi}_{L_y+1}(t)) \neq \operatorname{sign}(\hat{\phi}_{L_y+1}(0))$$

Remark 9.8 The FFDL data model captures the input–output characteristics of the unknown nonlinear plant and does not ignore any information about the plant given by (9.1). Hence, it can be claimed that it is an accurate enough equivalent of (9.1). Therefore, the adaptive controller based on the FFDL data model, although simple and practically implementable, enjoys model-free features and is robust with no un-modelled dynamics concerns.

Algorithm 9.3 FFDL–MFAC

Input Select $\mu, \lambda, \varepsilon, \rho, \eta$, initialise $\hat{\boldsymbol{\phi}}_{f,L_y,L_u}(t)$ and collect the measured system input–output data.

Step 1 Estimate $\hat{\boldsymbol{\phi}}_{f,L_y,L_u}(t)$ using (9.38).

Step 2 Reset $\hat{\boldsymbol{\phi}}_{f,L_y,L_u}(t)$ if the resetting condition occurred.

Step 3 Compute $u(t)$ by utilising $\hat{\boldsymbol{\phi}}_{f,L_y,L_u}(t)$ in ((9.36)).

Update $y(t)$ and go to step 1.

The closed-loop BIBO and internal stabilities and the monotonic convergence of the error dynamics of the plant shown in Figure 9.4 are discussed in [10], and the proof of the following convergence-stability theorem is provided. The presented stability analysis is based on the contraction mapping principle that provides a new framework for the analysis and design of adaptive control of unknown plants.[4]

4 Note that the main closed-loop stability analysis approaches in direct or indirect classical adaptive control techniques are Lyapunov or input-output (passivity) based techniques.

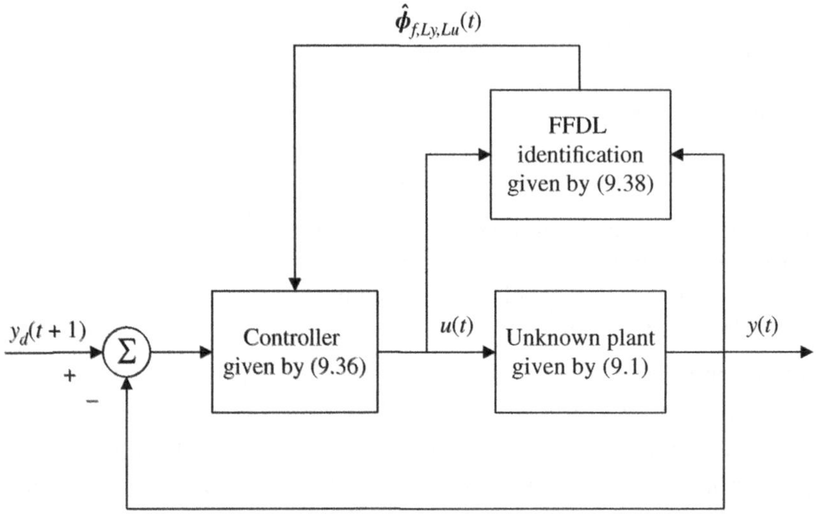

Figure 9.4 The MFAC closed-loop plant based on the FFDL data model identification.

Theorem 9.3 Consider the unknown nonlinear plant given by (9.1), satisfying Assumptions 1″, 2″, 5 and 6″. Then, there exists a constant $\lambda_{\min} > 0$, such that for all $\lambda > \lambda_{\min}$, the closed-loop control systems shown in Figure 9.4 has the following properties for a constant set-point y_d:

1. $\lim_{t \to \infty} |y_d - y(t)| \to 0$.
2. The closed-loop system is BIBO stable.
3. The closed-loop system is internally stable.

Example 9.2 Consider the following nonlinear single input single output system:

$$x_1(t) = x_2(t-2)$$
$$x_2(t) = -\frac{1}{64}\left((1.7\sin(t\pi/10))x_2(t-1) + (e^{-t}+3)x_2{}^3(t-1)\right) + \frac{1}{128}u(t-1)$$
$$y(t) = x_1(t)$$

In this example, the CFDL–MFAC, PFDL–MFAC and FFDL–MFAC techniques are employed to control the above nonlinear system utilising the input–output plant measured data. The following parameters and initial conditions are considered. For the CFDL–MFAC we have:

$$\mu = 0.5, \lambda = 1 \times 10^{-6}, \rho = 0.3, \eta = 0.1, \hat{\phi}_c(0) = 0.01, \varepsilon = 10^{-5},$$

for the PFDL–MFAC we have:

$$L = 3, \mu = 0.5, \lambda = 1 \times 10^{-6}, \boldsymbol{\rho} = [0.3, 0.3, 0.3], \eta = 0.1, \hat{\boldsymbol{\phi}}_{p,L}(0) = \begin{bmatrix} 0.01 & 0.01 & 0.01 \end{bmatrix}, \ \varepsilon = 10^{-5}.$$

and for the FFDL–MFAC we have:

$$L_u = 2, L_y = 2, \mu = 0.5, \lambda = 1 \times 10^{-6}, \boldsymbol{\rho} = [0.3, 0.3, 0.3, 0.3,], \eta = 0.1, \hat{\boldsymbol{\phi}}_{f,L_y,L_u}(0) = \begin{bmatrix} 0.1 & 0.01 & 0.01 & 0.01 \end{bmatrix}, \ \varepsilon = 10^{-5}.$$

The closed-loop time responses and the controller outputs of the three design approaches are shown in Figures 9.5 and 9.6, respectively. The desired closed-loop set-point tracking with an acceptable control signal is achieved for the following desired output:

$$y_d(t) = 0.4036 \sin\left(\frac{t\pi}{180}\right) + 0.12 \cos\left(\frac{t\pi}{50}\right)$$

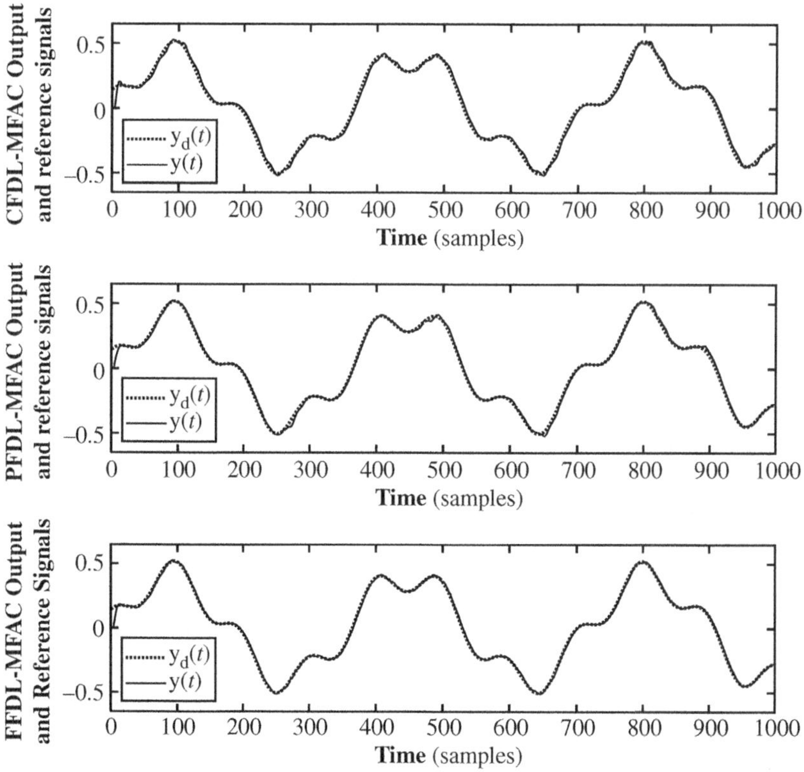

Figure 9.5 Closed-loop responses of the three MFAC approaches.

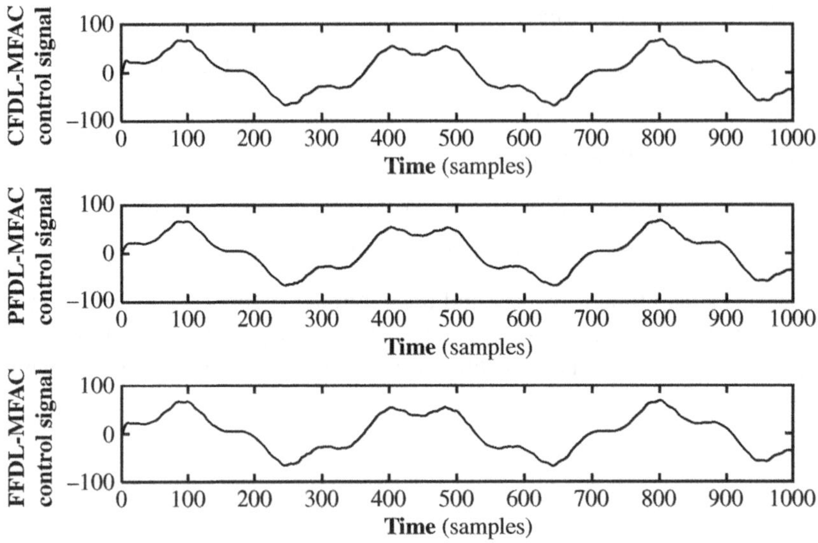

Figure 9.6 The MFAC control signals.

The PPD of the CFDL–MFAC scheme $\phi_c(t)$, the time-varying vector PG of the PFDL–MFAC scheme $\boldsymbol{\phi}_{p,L}(t)$ and the time-varying vector PG of the FFDL–MFAC scheme $\boldsymbol{\phi}_{f,L_y,L_u}(t)$, are shown if Figures 9.7 (a), (b) and (c), respectively.

9.5 Extensions of the Model-free Adaptive Control Methodologies to Multivariable Plants

The MFAC strategies presented in Section 9.4 are extended to unknown nonlinear multivariable plants. Consider the unknown nonlinear multivariable plant given by (9.21) satisfying the specified assumptions. It is assumed that the measured input–output vector sequences of (9.21) are available at $t=0,1,2,\ldots$. The main results on the multivariable MFAC are briefly reviewed. For detailed derivations, refer to [5].

9.5.1 MFAC Design Based on the CFDL Data Model for Nonlinear Multivariable Plants

For the nonlinear multivariable plant represented by (9.21) satisfying assumptions 3 and 4, there exists an equivalent $m \times m$ linear time-varying multivariable CFDL data model governed by (9.22). And $\boldsymbol{\Phi}_c(t)$ in (9.22) is the unknown bounded PJM given by $\boldsymbol{\Phi}_c(t)=[\phi_{ij}(t)]$, $i,j=1,\ldots,m$.

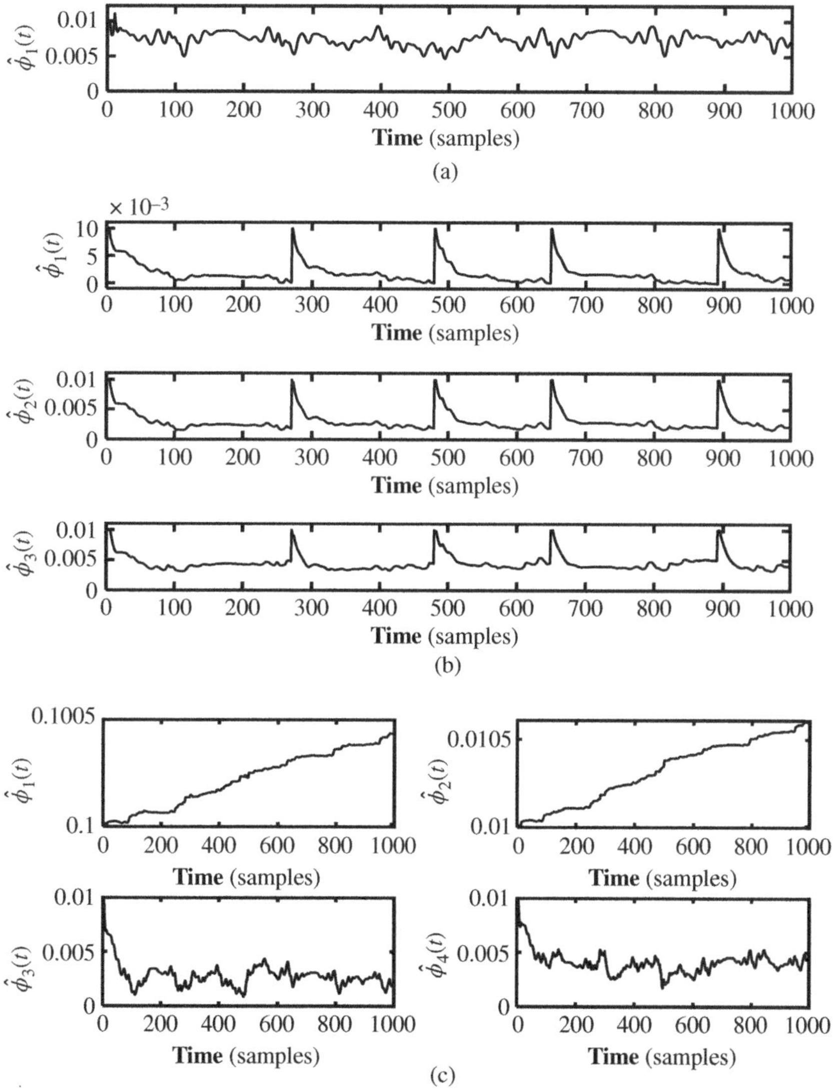

Figure 9.7 (a) PPD of CLFD-MFAC scheme, (b) PG of PFDL-MFAC scheme and (c) PG of FFDL–MFAC scheme.

Assumption 7. The PJM is *diagonally dominant*. Diagonal dominance is satisfied if the off-diagonal terms satisfy $|\phi_{ij}(t)| \leq b_1$ and the diagonal terms satisfy $b_2 \leq |\phi_{ii}(t)| \leq \alpha b_2$, where $i, j = 1, \ldots, m$, $i \neq j$, $\alpha \geq 1$, $b_2 > b_1(2\alpha + 1)(m - 1)$. It is also assumed that the signs of the elements of $\mathbf{\Phi}_c(t)$ are unchanged.

Remark 9.9 The concept of diagonal dominance was introduced in the 1970s by H. H. Rosenbrock to evaluate the interaction in multivariable plants. Diagonal dominance is defined in terms of the transfer function matrix of a linear multivariable plant. It was introduced to provide a more practical approach to multivariable control following the decoupling approach of Kalman in the 1960s. For a complete definition of Rosenbrock's diagonal dominance and its consequences in the design of multivariable control systems, refer to [11]. Rosenbrock's diagonal dominance condition is model-based. To provide a data-based definition of diagonal dominance in multivariable plants, Assumption 7 is proposed in Ref. [5].

Introduce the following *weighted one-step-ahead prediction cost function*, as in Section 9.4.1:

$$J(\mathbf{u}(t)) = \|\mathbf{y}_d(t+1) - \mathbf{y}(t+1)\|^2 + \lambda\|\mathbf{u}(t) - \mathbf{u}(t-1)\|^2 \tag{9.39}$$

where $\mathbf{y}_d(t)$ is the desired output vector, and $\lambda > 0$ is the controller weighting factor to restrict the changing rate of the control input vector. Inserting (9.22) in (9.39) and minimising (9.39) gives

$$\mathbf{u}(t) = \mathbf{u}(t-1) + \left(\lambda\mathbf{I} + \mathbf{\Phi}_c^T(t)\mathbf{\Phi}_c(t)\right)^{-1}\mathbf{\Phi}_c^T(t)(\mathbf{y}_d(t+1) - \mathbf{y}(t)) \tag{9.40}$$

Remark 9.10 The control law given by (9.40) requires an $m \times m$ matrix inversion at each sampling time. For plants with a large number of inputs and outputs, this can cause implementation problems. Hence, in Ref. [5], a simpler equation is given to avoid the matrix inversion:

$$\mathbf{u}(t) = \mathbf{u}(t-1) + \rho\left(\lambda + \|\mathbf{\Phi}_c(t)\|^2\right)^{-1}\mathbf{\Phi}_c^T(t)(\mathbf{y}_d(t+1) - \mathbf{y}(t)) \tag{9.41}$$

where a step factor $\rho \in (0,1]$ is added to provide a degree of freedom for controlling the effect of the correcting term (refer to Remark 9.7). Similar to the arguments in Section 9.4.1, $\mathbf{\Phi}_c(t)$ contains the only unknown variables in (9.40) or (9.41) that require recursive estimation at each sampling time. Let $\hat{\mathbf{\Phi}}_c(t) = [\hat{\phi}_{ij}]$, $i, j = 1, \dots, m$ be the estimate of $\mathbf{\Phi}_c(t)$, and consider the following cost function:

$$J(\mathbf{\Phi}_c(t)) = \|\Delta\mathbf{y}(t) - \mathbf{\Phi}_c(t)\Delta\mathbf{u}(t-1)\|^2 + \mu\|\mathbf{\Phi}_c(t) - \hat{\mathbf{\Phi}}_c(t-1)\|^2 \tag{9.42}$$

where $\mu > 0$ is a weighting factor. Minimisation of (9.42) with respect to $\mathbf{\Phi}_c(t)$, gives

$$\begin{aligned}\hat{\mathbf{\Phi}}_c(t) = {} & \hat{\mathbf{\Phi}}_c(t-1) + [\Delta\mathbf{y}(t) - \hat{\mathbf{\Phi}}_c(t-1)\Delta\mathbf{u}(t-1)]\Delta\mathbf{u}^T(t-1) \\ & \times [\mu\mathbf{I} + \Delta\mathbf{u}(t-1)\Delta\mathbf{u}^T(t-1)]^{-1}\end{aligned} \tag{9.43}$$

Or it can be equivalently written to reduce computational burden as follows [5]:

$$\begin{aligned}\hat{\mathbf{\Phi}}_c(t) = {} & \hat{\mathbf{\Phi}}_c(t-1) + \eta[\Delta\mathbf{y}(t) - \hat{\mathbf{\Phi}}_c(t-1)\Delta\mathbf{u}(t-1)]\Delta\mathbf{u}^T(t-1) \\ & \times [\mu + \|\Delta\mathbf{u}(t-1)\|^2]^{-1}\end{aligned} \tag{9.44}$$

Similar to the arguments in Remark 9.7, the step factor $\eta \in (0,2]$ is added as a free parameter to control the correcting term. Equations (9.40) or (9.41) and (9.43) or (9.44) comprise the CFDL-based MFAC algorithm, with the following reset mechanism, as previously discussed in Section 9.4.

$$\hat{\phi}_{ii}(t) = \hat{\phi}_{ii}(0)$$

in the case of

$$|\hat{\phi}_{ii}(t)| \leq b_2, \text{ or } |\hat{\phi}_{ii}(t)| \geq \alpha b_2, \text{ or } \operatorname{sign}(\hat{\phi}_{ii}(t)) \neq \operatorname{sign}(\hat{\phi}_{ii}(0))$$

And for $i \neq j$,

$$\hat{\phi}_{ij}(t) = \hat{\phi}_{ij}(0)$$

in the case of

$$|\hat{\phi}_{ij}(t)| \geq b_1, \text{ or } \operatorname{sign}(\hat{\phi}_{ij}(t)) \neq \operatorname{sign}(\hat{\phi}_{ij}(0))$$

The general structure for implementing the above algorithm is similar to Figure 9.2 within a matrix framework, and the corresponding equations are substituted.

Algorithm 9.4 Multivariable CFDL–MFAC

Input Select $\mu, \lambda, \rho, \eta, \alpha, b_1, b_2$ initialise $\hat{\mathbf{\Phi}}_c(t)$ and collect the measured system input–output data.

Step 1 Estimate $\hat{\mathbf{\Phi}}_c(t)$ using (9.44).

Step 2 Reset $\hat{\phi}_{ii}(t)$ if the resetting condition occurred.

Step 3 Reset $\hat{\phi}_{ij}(t)$ if the resetting condition occurred.

Step 4 Compute $\mathbf{u}(t)$ by utilising $\hat{\mathbf{\Phi}}_c(t)$ in (9.41).

Update $\mathbf{y}(t)$ and go to step 1.

The following theorem proved in Ref. [5] provides the steady-state constant set-point tracking and closed-loop stability guarantee of the proposed multivariable closed-loop system.

Theorem 9.4 Consider the unknown nonlinear plant given by (9.21), satisfying Assumptions 3, 4, 5 and 7 with the reset mechanism. Then, there exists a constant $\lambda_{\min} > 0$, such that for all $\lambda > \lambda_{\min}$, the multivariable closed-loop control system given by the CFDL-based MFAC has the following properties for a constant set-point vector $\mathbf{y}_d$:

1. $\lim_{t \to \infty} \|\mathbf{y}_d - \mathbf{y}(t)\| \to 0$.
2. The closed-loop system is BIBO stable.

9.5.2 MFAC Design Based on the PFDL Data Model for Nonlinear Multivariable Plants

Consider the nonlinear multivariable plant represented by (9.21), satisfying the required assumptions in Section 9.3. Then, a corresponding linear time-varying multivariable PFDL data model given by (9.23) exists for a given L, and the PPJM time-varying unknown bounded $m \times mL$ matrix $\mathbf{\Phi}_{p,L}(t) = [\mathbf{\Phi}_1(t), \ldots, \mathbf{\Phi}_L(t)]$ with $m \times m$ matrices as follows:

$$\mathbf{\Phi}_j(t) = \begin{bmatrix} \phi_{11j}(t) & \phi_{12j}(t) & \ldots & \phi_{1mj}(t) \\ \phi_{21j}(t) & \phi_{22j}(t) & \ldots & \phi_{2mj}(t) \\ \vdots & \vdots & \vdots & \vdots \\ \phi_{m1j}(t) & \phi_{m2j}(t) & \ldots & \phi_{mmj}(t) \end{bmatrix}, j = 1, \ldots, L.$$

Assumption 8. The matrix $\mathbf{\Phi}_1(t)$ of $\mathbf{\Phi}_{p,L}(t)$ is diagonally dominant. That is, the off-diagonal terms satisfy $|\phi_{ij1}(t)| \leq b_1$ and the diagonal terms satisfy $b_2 \leq |\phi_{ii1}(t)| \leq \alpha b_2$, where $i, j = 1, \ldots, m$, $i \neq j$, $\alpha \geq 1$, $b_2 > b_1(2\alpha + 1)(m-1)$ and the signs of the elements of $\mathbf{\Phi}_1(t)$ are unchanged.

Considering the weighted one-step-ahead prediction cost function given by (9.39) and inserting (9.23) in (9.39) and minimising (9.39), gives

$$\mathbf{u}(t) = \mathbf{u}(t-1) + \left(\lambda \mathbf{I} + \mathbf{\Phi}_1^T(t)\mathbf{\Phi}_1(t)\right)^{-1}\mathbf{\Phi}_1^T(t) \times \left[(\mathbf{y}_d(t+1) - \mathbf{y}(t)) - \sum_{i=2}^{L}\mathbf{\Phi}_i(t)\Delta\mathbf{u}(t-i+1)\right] \quad (9.45)$$

Similar to the arguments in Remark 9.10, a simpler version of the control law (9.45) is given as follows:

$$\mathbf{u}(t) = \mathbf{u}(t-1) + \left(\lambda + \|\mathbf{\Phi}_1(t)\|^2\right)^{-1}\mathbf{\Phi}_1^T(t) \times \left[\rho_1(\mathbf{y}_d(t+1) - \mathbf{y}(t)) - \sum_{i=2}^{L}\rho_i\mathbf{\Phi}_i(t)\Delta\mathbf{u}(t-i+1)\right] \quad (9.46)$$

where $\rho_i \in (0,1]$, $i = 1,2, \ldots, L$, and it is inserted in (9.46) in accordance with the arguments in Remark 9.7. Also, let $\hat{\mathbf{\Phi}}_{p,L}(t) = [\hat{\mathbf{\Phi}}_1(t), \ldots, \hat{\mathbf{\Phi}}_L(t)]$, be the estimate of $\mathbf{\Phi}_{p,L}(t)$, and consider the following cost function:

$$J(\mathbf{\Phi}_{p,L}(t)) = \|\Delta\mathbf{y}(t) - \mathbf{\Phi}_{p,L}(t)\Delta\overline{\boldsymbol{U}}_L(t-1)\|^2 + \mu\|\mathbf{\Phi}_{p,L}(t) - \hat{\mathbf{\Phi}}_{p,L}(t-1)\|^2 \quad (9.47)$$

where $\mu > 0$ is a weighting factor. Minimisation of (9.47) with respect to $\mathbf{\Phi}_{p,L}(t)$, gives

$$\hat{\mathbf{\Phi}}_{p,L}(t) = \hat{\mathbf{\Phi}}_{p,L}(t-1) + [\Delta\mathbf{y}(t) - \hat{\mathbf{\Phi}}_{p,L}(t-1)\Delta\overline{\boldsymbol{U}}_L(t-1)]\Delta\overline{\boldsymbol{U}}_L^T(t-1) \times \left[\mu\mathbf{I} + \Delta\overline{\boldsymbol{U}}_L(t-1)\Delta\overline{\boldsymbol{U}}_L^T(t-1)\right]^{-1} \quad (9.48)$$

Or in the following simplified form, we have:

$$\hat{\boldsymbol{\Phi}}_{p,L}(t) = \hat{\boldsymbol{\Phi}}_{p,L}(t-1) + \eta[\Delta \mathbf{y}(t) - \hat{\boldsymbol{\Phi}}_{p,L}(t-1)\Delta\overline{\boldsymbol{U}}_L(t-1)]\Delta\overline{\boldsymbol{U}}_L^T(t-1)$$
$$\times \left[\mu + \|\Delta\overline{\boldsymbol{U}}_L(t-1)\|^2\right]^{-1} \tag{9.49}$$

Similar to the arguments in Remark 9.7, the step factor $\eta \in (0,2]$ is added as a free parameter to control the correcting term. Equations (9.45) or (9.46) and (9.48) or (9.49) comprise the PFDL-based MFAC algorithm, with the following reset mechanism, as previously discussed in Section 9.4.

$$\hat{\phi}_{ii1}(t) = \hat{\phi}_{ii1}(0)$$

in the case of

$$|\hat{\phi}_{ii1}(t)| \leq b_2, \text{or } |\hat{\phi}_{ii1}(t)| \geq \alpha b_2, \text{or } \text{sign}(\hat{\phi}_{ii1}(t)) \neq \text{sign}(\hat{\phi}_{ii1}(0))$$

And for $i \neq j$,

$$\hat{\phi}_{ij1}(t) = \hat{\phi}_{ij1}(0)$$

in the case of

$$|\hat{\phi}_{ij1}(t)| \geq b_1, \text{or } \text{sign}(\hat{\phi}_{ij1}(t)) \neq \text{sign}(\hat{\phi}_{ij1}(0))$$

The general structure for implementing the above algorithm is similar to Figure 9.3 within a matrix framework, and the corresponding equations are substituted.

Algorithm 9.5 Multivariable PFDL–MFAC

Input Select μ, λ, $\rho_i(i=1, \ldots, L)$, η, α, b_1. b_2, initialise $\hat{\boldsymbol{\Phi}}_{p,L}(t)$ and collect the measured system input–output data.

Step 1 Estimate $\hat{\boldsymbol{\Phi}}_{p,L}(t)$ using (9.49).
Step 2 Reset $\hat{\phi}_{ii1}(t)$ if the resetting condition occurred.
Step 3 Reset $\hat{\phi}_{ij1}(t)$ if the resetting condition occurred.
Step 4 Compute $\mathbf{u}(t)$ by utilising $\hat{\boldsymbol{\Phi}}_{p,L}(t)$ in (9.46).
Update $\mathbf{y}(t)$ and go to step 1.

The following theorem proved in [5] provides the steady-state constant set-point tracking and closed-loop stability guarantee of the proposed multivariable closed-loop system.

Theorem 9.5 Consider the unknown nonlinear multivariable plant given by (9.21), satisfying Assumptions 3′, 4′, 5 and 8 with the reset mechanism. Then, there exists a constant $\lambda_{\min} > 0$, such that for all $\lambda > \lambda_{\min}$, the multivariable closed-loop control system given by the PFDL-based MFAC algorithm has the following properties for a constant set-point vector $\boldsymbol{y}_d$:

1. $\lim_{t\to\infty}\|\mathbf{y}_d - \mathbf{y}(t)\| \to 0$.
2. The closed-loop system is BIBO stable.

9.5.3 MFAC Design Based on the FFDL Data Model for Nonlinear Multivariable Plants

Consider the nonlinear multivariable plant represented by (9.21), satisfying the required assumptions in Section 9.3.3. Then, a corresponding linear time-varying multivariable FFDL data model given by (9.24) exists for a given L_y and L_u, and the PPJM time-varying unknown bounded $m \times m(L_u + L_y)$ matrix $\mathbf{\Phi}_{f,L_y,L_u}(t)$. Considering the weighted one-step-ahead prediction cost function given by (9.39) and inserting (9.24) in (9.39) and minimising (9.39) gives

$$\mathbf{u}(t) = \mathbf{u}(t-1) + \left(\lambda \mathbf{I} + \mathbf{\Phi}_{L_y+1}^T(t)\mathbf{\Phi}_{L_y+1}(t)\right)^{-1} \mathbf{\Phi}_{L_y+1}^T(t)$$
$$\times \left[(\mathbf{y}_d(t+1) - \mathbf{y}(t)) - \sum_{i=1}^{L_y} \mathbf{\Phi}_i(t)\Delta\mathbf{y}(t-i+1) \right.$$
$$\left. - \sum_{i=L_y+2}^{L_y+L_u} \mathbf{\Phi}_i(t)\Delta\mathbf{u}(t+L_y-i+1) \right] \tag{9.50}$$

Or in the following simplified form, we have:

$$\mathbf{u}(t) = \mathbf{u}(t-1) + \left(\lambda + \|\mathbf{\Phi}_{L_y+1}(t)\|^2\right)^{-1} \mathbf{\Phi}_{L_y+1}^T(t)$$
$$\times \left[\rho_{L_y+1}(\mathbf{y}_d(t+1) - \mathbf{y}(t)) - \sum_{i=1}^{L_y} \rho_i \mathbf{\Phi}_i(t)\Delta\mathbf{y}(t-i+1) \right.$$
$$\left. - \sum_{i=L_y+2}^{L_y+L_u} \rho_i \mathbf{\Phi}_i(t)\Delta\mathbf{u}(t+L_y-i+1) \right] \tag{9.51}$$

where $\rho_i \in (0,1]$, $i = 1, 2, \ldots, L_y + L_u$, and it is inserted in (9.51) in accordance with the arguments in Remark 9.7. Also, let $\hat{\mathbf{\Phi}}_{f,L_y,L_u}(t) = [\hat{\mathbf{\Phi}}_1(t), \ldots, \hat{\mathbf{\Phi}}_L(t)] \in \mathbb{R}^{m\times m(L_y+L_u)}$ be the estimate of $\mathbf{\Phi}_{f,L_y,L_u}(t)$, where

$$\hat{\mathbf{\Phi}}_j(t) = \begin{bmatrix} \hat{\phi}_{11j}(t) & \hat{\phi}_{12j}(t) & \ldots & \hat{\phi}_{1mj}(t) \\ \hat{\phi}_{21j}(t) & \hat{\phi}_{22j}(t) & \ldots & \hat{\phi}_{2mj}(t) \\ \vdots & \vdots & \vdots & \vdots \\ \hat{\phi}_{m1j}(t) & \hat{\phi}_{m2j}(t) & \ldots & \hat{\phi}_{mmj}(t) \end{bmatrix} \in \mathbb{R}^{m\times m}, j = 1, \ldots, L_y + L_u.$$

and consider the following cost function:

$$J(\boldsymbol{\Phi}_{f,L_y,L_u}(t)) = \|\Delta\mathbf{y}(t) - \boldsymbol{\Phi}_{f,L_y,L_u}(t)\Delta\overline{\boldsymbol{H}}_{L_y,L_u}(t-1)\|^2 + \mu\|\boldsymbol{\Phi}_{f,L_y,L_u}(t) - \widehat{\boldsymbol{\Phi}}_{f,L_y,L_u}(t-1)\|^2 \tag{9.52}$$

where $\mu > 0$ is a weighting factor. Minimisation of (9.52) with respect to $\boldsymbol{\Phi}_{f,L_y,L_u}(t)$, gives

$$\widehat{\boldsymbol{\Phi}}_{f,L_y,L_u}(t) = \widehat{\boldsymbol{\Phi}}_{f,L_y,L_u}(t-1) + [\Delta\mathbf{y}(t) - \widehat{\boldsymbol{\Phi}}_{f,L_y,L_u}(t-1)\Delta\overline{\boldsymbol{H}}_{L_y,L_u}(t-1)] \times \Delta\overline{\boldsymbol{H}}^T_{L_y,L_u}(t-1)\left[\mu\mathbf{I} + \Delta\overline{\boldsymbol{H}}_{L_y,L_u}(t-1)\Delta\overline{\boldsymbol{H}}^T_{L_y,L_u}(t-1)\right]^{-1} \tag{9.53}$$

Or, in the simplified form,

$$\widehat{\boldsymbol{\Phi}}_{f,L_y,L_u}(t) = \widehat{\boldsymbol{\Phi}}_{f,L_y,L_u}(t-1) + \eta[\Delta\mathbf{y}(t) - \widehat{\boldsymbol{\Phi}}_{f,L_y,L_u}(t-1)\Delta\overline{\boldsymbol{H}}_{L_y,L_u}(t-1)] \times \Delta\overline{\boldsymbol{H}}^T_{L_y,L_u}(t-1)\left[\mu + \|\Delta\overline{\boldsymbol{H}}_{L_y,L_u}(t-1)\|^2\right]^{-1} \tag{9.54}$$

Similar to the arguments in Remark 9.7, the step factor $\eta \in (0,2]$ is added as a free parameter to control the correcting term. Equations (9.50) or (9.51) and (9.53) or (9.54) comprise the FFDL-based MFAC algorithm, with the following reset mechanism, as previously discussed in Section 9.4.

$$\widehat{\phi}_{ii(L_y+1)}(t) = \widehat{\phi}_{ii(L_y+1)}(0)$$

in the case of

$$|\widehat{\phi}_{ii(L_y+1)}(t)| \le b_2, \text{or } |\widehat{\phi}_{ii(L_y+1)}(t)| \ge \alpha b_2, \text{or } \operatorname{sign}(\widehat{\phi}_{ii(L_y+1)}(t)) \ne \operatorname{sign}(\widehat{\phi}_{ii(L_y+1)}(0))$$

And for $i \ne j$,

$$\widehat{\phi}_{ij(L_y+1)}(t) = \widehat{\phi}_{ij(L_y+1)}(0)$$

in the case of

$$|\widehat{\phi}_{ij(L_y+1)}(t)| \ge b_1, \text{or } \operatorname{sign}(\widehat{\phi}_{ij(L_y+1)}(t)) \ne \operatorname{sign}(\widehat{\phi}_{ij(L_y+1)}(0))$$

The general structure for implementing the above algorithm is similar to Figure 9.4 within a matrix framework, and the corresponding equations are substituted.

Algorithm 9.6 Multivariable FFDL–MFAC

Input Select μ, λ, $\rho_i (i = 1, \ldots, L_u + L_y)$, η, α, b_1. b_2,initialise $\widehat{\boldsymbol{\Phi}}_{f,L_y,L_u}(t)$ and collect the measured system input–output data.

Step 1 Estimate $\hat{\boldsymbol{\Phi}}_{f,L_y,L_u}(t)$ using (9.54).
Step 2 Reset $\hat{\phi}_{ii(L_y+1)}(t)$ if the resetting condition occurred.
Step 3 Reset $\hat{\phi}_{ij(L_y+1)}(t)$ if the resetting condition occurred.
Step 4 Compute $\mathbf{u}(t)$ by utilising $\hat{\boldsymbol{\Phi}}_{f,L_y,L_u}(t)$ in (9.51).
Update $\mathbf{y}(t)$ and go to step 1.

The following theorem presented and proved in Ref. [12] provides the steady-state constant set-point tracking, boundedness of the input and output signals and closed-loop internal stability proof of the proposed multivariable closed-loop system in four parts. It first proves the boundedness of the estimated PPJM vector. Then, the regulator error is shown to be convergent and the system inputs and outputs are shown to be bounded. Finally, the closed-loop internal stability is proved.

Assumption 9. The matrix $\boldsymbol{\Phi}_{L_y+1}(t)$ in $\boldsymbol{\Phi}_{f,L_y,L_u}(t)$ is diagonally dominant. That is, the off-diagonal terms satisfy $|\phi_{ij(L_y+1)}(t)| \leq b_1$, $b_2 \leq |\phi_{ii(L_y+1)}(t)| \leq \alpha b_2$, $\alpha \geq 1$, $b_2 > b_1(2\alpha+1)(m-1)$, $i, j = 1, \ldots, m$, $i \neq j$, and the signs of all elements of $\boldsymbol{\Phi}_{L_y+1}(t)$ are unchanged.

Theorem 9.6 Consider the unknown nonlinear plant given by (9.21), satisfying Assumptions 3″, 4″, 5 and 9 with the reset mechanism. Then, for all real constants $\lambda \in (\lambda_{\min}, \lambda_{\max})$, $\lambda_{\min} > 0, \lambda_{\max} > 0$, the multivariable closed-loop control system given by the FFDL-based MFAC algorithm has the following properties for a constant set-point vector $\boldsymbol{y}_d$:

1. $\lim_{t\to\infty} \|\mathbf{y}_d - \mathbf{y}(t)\| \to 0$.
2. The input and output vector sequences, $\mathbf{u}(t)$, $\mathbf{y}(t)$, are bounded for all t.
3. The closed-loop system is internally stable.

Example 9.3 A linearised model of an automotive gas turbine with an operating point corresponding to 80% gas generator speed and 85% power turbine with the fuel pump excitation and nozzle actuator excitation as inputs, and gas generator speed and inter-turbine temperature as outputs are given in Ref. [11], as follows:

$$\mathbf{G}(s) = \begin{bmatrix} \dfrac{0.806s + 0.264}{s^2 + 1.15s + 0.202} & \dfrac{-(15s + 1.42)}{s^3 + 12.8s^2 + 13.6s + 2.36} \\ \dfrac{1.95s^2 + 2.12s + 0.49}{s^3 + 9.15s^2 + 9.39s + 1.62} & \dfrac{7.14s^2 + 25.8s + 9.35}{s^4 + 20.8s^3 + 116.4s^2 + 111.6s + 18.8} \end{bmatrix}$$

The schematic diagram of the gas turbine is shown in Figure 9.8. An inverse Nyquist diagram and Gershgorin band analysis showed that the gas turbine model is not diagonally dominant in the model-based sense defined in Ref. [11]. In this example, the CFDL–MFAC, PFDL–MFAC and FFDL–MFAC techniques are employed to control the above gas turbine model outputs utilising the

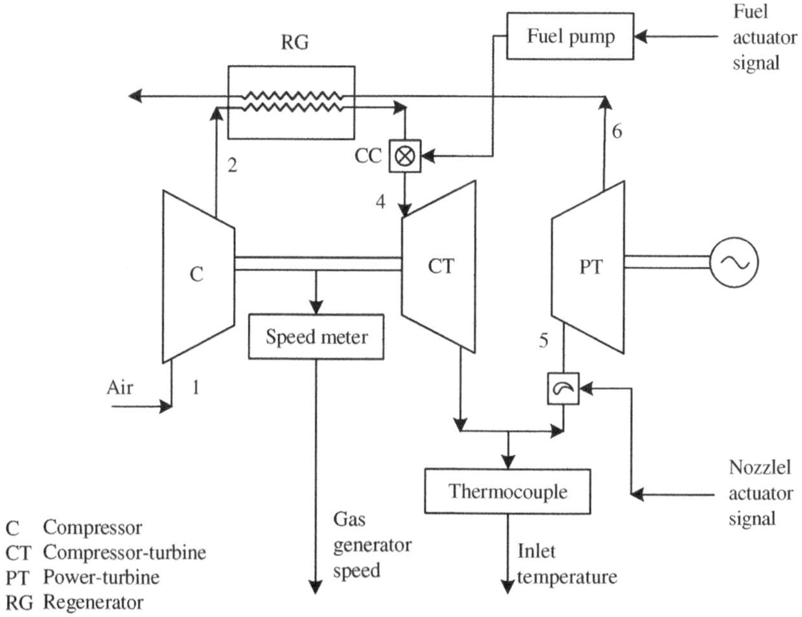

Figure 9.8 Schematic diagram of the gas turbine.

input–output plant measured data with a sampling time of $T = 0.1$ second. The following parameters and initial conditions are considered. For the CFDL–MFAC we have:

$$\mu = 0.5, \lambda = 3.7, \rho = 0.1, \eta = 0.5, \hat{\boldsymbol{\Phi}}_c(0) = \begin{bmatrix} 2 & -0.001 \\ 0.001 & 2 \end{bmatrix},$$

$$\alpha = 6, b_1 = 0.03. \ b_2 = 1,$$

for the PFDL–MFAC we have:

$$L = 3, \mu = 0.5, \lambda = 3.5, \boldsymbol{\rho} = 0.1 \times [1,1,1], \eta = 0.5,$$

$$\hat{\boldsymbol{\Phi}}_{p,L}(0) = \begin{bmatrix} 2 & -0.001 & 0.001 & 0.001 & 0.001 & 0.001 \\ 0.001 & 2 & 0.001 & 0.001 & 0.001 & 0.001 \end{bmatrix}, \alpha = 6,$$

$$b_1 = 0.03. \ b_2 = 1,$$

and for the FFDL–MFAC we have:

$$L_u = 2, L_y = 2, \mu = 0.5, \lambda = 2.5, \boldsymbol{\rho} = 0.1 \times [1, 1,1,1], \eta = 0.5,$$

$$\hat{\boldsymbol{\Phi}}_{f,L_y,L_u}(0) = \begin{bmatrix} 0.001 & \dots & 2 & -0.001 & 0.001 & \dots & 0.001 \\ 0.001 & \dots & 0.001 & 2 & 0.001 & \dots & 0.001 \end{bmatrix}_{2\times 8},$$

$$\alpha = 6, b_1 = 0.03. \ b_2 = 1.$$

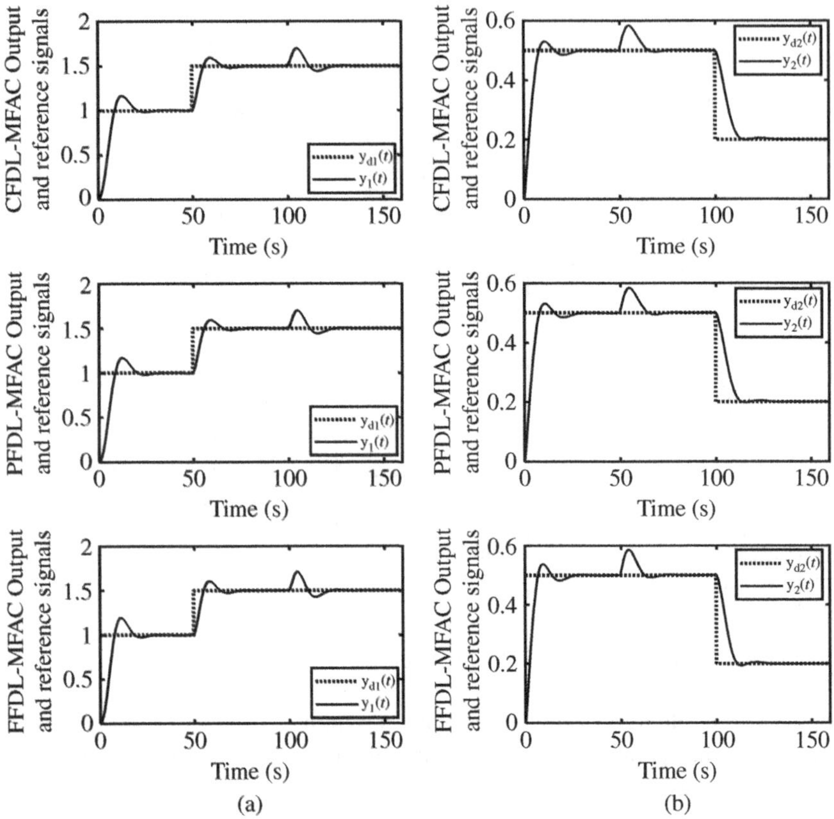

Figure 9.9 The closed-loop outputs.

The closed-loop time responses of the three design approaches are shown in Figure 9.9. The desired closed-loop set-point tracking with an acceptable interaction level is achieved for the following desired outputs:

$$\boldsymbol{y}_d(t) = \begin{cases} \begin{bmatrix} 1 \\ 0.5 \end{bmatrix} & 0 \leq t \times T < 50 \\ \begin{bmatrix} 1.5 \\ 0.5 \end{bmatrix} & 50 \leq t \times T < 100 \\ \begin{bmatrix} 1.5 \\ 0.2 \end{bmatrix} & 100 \leq t \times T < 160 \end{cases}$$

The corresponding control signals are shown in Figure 9.10. The results indicated that the control signals conform with the performance of the actual actuators in the gas turbine.

Figure 9.11 depicts the identification of the 2×2 time-varying PJM matrix $\boldsymbol{\Phi}_c(t)$.

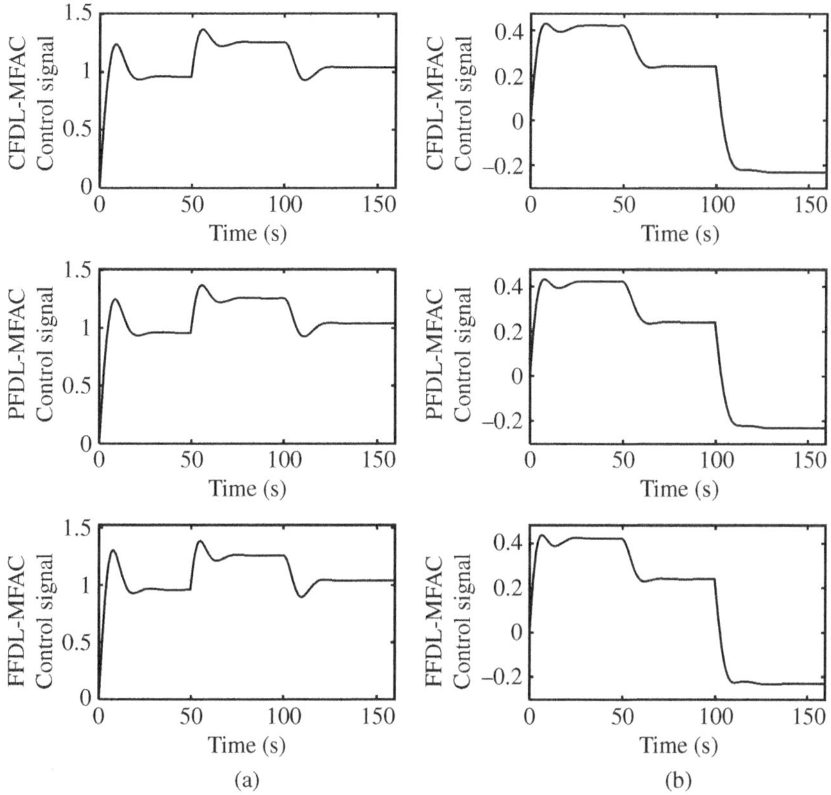

Figure 9.10 The control signals (a) first input and (b) second input.

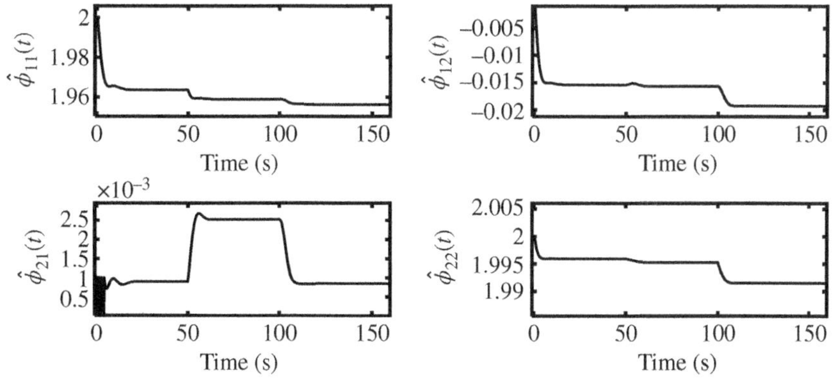

Figure 9.11 The PJM in the CFDL–MFAC approach.

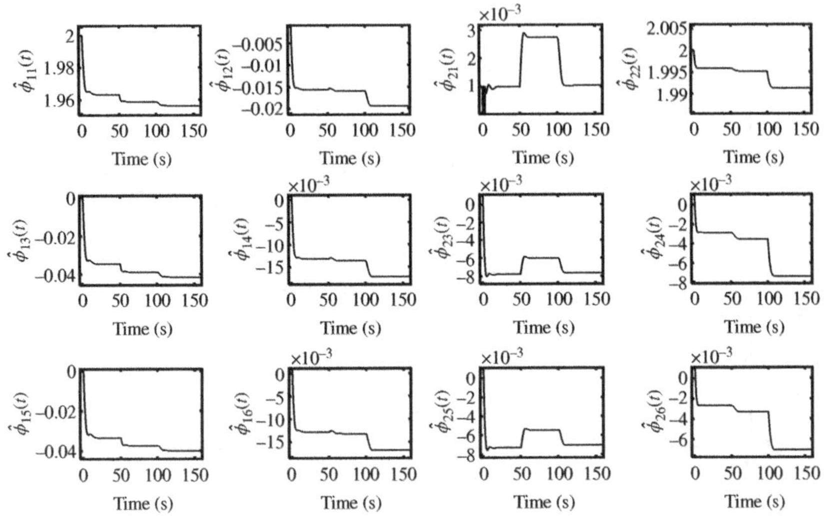

Figure 9.12 The PPJM in the PFDL–MFAC approach.

Figure 9.12 depicts the identification of the 2×6 time-varying PPJM matrix $\mathbf{\Phi}_{p,L}(t)$.

Figure 9.13 depicts the identification of the 2×8 time-varying PPJM matrix $\mathbf{\Phi}_{f,L_y,L_u}(t)$.

Example 9.4 A nonlinear model of a cascade of two continuous stirred tank reactors (CSTR) is given in [13], as follows:

$$\frac{dc_{A1}}{dt} = \frac{F}{V}(c_{A0} - c_{A1}) - k_0 e^{-E/RT_1} c_{A1}^2$$
$$\frac{dc_{A2}}{dt} = \frac{F}{V}(c_{A1} - c_{A2}) - k_0 e^{-E/RT_2} c_{A2}^2$$
$$\frac{dT_1}{dt} = \frac{F}{V}(T_0 - T_1) + \frac{-\Delta H}{\varrho c_p} k_0 e^{-E/RT_1} c_{A1}^2 + \frac{1}{V\varrho c_p} Q_1$$
$$\frac{dT_2}{dt} = \frac{F}{V}(T_1 - T_2) + \frac{-\Delta H}{\varrho c_p} k_0 e^{-E/RT_2} c_{A2}^2 + \frac{1}{V\varrho c_p} Q_2$$

where $c_{A0} = 0.75$ mol/dm^3, c_{A1}, c_{A2}, are the inlet, tank 1 and 2 molar concentrations, respectively, $T_0 = 300$ Kelvin is the inlet temperature, $k_0 = 0.00493$ dm^3/mol is the Arrhenius frequency factor, $\varrho = 55 + 3 \times \frac{(T-273.15)}{100}$ mol/dm^3 is the density, $R = 0.0019872$ kcal/(mol × kevin) is the universal gas constant, $c_p = 0.016$ Kelvin is the heat capacity, $E = 13.3$ kcal/mol is the activation energy, $F = 1$ dm^3/s is the volumetric flow rate, $-\Delta H = 10.5$ kcal/mol is the heat of reaction, $V = 50$ dm^3 is the volume, Q_1, Q_2, T_1, T_2 are the heat inputs and the temperatures of tanks 1 and 2, respectively. The cascade CSTR plant has a one-way interaction, that is the first output variations affect the second output, while the second output has no

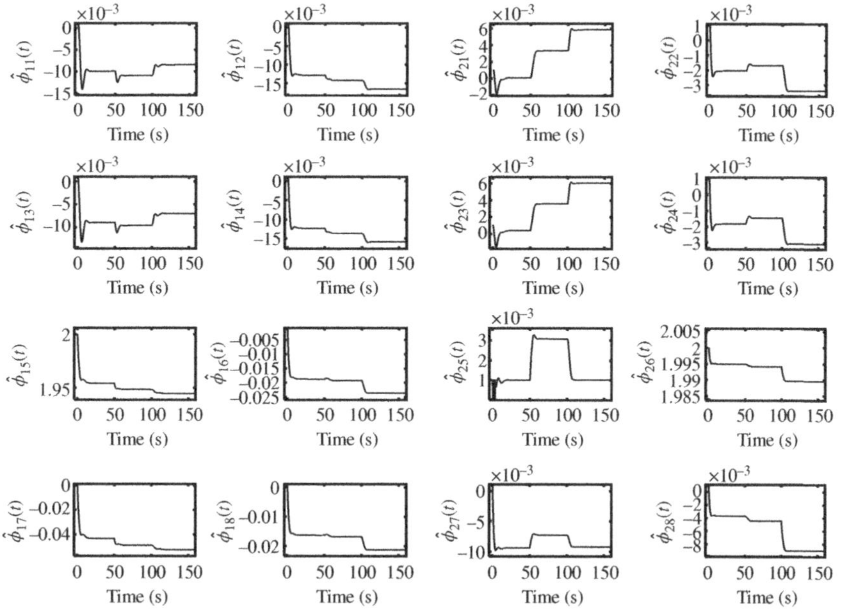

Figure 9.13 The PPJM in the FFDL–MFAC approach.

interaction effect on the first output. The schematic diagram of the cascade CSTR is shown in Figure 9.14.

In this example, the CFDL–MFAC, PFDL–MFAC and FFDL–MFAC techniques are employed to control the CSTR model outputs utilising the input–output plant measured data with a sampling time of $T = 10$ seconds. The following parameters

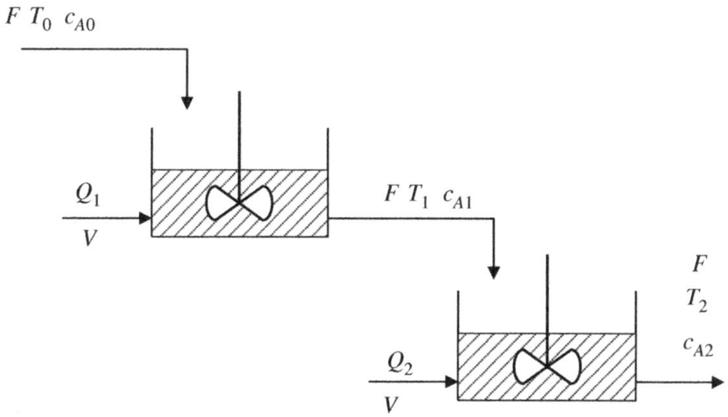

Figure 9.14 Schematic diagram of a cascade of two continuous stirred tank reactors.

and initial conditions are considered. For the CFDL–MFAC we have:

$$\mu = 1, \lambda = 1.8, \rho = 0.5, \eta = 1, \hat{\boldsymbol{\Phi}}_c(0) = \begin{bmatrix} 0.5 & 0.0001 \\ 0.0001 & 0.5 \end{bmatrix}, \alpha = 10,$$

$b_1 = 0.01.\ b_2 = 0.4$, for the PFDL–MFAC we have:

$$L = 2, \mu = 1, \lambda = 1.8, \boldsymbol{\rho} = [0.5{,}0.5], \eta = 1,$$

$$\hat{\boldsymbol{\Phi}}_{p,L}(0) = \begin{bmatrix} 0.5 & 0.0001 & 0.001 & 0.001 \\ 0.0001 & 0.5 & 0.001 & 0.001 \end{bmatrix}, \alpha = 10, b_1 = 0.01.\ b_2 = 0.4,$$

and for the FFDL–MFAC we have:

$$L_u = 2, L_y = 2, \mu = 1, \lambda = 1.8, \boldsymbol{\rho} = [0.5{,}0.5{,}0.5{,}0.5], \eta = 1,$$

$$\hat{\boldsymbol{\Phi}}_{f,L_y,L_u}(0) = \begin{bmatrix} 0.001 & \dots & 0.5 & 0.0001 & 0.001 & \dots & 0.001 \\ 0.001 & \dots & 0.0001 & 0.5 & 0.001 & \dots & 0.001 \end{bmatrix}_{2\times 8},$$

$$\alpha = 10, b_1 = 0.01.\ b_2 = 0.4.$$

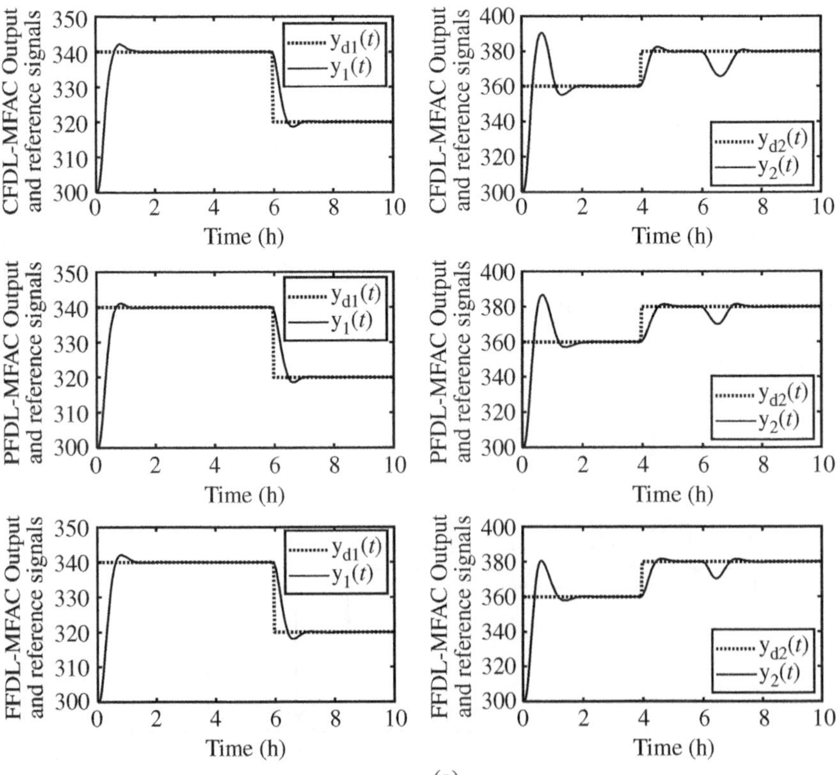

Figure 9.15 (a) Closed-loop responses and (b) control signals.

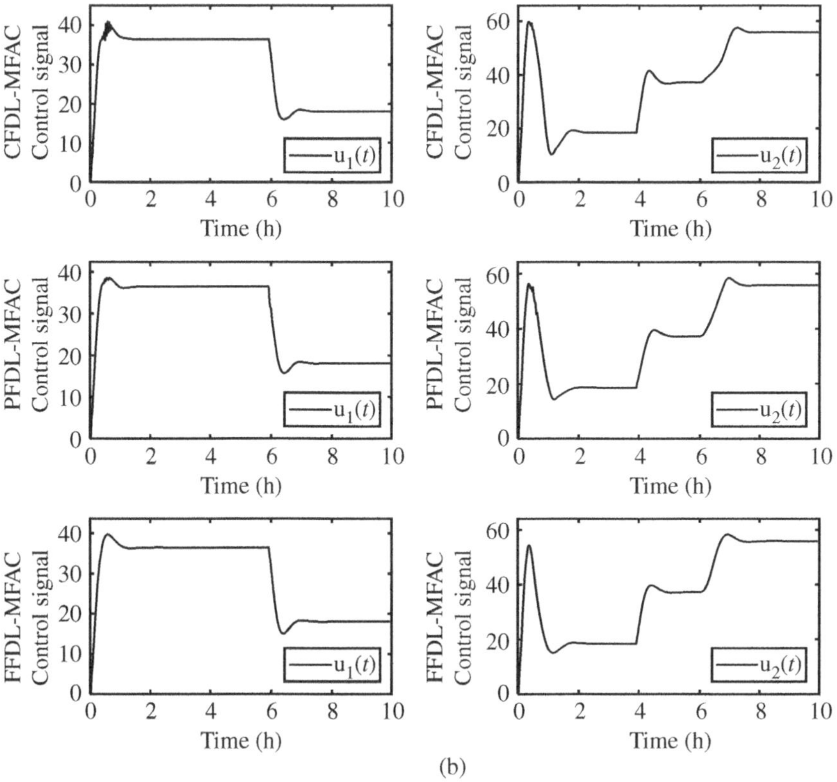

Figure 9.15 *(Continued)*

The closed-loop time responses of the three design approaches are shown in Figure 9.15 (a), and the corresponding control signals are shown in Figure 9.15 (b). The desired outputs are given as follows:

$$\mathbf{y}_d(t) = \begin{cases} \begin{bmatrix} 340 \\ 360 \end{bmatrix} & 0 \le \dfrac{t \times T}{360} < 4 \\ \begin{bmatrix} 340 \\ 380 \end{bmatrix} & 4 \le \dfrac{t \times T}{360} < 6 \\ \begin{bmatrix} 320 \\ 380 \end{bmatrix} & 6 \le \dfrac{t \times T}{360} < 10 \end{cases}$$

Figure 9.16 depicts the identification of the 2×2 time-varying PJM matrix $\mathbf{\Phi}_c(t)$.

Figure 9.17 depicts the identification of the 2×4 time-varying PPJM matrix $\mathbf{\Phi}_{p,L}(t)$.

Figure 9.18 depicts the identification of the 2×8 time-varying PPJM matrix $\mathbf{\Phi}_{f,L_y,L_u}(t)$.

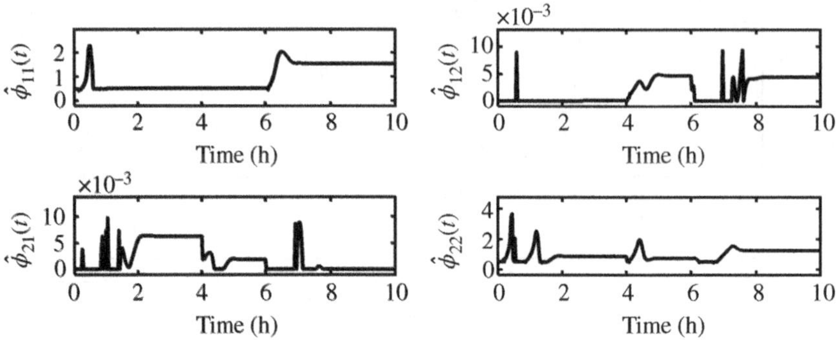

Figure 9.16 The PJM in the CFDL–MFAC approach.

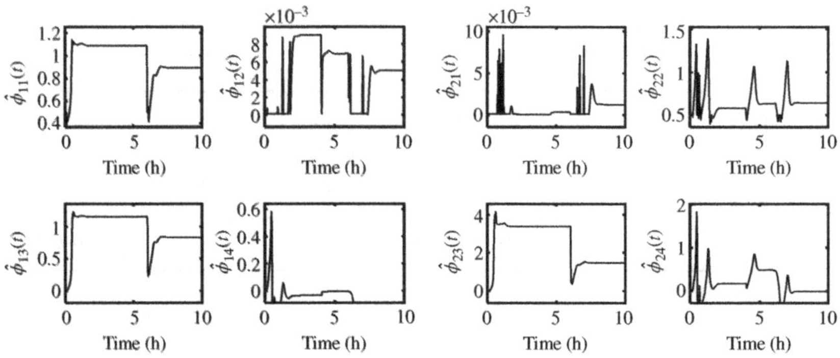

Figure 9.17 The PPJM in the PFDL–MFAC approach.

9.6 A Combined MFAC–SPSA Data-driven Control Strategy

To enhance the MFAC performance and widen its applicability, MFAC has been employed in conjunction with other design methodologies. See for example Refs. [14, 15]. This section uses MFAC with the simultaneous perturbation stochastic approximation (SPSA), presented in Chapter 6. In an inner loop, SPSA tunes internal control loops using the input–output data. The inner loop SPSA-designed controllers' main task is stabilising the open-loop unstable system. Then, the MFAC technique is used in the outer loop using the generated input–output data to ensure acceptable closed-loop performance. Closed-loop performance improvement and robustness of the combined methodology are investigated through an underactuated system described in Section 6.4.

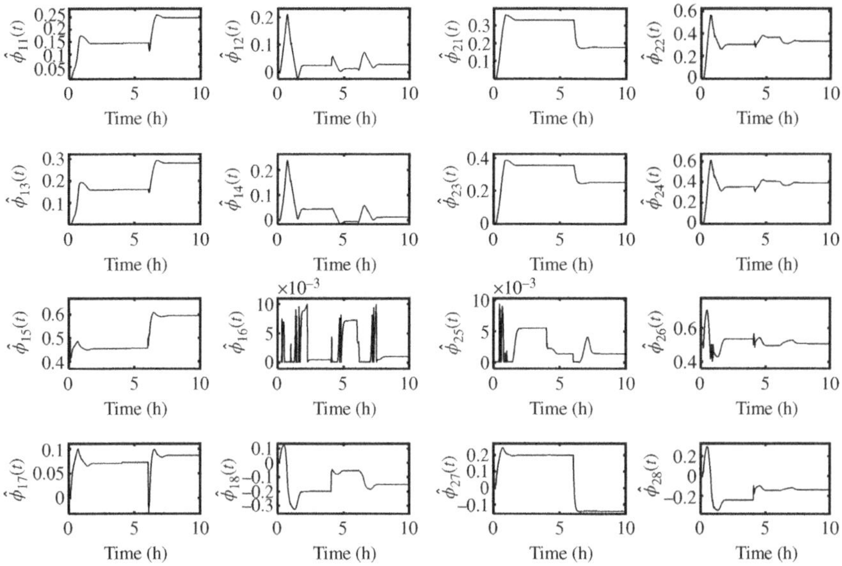

Figure 9.18 The PPJM in the FFDL–MFAC approach.

The combined MFAC–SPSA strategy is summarised as follows:

- The online data-driven SPSA method is used for tuning internal controllers, as discussed in Section 6.4. This online tuning delivers the optimal parameter values without prior plant assumptions.
- The data-driven MFAC uses the input–output data of the stabilised inner loop and provides enhanced closed-loop performance, as is shown in Figure 9.19.

Consider a closed-loop structure with a controller of pre-specified and known structure and a cost function that is a control performance criterion in the

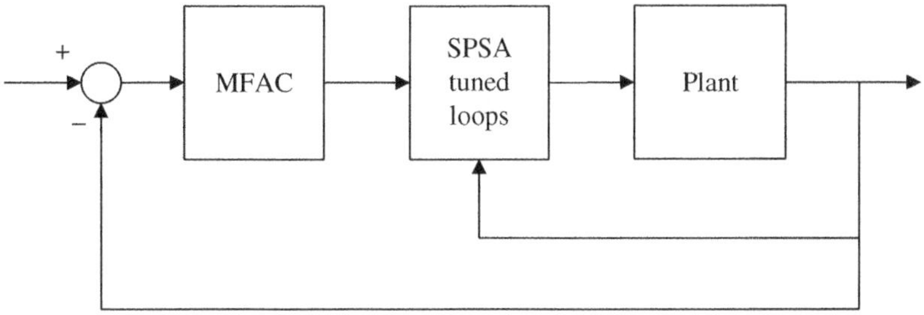

Figure 9.19 The general combined MFAC–SPSA control system.

inner loop. SPSA plays the role of a *tuning algorithm* and finds the controller parameter vector. The implementation issues of the inner loop design are discussed in Chapter 6, and the MFAC design principles are presented in previous sections. Hence, the following assumptions are briefly summarised for the combined MFAC–SPSA closed-loop system of Figure 9.19:

- The inner-loop controller structure is pre-specified and fixed.
- Input–output data of the unknown plant under control are available.
- The partial derivatives of the outer loop plant with respect to all the elements of the control input vector are continuous, and it satisfies the generalised Lipschitz condition.
- The open-loop unknown system is assumed to be output controllable from a data-driven viewpoint, and the controllability property is preserved in the inner loop compensation.
- The PJM of the closed loop inner loop system is diagonally dominant, and the sign of all its elements is unchanged.

The main advantage of SPSA is its stabilisation capability of a possibly unstable unknown open-loop system. MFAC improves the performance due to the use of an equivalent data model of the inner-loop closed-loop system and provides a satisfactory transient response for the time-varying reference values. However, MFAC cannot be directly applied to systems where the assumption on the fixed sign of the output changes with respect to input changes is violated.

Example 9.5 Consider the multivariable underactuated liquid slosh system in Section 6.4.1. Let $u(t) = \sin(t)$ be the system input for 20 seconds. The output responses are depicted in Figure 9.20. The sampling time is 0.01 second.

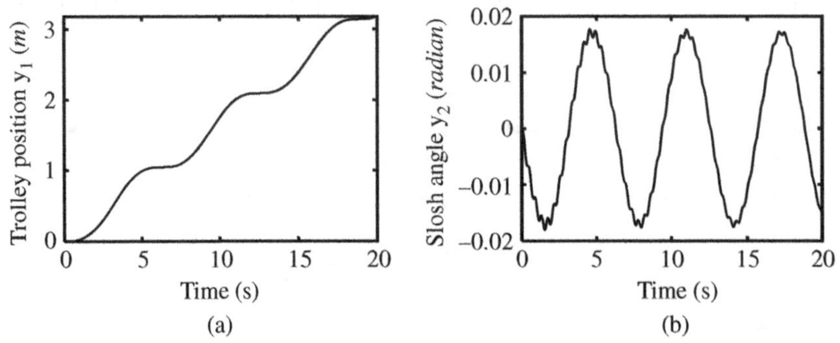

Figure 9.20 The open-loop responses (a) the trolley position and (b) the slosh angle of the liquid slosh.

It is clearly observed that the sign of the first output changes with respect to the input changes is not constant. Therefore, the above final assumption for MFAC does not hold, and MFAC cannot be directly applied to the plant. In Section 6.4.1, the plant is stabilised using an SPSA tuning strategy of two PID control loops. It is shown that the compensated plant satisfies the required assumption. Let the reference input be a sine signal $y_{r1}(t) = \sin(t)$. In addition, the PID controllers' gain vector is as $\boldsymbol{\zeta} = [2.41\ 0.31\ 2.01\ 1.96\ 4.05\ 0.3]^T$ that is the final gain vector of the SPSA tuning. The results are depicted in Figure 9.21. In this case, the sign of $\Delta\mathbf{y}$ with respect to $\Delta y_{r_{1_m}}$ remains constant. Hence, implementation of the MFAC becomes feasible in the combined framework. The proposed control structure is shown in Figure 9.22.

In Figure 9.22, the underactuated plant is assumed to have two degrees of freedom, where y_1 and y_2 represent the first and the second outputs, respectively,

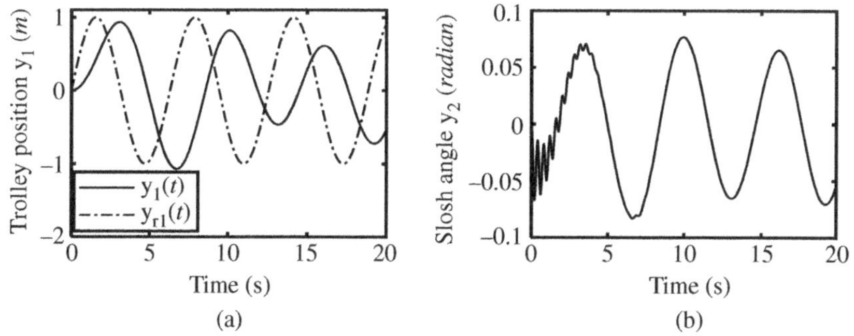

Figure 9.21 The closed-loop responses with two PIDs (a) the trolley position and (b) the slosh angle of the liquid slosh.

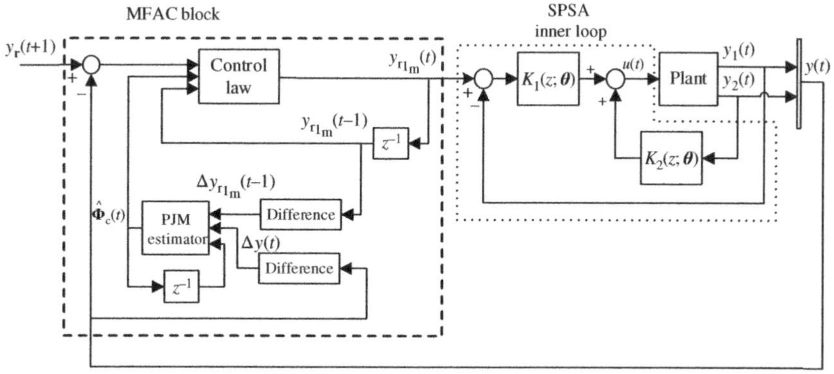

Figure 9.22 Combination of SPSA and CFDL-MFAC systems with two degrees of freedom.

and u is the only control input. In addition, $\mathbf{y} = \begin{bmatrix} y_1 \\ y_2 \end{bmatrix}$ is the system output vector, $\mathbf{y}_r = \begin{bmatrix} y_{r_1} \\ y_{r_2} \end{bmatrix}$ is the vector of reference signals and $y_{r_{1m}}$ is the modified reference signal of the output y_1.

The liquid slosh model equations and parameters are as given in Section 6.4.1 and are not repeated here. Initially, the step-by-step algorithm of SPSA is performed for the inner loop of Figure 9.22. After the system becomes stable, the tuning algorithm stops, and the gains of the PID controllers are fixed on their final values. Subsequently, the CFDL–MFAC algorithm in the outer loop starts operation. After 20 seconds, the set point of the first output y_1 changes from 50 to −50 cm. This performance is compared with the case that after the tuning algorithm stops, the PID gains become fixed on their last values and the system is controlled with the tuned PID controllers.

The SPSA parameters are as follows: $\alpha = 0.6$, $A = 1$, $a = 20$, $\gamma = 0.101$ and $c = 0.02$. The initial gain vector is chosen as $\zeta(0) = [k_{p_1}\ k_{i_1}\ k_{d_1}\ k_{p_2}\ k_{i_2}\ k_{d_2}]^T = [1\ 0.9\ 5\ 3\ 0.1\ 2]^T$. The MFAC parameters are as follows: $\lambda = 8$, $\rho = 0.01$, $\mu = 1$, $\eta = 1$. In addition, the initial value of the PJM estimate is considered as $\widehat{\mathbf{\Phi}}_c(0) = [16\ 5]^T$ and $b_1 = 5.1$, $b_2 = 15$, $\alpha b_2 = 17$, are the resetting condition values. Figure 9.23 shows the closed-loop responses of the MFAC–SPSA design methodology. The control signal and the cost function are shown in Figure 9.24 (a) and (b), respectively. These figures show the acceptable control signal performance for practical implementations and the reduction of the cost function. Also, the convergence of the PID controllers' parameters for the liquid slosh in the combined MFAC–SPSA scheme is shown in Figure 9.25. Figure 9.26 presents the estimated elements of the PJM estimate $\widehat{\mathbf{\Phi}}_c$ for the liquid slosh for the combined MFAC–SPSA scheme. According to Figure 9.27, when adding the outer-loop controller, the modified

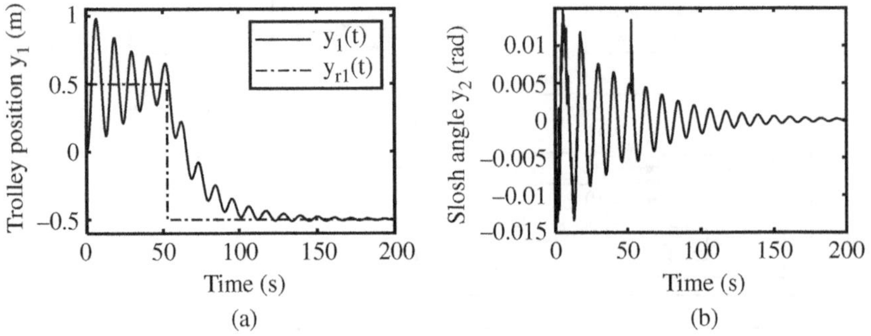

Figure 9.23 (a) y_1 the trolley position and (b) y_2 the slosh angle of the liquid slosh of the combined MFAC–SPSA scheme.

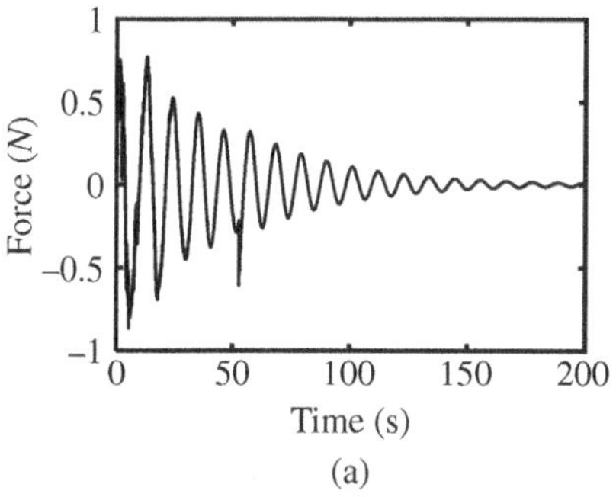

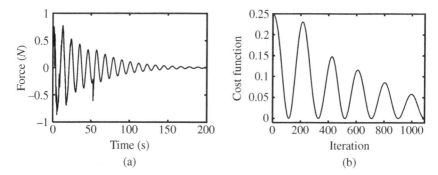

Figure 9.24 (a) Control input u and (b) the cost function J for the liquid slosh for the combined MFAC–SPSA scheme.

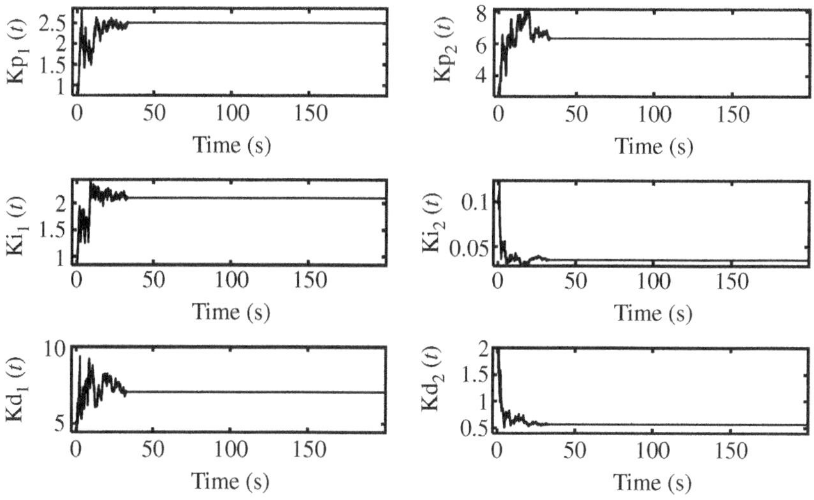

Figure 9.25 PID controllers' parameters for the combined MFAC–SPSA scheme.

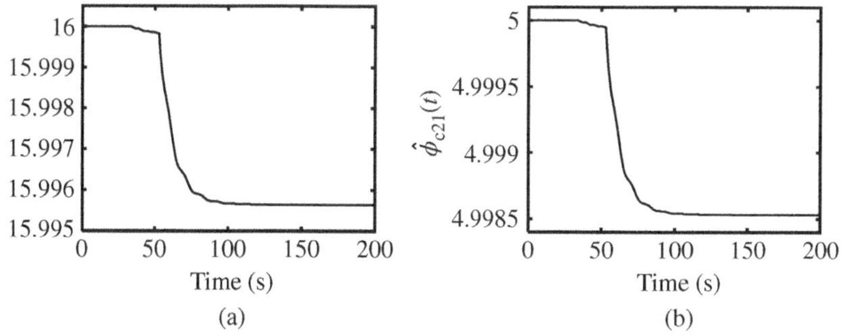

Figure 9.26 (a) The first and (b) the second element of the PJM estimate $\hat{\mathbf{\Phi}}_c$ for the liquid slosh for the combined MFAC–SPSA scheme.

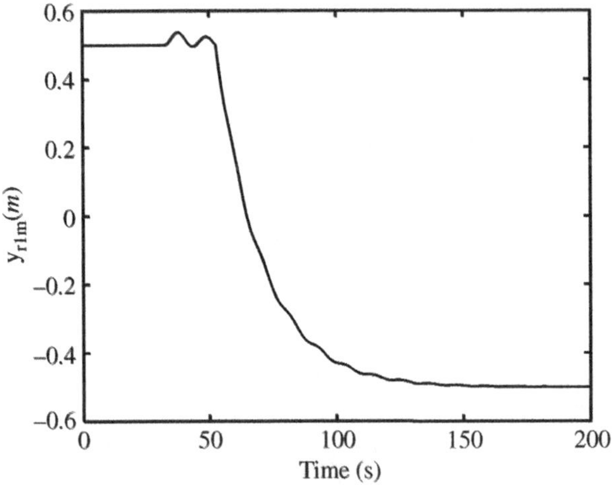

Figure 9.27 The liquid slosh's modified reference input $y_{r_{1m}}$ for the combined MFAC–SPSA scheme.

reference signal $y_{r_{1m}}$ has a slight variation and after the set point change it gradually converges to the new value −0.5 m.

To further show the effectiveness of the proposed combined methodology, after $t = 33$ seconds the PID controller gains are fixed and the outer control loop is disconnected. It is shown in Figures 9.28 and 9.29, representing the closed-loop outputs and control input, that up to the time $t = 53$ seconds, the performance of the two controllers is almost similar. However, after a set-point change at $t = 53$ seconds, the performance of the first output y_1 deteriorates with a larger settling time and overshoot. Also, the second output y_2 has oscillations with

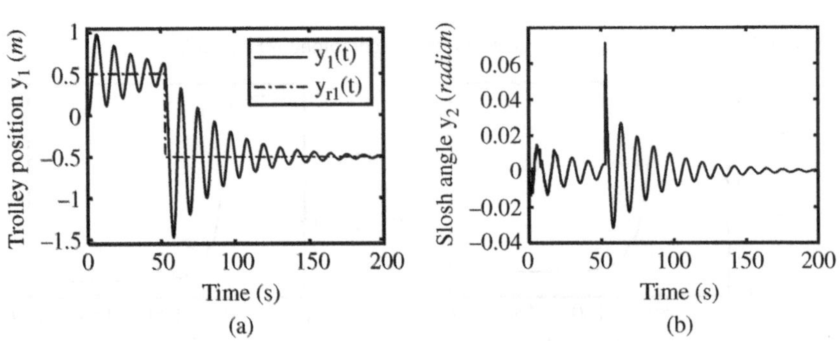

Figure 9.28 Performance of the inner loop PID controllers after disconnecting the outer loop MFAC at $t = 33$ seconds (a) y_1 the trolley position and (b) y_2 the slosh angle of the liquid slosh.

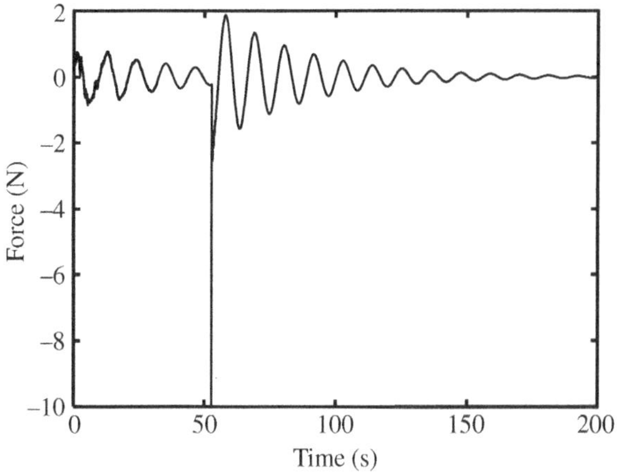

Figure 9.29 Control signal of the liquid slosh inner loop PID control after disconnecting the outer loop MFAC at $t = 33$ seconds.

larger amplitudes, as compared to the active MFAC, and the control input is considerably greater than the MFAC scheme.

9.7 Conclusions

The principles of MFAC are presented in this chapter. MFAC is a data-driven control systems design with minimum plant assumptions for unknown nonlinear plants with no classical system identification techniques. The MFAC concepts and design techniques are naturally extendable to multivariable plants. Rigorous mathematical proofs for error convergence and closed-loop stability are available in the literature. However, the corresponding theorems are presented in this chapter, avoiding detailed mathematical discussions. The main emphasis has been on implementation issues via providing algorithms and simulation studies on different linear and nonlinear plants. As the classical linearisation approaches require a mathematical model of the plant, MFAC employs the novel dynamical linearisation concept. The different forms of dynamic linearisation models utilised in the MFAC architecture are discussed, and it is shown that the virtual models capture the input–output dynamical system's behaviour by providing closed-loop stability and set-point tracking for unknown nonlinear plants. The three dynamic linearisation techniques and their corresponding MFAC designs provided are CFDL, PFDL and FFDL. Employing recursive estimation techniques, the initially unknown models are recursively updated, and a controller is designed utilising the estimated models at each sampling time. Finally, to widen the applicability of the MFAC design, a combined MFAC–SPSA strategy is presented, and simulation results are provided to show the effectiveness of the proposed methodology.

Problems

9.1 The data-driven model-free adaptive control (MFAC) and the model-based model reference adaptive control (MRAC) are prominent adaptive control strategies. Compare and discuss the underlying assumptions of these methodologies.

9.2 Compare and discuss the different data-driven control systems designs: MFAC, Koopman–Willems-based designs, VRFT, SPSA and UASC from a control structure perspective.

9.3 Consider the following discrete-time SISO nonlinear model that is an extension of (9.1) by incorporating the disturbance $d(t)$ and unknown n_d:

$$y(t+1) = f(y(t), \dots, y(t-n_y), u(t), \dots, u(t-n_u), d(t), \dots, d(t-n_d))$$

Also, extend Assumption 1 to include the continuity of $f(\cdot)$ with respect to $d(t)$ and the generalised Lipschitz condition in Assumption 2 as follows:

$$|y(t_1+1) - y(t_2+1)| \leq b_1 |u(t_1) - u(t_2)| \text{ for any } t_1 \neq t_2,$$
$$t_1, t_2 \geq 0, \text{and } u(t_1) \neq u(t_2)$$

$$|y(t_1+1) - y(t_2+1)| \leq b_2 |d(t_1) - d(t_2)| \text{ for any } t_1 \neq t_2,$$
$$t_1, t_2 \geq 0, \text{and } d(t_1) \neq d(t_2)$$

where b_1 and b_2 are positive constants. Show that the above nonlinear plant can be transformed in the following equivalent CFDL model:

$$\Delta y(t+1) = \phi_c(t)\Delta u(t) + \left.\frac{\partial f}{\partial d}\right|_c \Delta d(t)$$

where $\phi_c(t) = \overline{\eta}(t) + \left.\frac{\partial f}{\partial u}\right|_c \in \mathbb{R}$. All the variables and constants are as defined in Section 9.2.1.

9.4 Consider a power system power supply network given by the following nonlinear differential equations [16]:

$$\dot{x}_1(t) = x_2$$
$$\dot{x}_2(t) = (-1 + 1.3\cos 0.8t)\sin x_1 - 0.4x_2 + 0.2 + u$$
$$y(t) = x_1(t)$$

The initial conditions are assumed zero and the sampling time is 0.05.

a) Design a CFDL-based MFAC and study the effect of the different parameter tunings on the system's closed-loop performance.
b) Design a PFDL-based MFAC and study the effect of the different parameter tunings on the system's closed-loop performance.
c) Design an FFDL-based MFAC and study the effect of the different parameter tunings on the system's closed-loop performance.
d) Compare the results from the three MFAC design methodologies.

9.5 Consider a nonlinear chemical reactor given by the following nonlinear differential equations [17]:

$$\dot{x}_1(t) = -x_1 + 0.072(1 - x_1)e^{\frac{x_2}{1+0.05x_2}}$$

$$\dot{x}_2(t) = -0.7x_2 + 0.576(1 - x_1)e^{\frac{x_2}{1+0.05x_2}} + 0.3u$$

$$y(t) = x_2(t)$$

Where $x_1(t)$ and $x_2(t)$ are the dimensionless reactant concentration and reactor temperature, respectively. The initial conditions are assumed zero and the sampling time is 0.2.

The desired reference input is arbitrary steps of amplitudes between 6 and 10.

a) Design a CFDL-based MFAC and study the effect of the different parameter tunings on the system's closed-loop performance.
b) Design a PFDL-based MFAC and study the effect of the different parameter tunings on the system's closed-loop performance.
c) Design an FFDL-based MFAC and study the effect of the different parameter tunings on the system's closed-loop performance.
d) Compare the results from the three MFAC design methodologies.

9.6 Consider the following linear multivariable plant:

$$\mathbf{P}(s) = \frac{1}{s+\alpha}\begin{bmatrix} k_{11} & k_{12} \\ k_{21} & k_{22} \end{bmatrix}$$

where $1 \leq k_{11} \leq 2, 0.5 \leq k_{12} \leq 1, 0.25 \leq k_{21} \leq 0.5, 2 \leq k_{22} \leq 4$ and it is assumed that the parameters vary linearly from their minimum values to their maximum values. It is desired that the first input tracks a unit step reference input, and the second output regulates about zero.

a) For $\alpha = 0$, design MFACs based on the three CFDL, PFDL and FFDL strategies and discuss the results.
b) For $\alpha = 1$, design MFACs based on the three CFDL, PFDL and FFDL strategies and discuss the results.
c) Analyse the closed-loop performance for the above two cases.

References

1 M. Klein, T. Marczinkowsky, and M. Pandit, "An elementary pattern recognition self-tuning PI-controller," in *Intelligent Tuning and Adaptive Control*: Elsevier, 1991, pp. 17–21.

2 Z. Hou, "*The parameter identification, adaptive control and model free learning adaptive control for nonlinear systems*," Master thesis. Shenyang: North-eastern University, 1994.

3 Z. Hou and W. Huang, "The model-free learning adaptive control of a class of SISO nonlinear systems," in *Proceedings of the 1997 American Control Conference (Cat. No. 97CH36041)*, 1997, vol. 1: IEEE, pp. 343–344.

4 Z. Hou, R. Chi, and H. Gao, "An overview of dynamic-linearization-based data-driven control and applications," *IEEE Transactions on Industrial Electronics,* vol. 64, no. 5, pp. 4076–4090, 2016.

5 Z. Hou and S. Jin, *Model Free Adaptive Control: Theory and Applications.* CRC Press, 2019.

6 R.-E. Precup, R.-C. Roman, and A. Safaei, *Data-Driven Model-Free Controllers.* CRC Press, 2022.

7 W. Rudin, *Principles of Mathematical Analysis.* McGraw-Hill, New York, 1976.

8 K. J. Åström and B. Wittenmark, *Adaptive Control.* Courier Corporation, 2013.

9 K. S. Narendra and A. M. Annaswamy, *Stable Adaptive Systems.* Courier Corporation, 2012.

10 Z. Hou and S. Xiong, "On model-free adaptive control and its stability analysis," *IEEE Transactions on Automatic Control,* vol. 64, no. 11, pp. 4555–4569, 2019.

11 R. Patel and N. Munro, "*Multivariable System Theory and Design,*" Oxford, Pergamon Press *(International Series on Systems and Control),* vol. 4, 1982.

12 S. Xiong and Z. Hou, "Model-free adaptive control for unknown MIMO nonaffine nonlinear discrete-time systems with experimental validation," *IEEE Transactions on Neural Networks and Learning Systems,* vol. 33, no. 4, pp. 1727–1739, 2020.

13 A. Khaki-Sedigh and B. Moaveni, *Control Configuration Selection for Multivariable Plants.* Springer, 2009.

14 R.-C. Roman, R.-E. Precup, C.-A. Bojan-Dragos, and A.-I. Szedlak-Stinean, "Combined model-free adaptive control with fuzzy component by virtual reference feedback tuning for tower crane systems," *Procedia Computer Science,* vol. 162, pp. 267–274, 2019.

15 F. Zhang, "A new model-free method combined with neural networks for MIMO systems," *arXiv preprint arXiv:2010.15338,* 2020.

16 A. Bai, Y. Luo, and H. Zhang, "Model-free adaptive control based on neural network observer for the chaotic power supply system," in *2020 IEEE 9th Data Driven Control and Learning Systems Conference (DDCLS),* 2020: IEEE, pp. 766–771.

17 L. dos Santos Coelho, A. A. R. Coelho, and R. R. Sumar, "Model-free learning adaptive controller with neural network compensator and differential evolution optimization," in *2006 IEEE Conference on Computer Aided Control System Design, 2006 IEEE International Conference on Control Applications, 2006 IEEE International Symposium on Intelligent Control,* 2006: IEEE, pp. 2018–2023.

Appendix

This appendix provides supplementary information relevant to the materials presented in various chapters of the book that are not covered in the main text. It is organized as follows:

- Section A presents the different norm definitions used throughout the book chapters.
- In Section B, you will find the Lyapunov equations used in Chapter 7, Section 7.4.
- Section C discusses the incremental stability concept employed in Chapter 3.
- Section D defines and elaborates on the dwell-time notion used in Chapters 3 and 4.
- Section E delves into the inverse moment condition presented in Theorem 6.1 of Chapter 6.
- Section F covers the principles of least squares, a concept widely used in Chapters 5, 7, 8, and 9.
- In Section G, we present the linear matrix inequality (LMI) approach, which is extensively utilized in Chapter 7.
- Finally, Section H introduces Linear Fractional Transformations, a topic employed in Chapter 7, Section 7.5.

A Norms

Norms are utilised to describe and evaluate the performance of a control system. There are several definitions for norms, and they apply to signals, systems and matrices, respectively. However, all the different norms must satisfy the following four conditions for ζ and ϱ the corresponding operands of the norm:

1. $||\zeta|| \geq 0$
2. $||\zeta|| = 0 \Leftrightarrow \zeta(t) = 0 \quad \forall t$

An Introduction to Data-Driven Control Systems, First Edition. Ali Khaki-Sedigh.

3. $\|a\zeta\| = |a|\|\zeta\|, \quad \forall a \in \mathbb{R}$
4. $\|\zeta + \varrho\| \leq \|\zeta\| + \|\varrho\|$

   And in the case of a *matrix norm*, it must satisfy the above four conditions for given matrices and the added following condition:

5. $\|\mathbf{AB}\| \leq \|\mathbf{A}\| \cdot \|\mathbf{B}\|$

   for any two given matrices. This condition is called the *multiplicative property*.

The norms used in this book are the 2-norm, the ∞-norm and the Frobenius norm. These norms are subsequently defined.

The 2-Norm. This is defined for the signal $u(t)$ as follows:

$$\|u\|_2 \triangleq \left(\int_{-\infty}^{\infty} u(t)^2 dt\right)^{1/2}$$

It is also called the *Euclidean norm* and for the n-dimensional vector $\mathbf{x}$ is defined as follows:

$$\|\mathbf{x}\|_2 \triangleq \left(\sum_{i=1}^{n} x_i^{\,2}\right)^{1/2}$$

Also, the 2-norm for the transfer function $G(s)$ is defined as follows:

$$\|G\|_2 \triangleq \left(\frac{1}{2}\int_{-\infty}^{\infty} |G(j\omega)|^2 d\omega\right)^{1/2}$$

The ∞-Norm. This is defined for the signal $u(t)$ as

$$\|u\|_\infty \triangleq sup_t|u(t)|$$

It is also called the *Chebyshev vector norm* and for the n-dimensional vector $\mathbf{x}$ is defined as follows:

$$\|\mathbf{x}\|_\infty \triangleq \max |x_i| \quad 1 \leq i \leq n$$

Also, the ∞-norm for the transfer function $G(s)$ is defined as follows:

$$\|G\|_\infty \triangleq sup_\omega|G(j\omega)|$$

The Frobenius Norm. The *Frobenius* or the *Euclidean norm* is defined for the $n \times n$ matrix $\mathbf{A}$ as follows:

$$\|\mathbf{A}\|_F \triangleq \left(\sum_{i=1}^{n}\sum_{j=1}^{n} |a_{ij}|^2\right)^{1/2} = \sqrt{tr(\mathbf{A}^*\mathbf{A})}$$

where trace *tr* is the sum of the diagonal elements of the corresponding matrix, and * denotes the complex conjugate transpose.

B Lyapunov Equation

Consider the continuous-time LTI plant given as follows:

$$\dot{\mathbf{x}}(t) = \mathbf{A}\mathbf{x}(t) \tag{B.1}$$

The stability of (B.1) can be studied using the following candidate Lyapunov function:

$$V(\mathbf{x}) = \mathbf{x}^T\mathbf{P}\mathbf{x}$$

where $\mathbf{P}$ is a positive definite symmetric matrix. The time derivative of $V(\mathbf{x})$ gives the following:

$$\dot{V}(\mathbf{x}) = \mathbf{x}^T(\mathbf{A}^T\mathbf{P} + \mathbf{P}\mathbf{A})\mathbf{x} \tag{B.2}$$

The necessary and sufficient condition for the asymptotic stability of the plant given by (B.1) is the negative definiteness of (B.2). That is, the following equation must hold for a positive definite matrix $\mathbf{Q}$:

$$\mathbf{A}^T\mathbf{P} + \mathbf{P}\mathbf{A} = -\mathbf{Q} \tag{B.3}$$

Equation (B.3) is called the *Lyapunov equation* for continuous-time plants. Similarly, it can be shown that for a discrete-time LTI plant given as follows:

$$\mathbf{x}(t+1) = \mathbf{A}\mathbf{x}(t) \tag{B.4}$$

The necessary and sufficient condition for the asymptotic stability of (B.4) is the satisfaction of the following Lyapunov equation for the symmetric positive definite matrices $\mathbf{P}$ and $\mathbf{Q}$:

$$\mathbf{A}^T\mathbf{P}\mathbf{A} - \mathbf{P} = -\mathbf{Q} \tag{B.5}$$

The procedure for studying the stability of plants given by (B.1) or (B.4), assuming that $\mathbf{A}$ is known is to select a symmetric positive definite matrix $\mathbf{Q}$, solve the corresponding Lyapunov equations (B.3) or (B.5) and obtain the matrix $\mathbf{P}$. The sign of $\mathbf{P}$ will determine the stability or instability of the respective plants.

C Incremental Stability

Consider a nonlinear system $S: l_{2e} \to l_{2e}$, described by the following equations:

$$\dot{\mathbf{x}}(t) = \boldsymbol{f}(t, \mathbf{x}(t), \mathbf{u}(t))$$

$$\mathbf{y}(t) = \boldsymbol{h}(t, \mathbf{x}(t), \mathbf{u}(t))$$

where l_{2e} is as defined in Section 3.3.1, $\mathbf{x}(t) \in \mathbb{R}^n$, $\mathbf{u}(t) \in \mathbb{R}^m$, $\mathbf{y}(t) \in \mathbb{R}^l$, $\boldsymbol{f}$ and $\boldsymbol{h}$ are nonlinear functions of appropriate dimensions. In the case of Lyapunov stability analysis, the behaviour of the state trajectories is studied for an equilibrium point. However, a stronger stability notion is *incremental stability* that compares arbitrary trajectories with each other. Incremental stability requires that any two trajectories of S converge to each other. It should be noted that incremental stability is different from the other stability concept of convergent systems. The relationship between these two stability definitions is discussed in Ref. [1]. Also, incremental stability implies that the steady-state properties of S are suitable and the effect of a non-zero initial condition is guaranteed to decay asymptotically to zero [2].

Definition C.1 The nonlinear system S has a *finite incremental gain* if there exits $\eta \geq 0$ such that for a given initial condition $\mathbf{x}(0)$ we have $\|Sv - Sw\|_2 \leq \eta \|v - w\|_2$ for any two $v, w \in l_2$. The *incremental gain* of S is defined as the minimum value of η. Then, S is *incrementally stable* if it is stable, that is it maps l_2 to l_2, and if it has a finite incremental gain.

Note that for linear systems, incremental stability is equivalent to stability. The incremental stability of switched systems is presented in Ref. [3].

D Switching and the Dwell-time

Consider a *switched linear system* described as follows:

$$\dot{\mathbf{x}}(t) = \mathbf{A}_{\sigma(t)}\mathbf{x}(t) \tag{D.1}$$

In this case, all the subsystems are assumed linear, $\mathbf{x}(t) \in \mathbb{R}^n$, $\mathbf{A}_{\sigma(t)}$ is the corresponding state matrix and $\sigma(t)$ is a piecewise constant function called the *switching signal* with a finite number of discontinuities that are the switching times denoted by $t_1, t_2, \ldots$. The switching signal $\sigma(t)$ is constant on the intervals between any two consecutive switching times. That is,

$$\sigma(t) = \sigma_k \text{ for } t \in \left[t_i, t_{i+1}\right), \qquad i = 0, 1, \ldots$$

where k corresponds to the selected plant. A typical switching signal is shown in Figure D.1.

The hysteresis switching algorithm introduced in Chapter 3 has many attractive features. However, this switching strategy can have shortcomings in the face of unmodelled dynamics and disturbances. To overcome these shortcomings, the *dwell-time switching* concept is introduced. Also, a well-known property of a switched system is that it is stable if all the individual subsystems are stable and the switching is sufficiently slow. Slow switching provides sufficient time for the switching transient effects to dissipate. Slow switching can be formulated in terms of the *dwell time* concept as follows. It can be alternatively stated that we require a minimum time interval between two successive switching times

Figure D.1

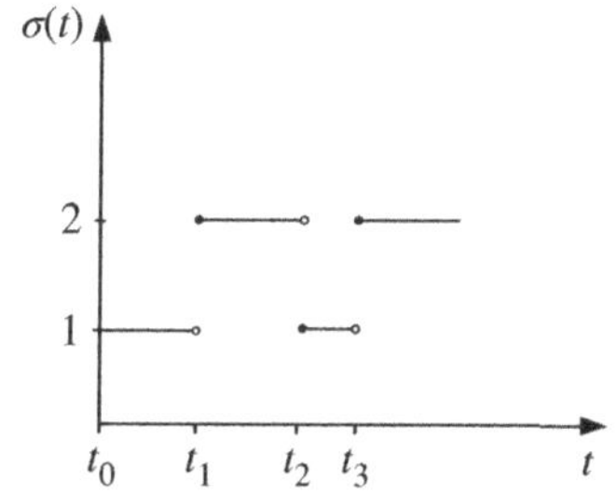

to avoid fast switching and its consequences. Let $\tau_d > 0$ and $t_1, t_2, \ldots$ be the switching times. Then, consider the class of admissible switching signals with the property that $t_{i+1} - t_i \geq \tau_d$ and τ_d is defined as the *dwell time*. In this case, if all the linear systems are asymptotically stable, and τ_d is *sufficiently* large, the switched system (D.1) is asymptotically stable. A lower bound on τ_d can be obtained from the exponential decay bounds on the transition matrices of each subsystem [4]. An interesting point to note is that under certain assumptions, the asymptotic stability of the switching control strategy for nonlinear systems can also be proved for a sufficiently large dwell time. However, the dwelling time approach can be conservative and undermine the effectiveness of the switching control methodology by delaying the necessary switching in response to the parameter or operating point changes, and possible faults occurring or deteriorating performance for whatever cause. To handle this challenge, the *average dwell time* notion is introduced to facilitate faster switching when required and to compensate for the fast switching by slow switching in a later stage. Consider a time interval (t, T) and denote the number of discontinuities of the switching signal $\sigma(t)$ on that interval by $N_\sigma(t, T)$. Then, $\sigma(t)$ has average dwell time τ_a if there exists $N_0, \tau_a \in \mathbb{R}^+$ such that

$$N_\sigma(t, T) \leq N_0 + \frac{T - t}{\tau_a} \qquad \forall T \geq t \geq 0 \tag{D.2}$$

Note that for $N_0 = 0$, Eq. (D.2) implies that if $\tau_a > T - t$, no switching will occur, and for $N_0 = 1$, Eq. (D.2) implies that if $\tau_a > T - t$, only one switching will occur, that is $\sigma(t)$ cannot switch twice on any interval of length smaller than τ_a. Hence, in general, discarding the first N_0 switches, the average time between consecutive switches is at least τ_a [4]. Estimating the upper bounds on dwell time and average dwell time is an important issue in model-based switching control systems, for example, refer to Ref. [5]. The main result concerning the average dwell time is that for any family of plants characterised by its state matrices, there always exists an average dwell time τ_a such that (D.1) is uniformly exponentially stable for any given N_0.

E Inverse Moments

To randomly perturb the elements of a tuning parameter vector, a mean-zero random perturbation vector is required. Denote this perturbation vector by the column vector $\boldsymbol{\delta}_j = [\delta_{j_i}],\ \ i = 1, \ldots, p$. This is a user-specified distribution satisfying condition (ii) given in Theorem 6.1 and the discussions provided in Remark 6.2. This is also discussed in detail in Ref. [6].

A key condition to be satisfied by the employed distributions in the *bounded inverse moment* condition given by $E\left[(\delta_{j_i})^{-2}\right] \leq C$ for some positive C. In what follows, the bounded inverse moment condition for some commonly used distributions is derived. Note that for a random variable δ_{j_i} with a probability density function given by $f(\delta)$, we have

$$E\left(\left|\frac{1}{\delta_{j_i}}\right|^{2(1+\tau)}\right) = \int_{-\infty}^{\infty} \left(\frac{1}{\delta}\right)^{2(1+\tau)} f(\delta)\, d\delta$$

for some positive τ. And for random variables with countably many outcomes with possible δ_i outcomes of the random variable δ_{j_i} with the corresponding probability $p(\delta_i)$, we have

$$E\left(\left|\frac{1}{\delta_{j_i}}\right|^{2(\tau+1)}\right) = \sum_{i=1}^{N} \left(\left|\frac{1}{\delta_i}\right|^{2(\tau+1)} p(\delta_i)\right)$$

The uniform distribution: This distribution is shown in Figure E.1. We have,

$$E\left(\left|\frac{1}{\delta_{j_i}}\right|^{2(1+\tau)}\right) = \int_{-m}^{m} \left(\frac{1}{\delta}\right)^{2(1+\tau)} \frac{1}{2m} d\delta = \frac{1}{m} \int_{0}^{m} \left(\frac{1}{\delta}\right)^{2(1+\tau)}$$

$$d\delta = \left.\frac{-1}{m(2\tau+1)\delta^{2\tau+1}}\right|_0^m = \infty$$

Hence, the inverse moment is unbounded and the uniform distribution does not satisfy this condition.

The segmented uniform distribution: This distribution is shown in Figure E.2. We have,

$$\begin{aligned}E\left(\left|\frac{1}{\delta_{j_i}}\right|^{2(\tau+1)}\right) &= \frac{1}{2(\alpha-\beta)}\left[\int_{-\alpha}^{-\beta} \left(\frac{1}{\delta}\right)^{2(\tau+1)} d\delta + \int_{\beta}^{\alpha} \left(\frac{1}{\delta}\right)^{2(\tau+1)} d\delta\right] \\ &= \frac{-1}{2(\alpha-\beta)}\left[\left.\frac{1}{(2\tau+1)\delta^{2\tau+1}}\right|_{-\alpha}^{-\beta} + \left.\frac{1}{(2\tau+1)\delta^{2\tau+1}}\right|_{\beta}^{\alpha}\right] \\ &= \frac{1}{2(\alpha-\beta)(2\tau+1)}\left[\frac{2}{\beta^{2\tau+1}} - \frac{2}{\alpha^{2\tau+1}}\right]\end{aligned}$$

Hence, the inverse moment is bounded and the segmented uniform distribution satisfies this condition.

Figure E.1

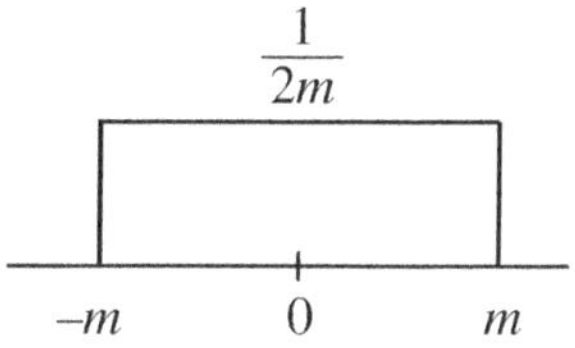

Figure E.2

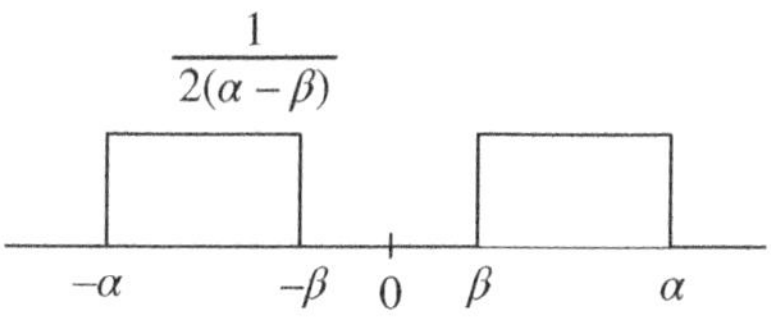

The Bernoulli distribution: This distribution is shown in Figure E.3. We have,

$$E\left(\left|\frac{1}{\delta_{j_i}}\right|^{2(\tau+1)}\right)=\sum_{i=1}^{N}\left(\left|\frac{1}{\delta_i}\right|^{2(\tau+1)}p(\delta_i)\right)=\left(\frac{1}{-\alpha}\right)^{2(\tau+1)}\left(\frac{1}{2}\right)$$

$$+\left(\frac{1}{\alpha}\right)^{2(\tau+1)}\left(\frac{1}{2}\right)=\left(\frac{1}{\alpha}\right)^{2(\tau+1)}$$

Hence, the inverse moment is bounded and the Bernoulli distribution satisfies this condition.

The normal distribution: This distribution is shown in Figure E.4. We have,

$$E\left(\left|\frac{1}{\delta_{j_i}}\right|^{2(\tau+1)}\right)=\int_{-\infty}^{\infty}\left(\frac{1}{\delta}\right)^{2(\tau+1)}\frac{1}{\sigma\sqrt{2\pi}}e^{-\frac{1}{2}\left(\frac{\delta}{\sigma}\right)^2}d\delta$$

$$=\frac{2}{\sigma\sqrt{2\pi}}\int_{0}^{\infty}\left(\frac{1}{\delta}\right)^{2(\tau+1)}e^{-\frac{1}{2}\left(\frac{\delta}{\sigma}\right)^2}d\delta$$

Figure E.3

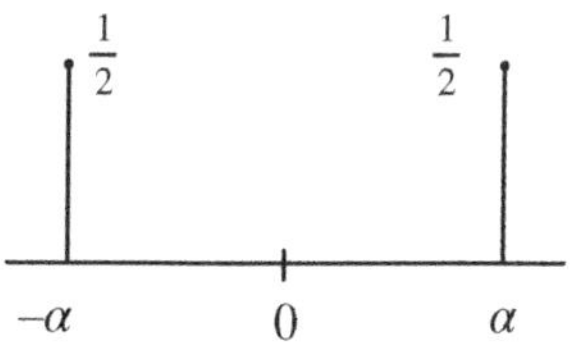

Figure E.4

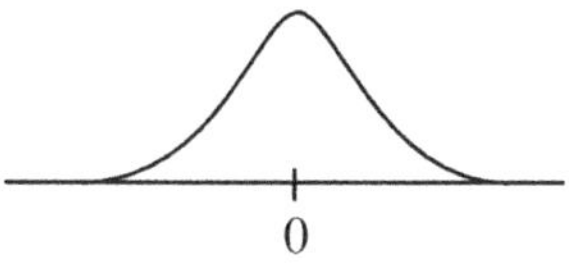

Let $f = e^{-\frac{1}{2}\left(\frac{\delta}{\sigma}\right)^2}, g' = \left(\frac{1}{\delta}\right)^{2(\tau+1)}$, and integration by parts gives

$$E\left(\left|\frac{1}{\delta_{j_i}}\right|^{2(\tau+1)}\right) = \frac{2}{\sigma\sqrt{2\pi}}$$
$$\times\left[-e^{-\frac{1}{2}\left(\frac{\delta}{\sigma}\right)^2}\frac{1}{2\tau+1}\left(\frac{1}{\delta}\right)^{2\tau+1}\Bigg|_0^{\infty} - \int_0^{\infty}\frac{1}{2\tau+1}\left(\frac{1}{\delta}\right)^{2\tau+1}\frac{\delta e^{-\frac{1}{2}\left(\frac{\delta}{\sigma}\right)^2}}{\sigma^2}d\delta\right]$$

Hence, without any further calculation, it is clear from the first term that the inverse moment is unbounded and the normal distribution does not satisfy this condition.

The V-shaped distribution: This distribution is shown in Figure E.5. We have,

$$E\left(\left|\frac{1}{\delta_{j_i}}\right|^{2(\tau+1)}\right) = \left[\int_{-m}^{0}\left(\frac{1}{\delta}\right)^{2(\tau+1)}\frac{-\delta}{m^2}d\delta + \int_0^m\left(\frac{1}{\delta}\right)^{2(\tau+1)}\frac{\delta}{m^2}d\delta\right]$$
$$2\int_0^m\left(\frac{1}{\delta}\right)^{2(\tau+1)}\frac{\delta}{m^2}d\delta = \infty$$

Hence, the inverse moment is unbounded and the V-shaped distribution does not satisfy this condition.

The U-shaped distribution: This distribution is shown in Figure E.6. We have,

$$E\left(\left|\frac{1}{\delta_{j_i}}\right|^{2(\tau+1)}\right) = \left[\int_{-\beta}^{0}\left(\frac{1}{\delta}\right)^{2(\tau+1)}\delta^n d\delta + \int_0^{\beta}\left(\frac{1}{\delta}\right)^{2(\tau+1)}\delta^n d\delta\right]$$
$$= 2\int_0^{\beta}\delta^{n-2(\tau+1)}d\delta = \frac{2}{n-(2\tau+1)}\delta^{n-(2\tau+1)}\Bigg|_0^{\beta}, n \geq 2(1+\tau)$$

Hence, for $n \geq 2(1+\tau)$, the inverse moment is bounded and the U-shaped distribution satisfies this condition.

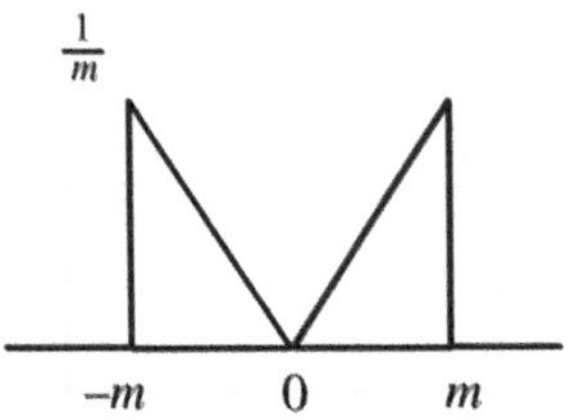

Figure E.5

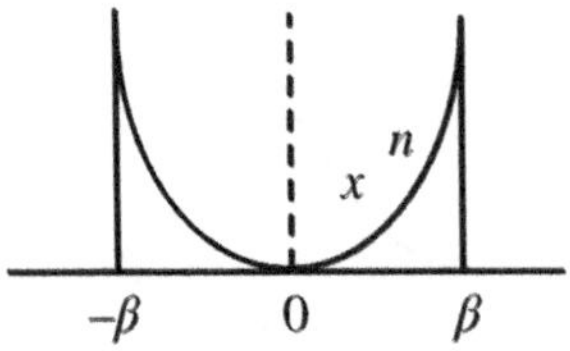

Figure E.6

F Least Squares Estimation

Consider an unknown single-input single-output linear time-invariant (LTI) plant given by the input–output relationship as $Y(s) = G(s)U(s)$. The equivalent discrete-time LTI model for a specified sampling time is given as follows:

$$A(q)y(t) = B(q)u(t) \tag{F.1}$$

where q is the shift operator, $A(q) = q^n + a_1 q^{n-1} + \cdots + a_n$, and $B(q) = b_1 q^{m-1} + b_2 q^{m-2} + \cdots + b_m$. Rewriting (F.1) in terms of q^{-1} gives

$$y(t) = -a_1 y(t-1) - \cdots - a_n y(t-n) + b_1 u(t+m-n-1) + \cdots + b_m u(t-n)$$

and

$$y(t) = [-y(t-1) \cdots -y(t-n)\ u(t+m-n-1) \cdots u(t-n)] \begin{bmatrix} a_1 \\ \vdots \\ a_n \\ b_1 \\ \vdots \\ b_m \end{bmatrix}$$

Or in the more compact *regression* form as follows:

$$y(t) = \boldsymbol{\phi}^T(t-1)\theta \tag{F.2}$$

where $\boldsymbol{\phi}^T(t-1)$ is the *regressor vector* and θ is the unknown *parameter vector*. Least squares (LS) estimation is an offline identification technique to estimate the unknown parameter vector θ in (F.2). Define the following cost function:

$$V(\theta, t) = \frac{1}{2}\sum_{i=1}^{t}(y(i) - \hat{y}(i))^2$$

where $y(i)$ is the measured plant output at the ith sample and $\hat{y}(i) = \boldsymbol{\phi}^T(i-1)\theta$ is the corresponding measured model output and $y(i) - \hat{y}(i)$ is the *estimation error*, also called the *residual* at the ith sample. The LS problem minimises $V(\theta, t)$ to obtain the estimated parameter vector $\hat{\theta}$. By defining $\mathbf{Y}(t) = [y(1) \cdots y(t)]^T$ and $\boldsymbol{\Phi}^T(t) = [\boldsymbol{\phi}(1) \ \cdots \ \boldsymbol{\phi}(t)]$, minimising $V(\theta, t)$ for θ leads to the following equation:

$$\boldsymbol{\Phi}^T(t)\boldsymbol{\Phi}(t)\hat{\theta} = \boldsymbol{\Phi}^T(t)\mathbf{Y}(t) \tag{F.3}$$

which is called the *normal equation* and provided that $\boldsymbol{\Phi}^T(t)\boldsymbol{\Phi}(t)$ is non-singular, Eq. (F.3) has a *unique* solution given as follows:

$$\hat{\theta} = (\boldsymbol{\Phi}^T(t)\boldsymbol{\Phi}(t))^{-1}\boldsymbol{\Phi}^T(t)\mathbf{Y}(t) \tag{F.4}$$

This is the *LS estimation solution*. The non-singularity condition of $\boldsymbol{\Phi}^T(t)\boldsymbol{\Phi}(t)$ is satisfied if the input signal is *persistently exciting*. For further theoretical properties of LS estimation and discussions on the persistent excitation of input signals refer to [7].

The LS parameter estimation can be rendered *online* by simple mathematical manipulations. Let $\widehat{\theta}(t-1)$ be the LS estimated parameter up to time $t-1$. Define $\mathbf{P}(t) = (\mathbf{\Phi}^T(t)\mathbf{\Phi}(t))^{-1}$ as the *covariance matrix*. We have, $\mathbf{P}(t)^{-1} = \mathbf{\Phi}^T(t)\mathbf{\Phi}(t)$ and using the definition of $\mathbf{\Phi}(t)$ this can be readily written as follows:

$$\mathbf{P}(t)^{-1} = \mathbf{P}(t-1)^{-1} + \boldsymbol{\phi}(t)\boldsymbol{\phi}^T(t) \tag{F.5}$$

Equation (F.4) gives

$$\widehat{\theta}(t) = \mathbf{P}(t)\mathbf{\Phi}^T(t)\boldsymbol{Y}(t) = \mathbf{P}(t)\left[\sum_{i=1}^{t}\boldsymbol{\phi}(i)y(i)\right] = \mathbf{P}(t)\left[\sum_{i=1}^{t-1}\boldsymbol{\phi}(i)y(i) + \boldsymbol{\phi}(t)y(t)\right] \tag{F.6}$$

On the other hand

$$\widehat{\theta}(t-1) = (\mathbf{\Phi}^T(t-1)\mathbf{\Phi}(t-1))^{-1}\mathbf{\Phi}^T(t-1)\boldsymbol{Y}(t-1)$$

which gives

$$\boldsymbol{P}(t-1)^{-1}\widehat{\theta}(t-1) = \mathbf{\Phi}^T(t-1)\boldsymbol{Y}(t-1) = \sum_{i=1}^{t-1}\boldsymbol{\phi}(i)y(i) \tag{F.7}$$

Substitution of (F.7) in (F.6) and using (F.5) gives

$$\begin{aligned}\widehat{\theta}(t) &= \mathbf{P}(t)[(\mathbf{P}(t)^{-1} - \boldsymbol{\phi}(t)\boldsymbol{\phi}^T(t))\widehat{\theta}(t-1) + \boldsymbol{\phi}(t)y(t)] \\ &= \widehat{\theta}(t-1) - \boldsymbol{P}(t)\boldsymbol{\phi}(t)\boldsymbol{\phi}^T(t)\widehat{\theta}(t-1) + \boldsymbol{P}(t)\boldsymbol{\phi}(t)y(t)\end{aligned}$$

As the estimated model output is $\widehat{y}(t) = \boldsymbol{\phi}^T(t)\widehat{\theta}(t-1)$, we have

$$\widehat{\theta}(t) = \widehat{\theta}(t-1) + \boldsymbol{P}(t)\boldsymbol{\phi}(t)[y(t) - \widehat{y}(t)] = \widehat{\theta}(t-1) + \boldsymbol{K}(t)e(t) \tag{F.8}$$

where $\boldsymbol{K}(t) = \boldsymbol{P}(t)\boldsymbol{\phi}(t)$ is defined as the *Kalman gain* and $e(t)$ is the *prediction* or *estimation error*. Eqaution (F.8) provides a formula for recursively updating the parameter estimation vector $\widehat{\theta}(t)$. However, to circumvent the need for inverting the matrix $\mathbf{\Phi}^T(t)\mathbf{\Phi}(t)$ at each sampling time, the matrix inversion lemma (see the footnote for Eq. (9.34)) is utilised as follows:

$$\begin{aligned}\mathbf{P}(t) &= [\mathbf{P}(t-1)^{-1} + \boldsymbol{\phi}(t)\boldsymbol{\phi}^T(t)]^{-1} \\ &= \mathbf{P}(t-1) - \mathbf{P}(t-1)\boldsymbol{\phi}(t)[\mathbf{I} + \boldsymbol{\phi}^T(t)\boldsymbol{P}(t-1)\boldsymbol{\phi}(t)]^{-1}\boldsymbol{\phi}^T(t)\boldsymbol{P}(t-1)\end{aligned}$$

which reduces the dimension of the matrix inversion and also gives

$$\boldsymbol{K}(t) = \mathbf{P}(t-1)\boldsymbol{\phi}(t)[\mathbf{I} + \boldsymbol{\phi}^T(t)\boldsymbol{P}(t-1)\boldsymbol{\phi}(t)]^{-1} \tag{F.9}$$

Equations (F.8) and (F.9) comprise the recursive least squares (RLS) procedure for the online parameter identification of an unknown system described by the regressor model (F.2). For further theoretical properties of RLS estimation and discussions on practical implementation issues, time-varying systems, forgetting factor and covariance resetting refer to [7].

G Linear Matrix Inequalities

The linear matrix inequality (LMI) approach is an effective tool that solves a wide range of problems in control systems analysis and design. Such problems can be reduced to certain standard convex optimisation problems involving LMIs. Subsequently, by employing effective optimisation toolboxes, the LMIs are numerically and efficiently solved. An LMI has the following general form:

$$\mathbf{F}(\mathbf{x}) \triangleq \mathbf{F}_0 + \sum_{i=1}^{n} x_i \mathbf{F}_i < \mathbf{0} \tag{G.1}$$

where $\mathbf{x} = \begin{bmatrix} x_1 & \cdots & x_n \end{bmatrix}$ is a vector of n real numbers that are the *decision variables*, $\mathbf{F}_i$, $i = 0,\ldots, n$, are $n \times n$ real symmetric matrices and $\mathbf{F}(\mathbf{x})$ is a negative definite matrix. Often in control systems analysis and design, the decision variables are matrices. In this case, Eq. (G.1) is written as a function of the matrix $\mathbf{X}$. As an example, consider the Lyapunov matrix inequality defined as follows:

$$\mathbf{F}(\mathbf{X}) = \mathbf{A}^T\mathbf{X} + \mathbf{X}^T\mathbf{A} + \mathbf{Q} < \mathbf{0} \tag{G.2}$$

where the known matrices $\mathbf{A}$, $\mathbf{Q} \in \mathbb{R}^{n\times n}$, and $\mathbf{X} \in \mathbb{R}^{n\times n}$ is the unknown matrix. Equation (G.2) is an LMI if $\mathbf{Q}$ is a symmetric matrix. Note that LMIs with $\mathbf{X}$ can be written in the form of (G.1), if the matrix $\mathbf{X}$ is expressed in terms of the basis matrices of the $n \times n$ matrix space given by $\mathbf{E}_1, \ldots, \mathbf{E}_N$, as $\mathbf{X} = \sum_{i=1}^{N} x_i \mathbf{E}_i$, and $N \leq n^2$ depending on the structure of $\mathbf{X}$. Then, for the LMI (G.2), we have

$$\begin{aligned}\mathbf{F}(\mathbf{X}) &= \mathbf{F}\left(\sum_{i=1}^{N} x_i \mathbf{E}_i\right) = \mathbf{A}^T \sum_{i=1}^{N} x_i \mathbf{E}_i + \left(\sum_{i=1}^{N} x_i {\mathbf{E}_i}^T\right)\mathbf{A} + \mathbf{Q} = \sum_{i=1}^{N} x_i \mathbf{A}^T \mathbf{E}_i \\ &\quad + \sum_{i=1}^{N} x_i {\mathbf{E}_i}^T \mathbf{A} + \mathbf{Q} = \sum_{i=1}^{N} x_i \left(\mathbf{A}^T\mathbf{E}_i + {\mathbf{E}_i}^T\mathbf{A} + \mathbf{Q}\right) - \sum_{i=1}^{N} x_i \mathbf{Q} + \mathbf{Q} \\ &= \sum_{i=1}^{N} x_i \mathbf{F}(\mathbf{E}_i) + \mathbf{F}_0 = \mathbf{F}_0 + \sum_{i=1}^{N} x_i \mathbf{F}_i < \mathbf{0}\end{aligned}$$

where x_i would be the decision variables and $\mathbf{F}_0$ is appropriately defined.

LMIs have the following interesting properties from Ref. [8]. It is shown that the LMI (G.1) is a *convex constraint* on $\mathbf{x}$ and its set of solutions is also convex. Moreover, for a system of LMIs given by $\mathbf{F}_1(\mathbf{x}) < \mathbf{0}$, …, $\mathbf{F}_K(\mathbf{x}) < \mathbf{0}$, the corresponding solution set, that is the set of solution vectors satisfying the given LMIs, is also convex. The system of LMIs can be written in the following compact form $\mathbf{F}(\mathbf{x}) = \text{diagonal}[\mathbf{F}_1(\mathbf{x}), \ldots, \mathbf{F}_K(\mathbf{x})] < \mathbf{0}$, and the corresponding solution satisfies the system of LMIs, and vice versa. Also, LMIs can handle affine constraints of $\mathbf{x}$ by grouping the LMI and constraints in a new LMI. Finally, if we partition $\mathbf{F}(\mathbf{x})$ as follows:

$$\mathbf{F}(\mathbf{x}) = \begin{bmatrix} \mathbf{F}_{11}(\mathbf{x}) & \mathbf{F}_{12}(\mathbf{x}) \\ \mathbf{F}_{21}(\mathbf{x}) & \mathbf{F}_{22}(\mathbf{x}) \end{bmatrix}$$

where $\mathbf{F}_{11}(\mathbf{x})$ is square and invertible, then $\mathbf{F}(\mathbf{x}) < \mathbf{0}$ if and only if

$$\begin{cases} \mathbf{F}_{11}(\mathbf{x}) < \mathbf{0} \\ \mathbf{F}_{22}(\mathbf{x}) - \mathbf{F}_{21}(\mathbf{x})[\mathbf{F}_{11}(\mathbf{x})]^{-1}\mathbf{F}_{12}(\mathbf{x}) < \mathbf{0} \end{cases}$$

Similarly, if $\mathbf{F}_{22}(\mathbf{x})$ is square and invertible, then $\mathbf{F}(\mathbf{x}) < \mathbf{0}$ if and only if

$$\begin{cases} \mathbf{F}_{22}(\mathbf{x}) < \mathbf{0} \\ \mathbf{F}_{11}(\mathbf{x}) - \mathbf{F}_{12}(\mathbf{x})[\mathbf{F}_{22}(\mathbf{x})]^{-1}\mathbf{F}_{21}(\mathbf{x}) < \mathbf{0} \end{cases}$$

The second inequality in the brackets is called the *Schur complement* of $\mathbf{F}_{11}(\mathbf{x})$ or $\mathbf{F}_{22}(\mathbf{x})$ in $\mathbf{F}(\mathbf{x})$, respectively.

An LMI is feasible if there exists an $\mathbf{x}$ that satisfies $\mathbf{F}(\mathbf{x}) < \mathbf{0}$. In what follows, simple examples of the applications of LMI are presented.

Example G.1 Let $\dot{\mathbf{x}}(t) = \mathbf{A}\mathbf{x}(t)$, as discussed in Section B, Lyapunov theory indicates that the given system is stable if and only if there exists a positive definite symmetric matrix $\mathbf{P}$ for which $\mathbf{A}^T\mathbf{P} + \mathbf{P}\mathbf{A} < \mathbf{0}$. This can be equivalently stated in terms of the feasibility of the following LMI:

$$\begin{bmatrix} -\mathbf{P} & 0 \\ 0 & \mathbf{A}^T\mathbf{P} + \mathbf{P}\mathbf{A} \end{bmatrix} < \mathbf{0}$$

Example G.2 Let $\dot{\mathbf{x}}(t) = \mathbf{A}_i\mathbf{x}(t) + \mathbf{B}_i\mathbf{u}(t)$, where $i = 1, \ldots, k$ and $\mathbf{A}_i \in \mathbb{R}^{n\times n}$ and $\mathbf{B}_i \in \mathbb{R}^{n\times m}$ are the state and input matrices, respectively. The given family of k systems is *simultaneously stabilised* if there exists a state feedback matrix $\mathbf{K} \in \mathbb{R}^{m\times n}$ such that the state feedback law $\mathbf{u}(t) = -\mathbf{K}\mathbf{x}(t)$ stabilises all the k systems. That is, all the closed-loop matrices $\mathbf{A}_i + \mathbf{B}_i\mathbf{K}$, $i = 1, \ldots, k$ are stable. Using the result of Example G.1, this is equivalent to finding matrices $\mathbf{K}$ and $\mathbf{P}_i$ for $i = 1, \ldots, k$, satisfying the following matrix inequalities:

$$\begin{cases} \mathbf{P}_i > \mathbf{0} \\ (\mathbf{A}_i + \mathbf{B}_i\mathbf{K})^T\mathbf{P}_i + \mathbf{P}_i(\mathbf{A}_i + \mathbf{B}_i\mathbf{K}) < \mathbf{0} \end{cases} \tag{G.3}$$

However, as both $\mathbf{K}$ and $\mathbf{P}_i$ are unknown, this is not in the conventional form of an LMI. To transform (G.3) into an LMI, let $\mathbf{P} \triangleq \mathbf{P}_1 = \mathbf{P}_2 = \cdots = \mathbf{P}_k$, and $\mathbf{Y} = \mathbf{P}^{-1}$ and $\mathbf{X} = \mathbf{K}\mathbf{Y}$, then (G.3) becomes

$$\begin{cases} \mathbf{Y} > \mathbf{0} \\ \mathbf{A}_i\mathbf{Y} + \mathbf{Y}\mathbf{A}_i^T + \mathbf{B}_i\mathbf{X} + \mathbf{X}^T\mathbf{B}_i^T < \mathbf{0} \end{cases} \tag{G.4}$$

Note that (G.4) is in the form of a system of LMIs in $\mathbf{X}$ and $\mathbf{Y}$ and the feasibility of (G.4) would imply a solution $\mathbf{K} = \mathbf{X}\mathbf{Y}^{-1}$ for simultaneous stabilisation of the given family of systems.

H Linear Fractional Transformations

Linear fractional transformations (LFTs) also known as *bilinear transformations* provide a framework where many control design problems can be treated in a unified manner once stated in this framework. In the scalar-valued case, $F(s) = \frac{a+bs}{c+ds}$ with a, b, c and $d \in \mathbb{R}$, is called an LFT and if $c \neq 0$, then we can write $F(s) = \alpha + \beta s(1 - \gamma s)^{-1}$ for some α, β and $\gamma \in \mathbb{R}$. This LFT can be generalised to the matrix-valued case as follows [9]. Let

$$\mathbf{M} = \begin{bmatrix} \boldsymbol{M}_{11} & \boldsymbol{M}_{12} \\ \boldsymbol{M}_{21} & \boldsymbol{M}_{22} \end{bmatrix} \in \mathbb{R}^{(p_1+p_2)\times(q_1+q_2)} \tag{H.1}$$

and $\Delta_l \in \mathbb{R}^{q_2 \times p_2}$ and $\Delta_u \in \mathbb{R}^{q_1 \times p_1}$. Then, a *lower LFT* for Δ_l is defined as

$$\mathrm{F}_l(\mathbf{M}, \Delta_l) \triangleq \boldsymbol{M}_{11} + \boldsymbol{M}_{12}\,\Delta_l(\mathbf{I} - \boldsymbol{M}_{22}\Delta_l)^{-1}\boldsymbol{M}_{21} \tag{H.2}$$

provided that $(\mathbf{I} - \boldsymbol{M}_{22}\Delta_l)^{-1}$ exists. Also, an *upper LFT* for Δ_u is defined as

$$\mathrm{F}_u(\mathbf{M}, \Delta_u) \triangleq \boldsymbol{M}_{22} + \boldsymbol{M}_{21}\,\Delta_u(\mathbf{I} - \boldsymbol{M}_{11}\Delta_u)^{-1}\boldsymbol{M}_{12} \tag{H.3}$$

provided that $(\mathbf{I} - \boldsymbol{M}_{11}\Delta_u)^{-1}$ exists. These are the widely used LFT representations of uncertain plants, where $\mathbf{M}$ denotes the nominal part of an uncertain plant.

Based on the LFT framework, in a slightly different representational form, the following results from [10] ensure closed-loop stability for uncertain systems. Consider the following open-loop LTI system in which the uncertainty enters in an LFT form:

$$\begin{bmatrix} \dot{\mathbf{x}} \\ \mathbf{z}_1 \\ \mathbf{z}_2 \\ \mathbf{y} \end{bmatrix} = \begin{bmatrix} \mathbf{A} & \mathbf{B}_1 & \mathbf{B}_2 & \mathbf{B} \\ \mathbf{C}_1 & \mathbf{D}_{11} & \mathbf{D}_{12} & \mathbf{E}_1 \\ \mathbf{C}_1 & \mathbf{D}_{21} & \mathbf{D}_{22} & \mathbf{E}_2 \\ \mathbf{C}_1 & \mathbf{F}_1 & \mathbf{F}_2 & \mathbf{0} \end{bmatrix} \begin{bmatrix} \mathbf{x} \\ \mathbf{w}_1 \\ \mathbf{w}_2 \\ \mathbf{u} \end{bmatrix}, \mathbf{w}_1 = \Delta\mathbf{z}_1$$

It is observed that $\mathbf{w}_1 \rightarrow \mathbf{z}_1$ represents the *uncertainty channel*, and $\mathbf{w}_2 \rightarrow \mathbf{z}_2$ represents the *performance channel* that describes the desired performance specifications, and $\mathbf{u} \rightarrow \mathbf{y}$ is the *control channel*. Assume that an LTI controller is designed

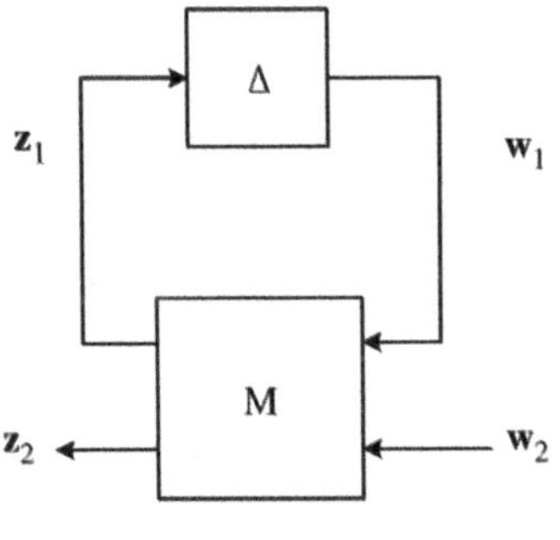

Figure H.1

and the resulting closed-loop system, shown in the nominal and uncertainty parts with a lower LFT depicted in Figure H.1, is given as follows:

$$\begin{bmatrix} \dot{\xi} \\ \mathbf{z}_1 \\ \mathbf{z}_2 \end{bmatrix} = \begin{bmatrix} A & B_1 & B_2 \\ C_1 & D_{11} & D_{12} \\ C_2 & D_{21} & D_{22} \end{bmatrix} \begin{bmatrix} \xi \\ \mathbf{w}_1 \\ \mathbf{w}_2 \end{bmatrix}, \mathbf{w}_1 = \Delta \mathbf{z}_1 \tag{H.4}$$

It is assumed that (H.4) is *well-posed*, that is $\mathbf{I} - \Delta D_{11}$ is nonsingular for all Δ in the uncertainty set. Then, we have

$$\mathbf{w}_1 = \Delta(C_1\xi + D_{11}\mathbf{w}_1 + D_{12}\mathbf{w}_2)$$

which gives

$$\mathbf{w}_1 = (\mathbf{I} - \Delta D_{11})^{-1}\Delta(C_1\xi + D_{12}\mathbf{w}_2) = (\mathbf{I} - \Delta D_{11})^{-1}\Delta C_1\xi \\ + (\mathbf{I} - \Delta D_{11})^{-1}\Delta D_{12}\mathbf{w}_2$$

And we finally have

$$\dot{\xi} = \left(A + B_1(\mathbf{I} - \Delta D_{11})^{-1}\Delta C_1\right)\xi + \left(B_2 + B_1(\mathbf{I} - \Delta D_{11})^{-1}\Delta D_{12}\right)\mathbf{w}_2 \tag{H.5}$$

$$\mathbf{z}_2 = \left(C_2 + D_{21}(\mathbf{I} - \Delta D_{11})^{-1}\Delta C_1\right)\xi + \left(D_{22} + D_{21}(\mathbf{I} - \Delta D_{11})^{-1}\Delta D_{12}\right)\mathbf{w}_2$$

The closed-loop state matrix is therefore $A(\Delta) = A + B_1(\mathbf{I} - \Delta \mathrm{D}_{11})^{-1}\Delta C_1$. Comparing (H.1) and (H.4), and (H.2) and (H.5), shows a slight difference in the LFT representations.

Theorem H.1 Let (H.4) be well-posed and suppose that there exists an $\chi > \mathbf{0}$ satisfying

$$\begin{bmatrix} \mathbf{I} \\ A(\Delta) \end{bmatrix}^T \begin{bmatrix} \mathbf{0} & \chi \\ \chi & \mathbf{0} \end{bmatrix} \begin{bmatrix} \mathbf{I} \\ A(\Delta) \end{bmatrix} < \mathbf{0} \quad \forall \boldsymbol{\Delta} \text{ in the uncertainty set} \tag{H.6}$$

Then the uncertain system (H.1) is uniformly exponentially stable.

The stability condition in Theorem H.1 is given in terms of $A(\Delta)$ on the entire uncertainty set and is therefore difficult to apply. It is shown in Ref. [10] that this condition can be equivalently expressed in terms of a more tractable condition by defining so-called *multiplier symmetric matrices* of appropriate dimensions satisfying the following inequality for all admissible uncertainties Δ

$$\begin{bmatrix} \Delta \\ \mathrm{I} \end{bmatrix}^T \boldsymbol{P} \begin{bmatrix} \Delta \\ \mathrm{I} \end{bmatrix} > 0$$

where the multipliers are assumed to be partitioned as follows:

$$\boldsymbol{P} = \begin{bmatrix} \boldsymbol{Q} & \boldsymbol{S} \\ \boldsymbol{S}^T & \boldsymbol{R} \end{bmatrix}$$

Theorem H.2 The representation (H.4) is well-posed and χ satisfies (H.6) *if and only if* there exists a multiplier $\boldsymbol{P}$ such that

$$\begin{bmatrix} \mathrm{I} & 0 \\ A & B \\ 0 & \mathrm{I} \\ C & D \end{bmatrix}^T \begin{bmatrix} 0 & \chi & 0 & 0 \\ \chi & 0 & 0 & 0 \\ 0 & 0 & Q & S \\ 0 & 0 & S^T & R \end{bmatrix} \begin{bmatrix} \mathrm{I} & 0 \\ A & B \\ 0 & \mathrm{I} \\ C & D \end{bmatrix} < 0$$

References

1 B. S. Rüffer, N. Van De Wouw, and M. Mueller, "Convergent systems vs. incremental stability," *Systems & Control Letters,* vol. 62, no. 3, pp. 277–285, 2013.
2 V. Fromion, G. Scorletti, and G. Ferreres, "Nonlinear performance of a PI controlled missile: an explanation," *International Journal of Robust and Nonlinear Control: IFAC-Affiliated Journal,* vol. 9, no. 8, pp. 485–518, 1999.
3 H. Yang and B. Jiang, "Incremental stability of switched non-linear systems," *IET Control Theory and Applications,* vol. 10, no. 2, pp. 220–225, 2016.
4 D. Liberzon, *Switching in Systems and Control.* Springer, 2003.
5 Ö. Karabacak, "Dwell time and average dwell time methods based on the cycle ratio of the switching graph," *Systems & Control Letters,* vol. 62, no. 11, pp. 1032–1037, 2013.
6 J. C. Spall, *Introduction to Stochastic Search and Optimization: Estimation, Simulation, and Control.* John Wiley & Sons, 2005.
7 K. J. Åström and B. Wittenmark, *Adaptive Control.* Courier Corporation, 2013.
8 C. Scherer and S. Weiland, "Linear matrix inequalities in control," *Lecture Notes, Dutch Institute for Systems and Control, Delft, The Netherlands,* vol. 3, no. 2, 2000.
9 K. Zhou and J. C. Doyle, *Essentials of Robust Control.* Prentice Hall, Upper Saddle River, NJ, 1998.
10 C. W. Scherer, "Robust mixed control and linear parameter-varying control with full block scalings," in *Advances in Linear Matrix Inequality Methods in Control,* 2000: SIAM, pp. 187–207.

Index

Please note that the page numbers referring to figures are in *italics* and those referring to tables are in **bold**.

d

t

Printed and bound by CPI Group (UK) Ltd, Croydon, CR0 4YY
16/04/2025
14658346-0003